Medical Minds and Machine Learning

Transforming Care with AI

Editors:

Sandeep Kumar Mathivanan
Assistant Professor, School of Computing Science and Engineering
Galgotias University, Greater Noida, India

Saurav Mallik
Research Scientist
Department of Pharmacology & Toxicology
R Ken Coit College of Pharmacy, University of Arizona, Tucson, AZ, USA

Basu Dev Shivahare
Associate Professor, School of Computing Science and Engineering
Galgotias University, Greater Noida, India

Sangeetha S.K.B.
Manipal Institute of Technology Bengaluru
Manipal Academy of Higher Education, Manipal, India

Shakila Basheer
Associate Professor
Department of Information Systems
College of Computer and Information Science
Princess Nourah Bint Abdulrahman University, Riyadh Saudi Arabia

CRC Press
Taylor & Francis Group
Boca Raton London New York

CRC Press is an imprint of the
Taylor & Francis Group, an **informa** business
A SCIENCE PUBLISHERS BOOK

First edition published 2027
by CRC Press
2385 NW Executive Center Drive, Suite 320, Boca Raton FL 33431

and by CRC Press
4 Park Square, Milton Park, Abingdon, Oxon, OX14 4RN

CRC Press is an imprint of Taylor & Francis Group, LLC

Library of Congress Cataloging-in-Publication Data (applied for)

ISBN: 978-1-041-21286-7 (hbk)
ISBN: 978-1-041-21378-9 (pbk)
ISBN: 978-1-003-72691-3 (ebk)

DOI: 10.1201/9781003726913

Typeset in Times New Roman
by Prime Publishing Services

Preface

The convergence of medical science and machine learning marks one of the most transformative phases in the evolution of modern healthcare. Advances in artificial intelligence (AI), data-driven modeling, and computational intelligence are reshaping how diseases are detected, diagnosed, managed, and prevented. *Medical Minds and Machine Learning: Transforming Care with AI* is conceived in response to this paradigm shift, offering a comprehensive and forward-looking exploration of intelligent technologies that augment clinical decision-making and personalized healthcare delivery.

Healthcare systems today generate vast volumes of heterogeneous data, including electronic health records, medical imaging, genomic profiles, wearable sensor streams, and unstructured clinical text. Effectively harnessing this data requires robust AI frameworks that are not only accurate and scalable but also interpretable, ethical, and clinically meaningful. This book addresses these challenges by presenting state-of-the-art research contributions that integrate machine learning, deep learning, federated learning, multimodal data fusion, and emerging paradigms such as generative AI, edge intelligence, and quantum-enhanced computing.

The chapters in this volume are carefully curated to cover the full spectrum of AI-enabled healthcare, ranging from intelligent clinical decision support systems and explainable diagnostic models to privacy-preserving federated architectures, wearable-based personalized medicine, ethical AI frameworks, and conversational medical assistants. Several chapters emphasize real-world applicability, scalability, and deployment readiness, reflecting the growing need to translate AI innovations from laboratory research into clinical practice. Additionally, the inclusion of topics such as bias mitigation, transparency, accountability, and patient-centric AI underscores the importance of responsible and trustworthy AI adoption in healthcare.

This book is intended for a broad audience, including researchers, clinicians, healthcare technologists, graduate students, and industry practitioners who seek to understand both the theoretical foundations and practical implementations of AI in medicine. By bridging computational intelligence with clinical insight, the volume aims to foster interdisciplinary collaboration and inspire innovative solutions to pressing healthcare challenges.

It is our hope that *Medical Minds and Machine Learning: Transforming Care with AI* will serve as a valuable reference and a catalyst for future research, innovation, and ethical deployment of artificial intelligence in global healthcare systems.

Contents

Chapter 1

OptiCareNet

An Optimization-Driven AI Framework for Intelligent Clinical Decision Support

Hari Prasath K.,[1,*] *Mili Dhar*,[2] *Vartika Mishra*[3] and *Sathishkumar V.E.*[4]

Due to computerized diagnosis, risk prediction, and clinical decision support, Artificial Intelligence (AI) has been a central part of modern healthcare. Most AI systems are built upon static models and hand-tuned parameters, which limit efficiency, accuracy, and flexibility. The study suggests OptiCareNet, an optimization-augmented AI model integrating Reinforcement Learning-based Clinical Decision Routing (RL-CDR) for dynamic decision and Particle Swarm Optimization (PSO) for hyperparameter optimization to bridge these limitations. To minimize training loss and enhance diagnostic accuracy, OptiCareNet systematically optimizes neural network architectural parameters such as learning rate, dropout rate, and layer structure. Meanwhile, RL-CDR agents boost real-time therapy planning and response ordering by optimizing treatment trajectories depending on patient state alterations.

[1] School of Computing Science and Engineering, Galgotias University, Greater Noida, 203201, India.
[2] Department of Computer Science and Engineering, Pranveer Singh Institute of Technology, Kanpur, Uttar Pradesh, India.
[3] Department of Computer Science and Engineering, Alliance School of Advanced Computing, Alliance University, Bangalore, India, 562106.
[4] School of Computing and Artificial Intelligence, Faculty of Engineering and Technology, Sunway University, No. 5, Jalan Universiti, Bandar Sunway, 47500 Selangor Darul Ehsan, Malaysia.
Emails: milidhar28@gmail.com; vartika.mishra@alliance.edu.in; sathishv@sunway.edu.my
* Corresponding author: khariprasathit@gmail.com

DOI: 10.1201/9781003726913-1

The MIMIC-IV and UCI Heart Disease datasets were employed to test the framework. Experimental outcomes report improved diagnostic accuracy (98.2%), decreased convergence time (42%), and 35.7% decrease in the rate of erroneous diagnoses compared to conventional deep learning models. The strength of integrating AI with optimization algorithms to provide versatile, efficient, and clinically relevant healthcare solutions is exemplified by OptiCareNet. This approach is scalable and applicable across diagnostic areas such as oncology, critical care, and cardiology.

1. Introduction

With its capacity to enhance clinical diagnosis, drive data-driven decision-making, and enhance diagnostic accuracy, Artificial Intelligence (AI) is now a central disruptor across the healthcare industry. AI technologies have enabled the deployment of automated solutions to support patient outcomes and efficiency, from image-based diagnosis in radiology and pathology to predictive analytics in intensive care units. Historically, AI models in healthcare are implemented with machine learning and deep learning models that work mostly with historic data and static parameter settings. Deterministic decision-making pipelines, rule-based inference systems, and static hyperparameters are generally employed with such approaches. These systems work well in structured settings, but they tend not to generalize to real-world, complex clinical settings where data distributions shift, patient states vary, and choices need to be made on the fly.

Existing methods also have some notable limitations. First, the performance and scalability of machine learning models are constrained by the manually adjusted hyperparameters. Second, traditional training processes are prone to overfitting and local minima, and have prolonged convergence times. Third, the majority of AI systems used in healthcare do not have adaptive abilities that allow them to modify their decisions in response to evolving data or ongoing patient monitoring. Moreover, in serious applications like emergency diagnosis, treatment triage, or critical care therapies, the lack of optimization-based learning results in poor model performance. Uncertainty, time-varying variables, and the requirement for adapted decision steps are all neglected by non-dynamic models.

The inability of traditional systems to maximize therapeutic outcomes while simultaneously improving computational effectiveness is another important limitation. Since medical information tends to be high-dimensional and heterogeneous in character, probing challenging solution spaces requires sophisticated optimization techniques. Most models rely on time-inefficient and computationally intensive grid search or random search for parameter tuning. These limitations make broad application of AI in the field of many clinical disciplines challenging. Decision support systems also tend to be lacking in dynamically adjusting to patient input, leading to delayed diagnosis, inappropriate testing, or incorrect recommendations for treatment. Healthcare sophistication and the amount of clinical information call for intelligent systems that can self-refresh to ensure accuracy, speed, and responsiveness as well as learning from data. AI systems that incorporate strong optimization algorithms are obviously required to improve model performance and strengthen decision paths in real-time. Without intervention by man, these systems would be able to discover the best hyperparameters and strategies and thus become more trustworthy and clinically meaningful.

These significant weaknesses are the impetus to create an optimization-based artificial intelligence system for medicine. Errors in decision-making can be disastrous in high-risk, high-speed clinical settings. Therefore, AI has to use techniques like swarm intelligence, reinforcement learning (RL), and adaptive policy optimization (APO) alongside static learning. These techniques may be able to improve clinical decision-making under uncertainty, facilitate convergence, and give intelligent feedback loops. Building trustable, precise, and context-sensitive next-generation healthcare systems requires a convergence of global optimization and learning in real-time. In order to advance from legacy AI deployments to smart, self-tuning, and clinically reliable solutions, these disparities must be filled.

The main objectives are:

1. To design an AI-based healthcare framework optimized using swarm intelligence.
2. To develop an RL mechanism for adaptive clinical decision-making.

2. Related Study

The need for intelligent, effective, and accurate clinical decision-support and diagnostic software has also brought a significant step forward in AI integration with health systems. In optimization of healthcare decision-making to its fullest potential, optimization techniques are being used more in AI systems, as seen in the study. Retaining the importance of synergy between data mining, deep learning, and optimization algorithms towards developing smart and adaptive systems that can analyze complex clinical data and maximize patient outcomes lies the platform of this revolution [1, 2].

Research studies based on laboratory data, real-time sensor streams, and Electronic Health Records (EHRs) have been utilized in a series of studies to improve data-driven clinical procedures using predictive models and neural networks. While such systems are promising, they have fixed rule-based models, hard-coded parameter settings, and limited flexibility in responding to varying clinical situations [3, 4]. Swarm intelligence, evolutionary methods, and RL are a few of the optimization processes that have emerged lately as useful tools for optimizing AI models, particularly in the sphere of healthcare diagnosis, resource allocation, and treatment planning [5, 6].

Traditional and bio-inspired optimization algorithms have both been employed in recent studies to attain various goals of healthcare. A healthcare monitoring system based on Internet of Things (IoT)-cloud that was intelligent, employed mutant convolutional neural networks (CNNs) with seahorse tuning to enhance the quality of the signal and monitor patients in real-time [7]. Similarly, the hybrid frameworks that combine Bayesian optimization and extreme learning machines have been suggested to identify cyberattacks in the Internet of Medical Things (IoMT) domain with a guarantee of security and acceleration [8]. To facilitate increased personalization of care, as well as efficiency of flow, decision trees and optimization through crowdsourcing have also been investigated for patient-centered clinical pathways in cardiology [9]. Furthermore, to improve the accuracy of time-series prediction in public health analytics, ant colony optimization algorithms have been combined with Long Short-Term Memory (LSTM) networks to predict trends in disease [10].

Many systems are aimed at using deep learning and optimization to extract insights from unstructured as well as structured data within EHRs. It has been shown that deep learning networks supplemented by attention and RL components greatly improve the prediction of disease progression and treatment outcomes [11, 12]. Explainable AI (XAI) models have also been increasingly applied to risk analysis and health finance, where optimization algorithms like neural networks and Tabu Search support strategic decision-making as well as vendor picking [13, 14]. Large-scale digital twin simulation and energy-conserving clinical decision-making systems are made possible through the advances in AI-designed network architecture and smart monitoring systems, which also remain mainly reliant on optimization-based learning [15, 16].

Multi-objective optimization approaches have been utilized to improve consensus, fairness, and stakeholder agreement in collaborative medical opinion and decision-making formation [17]. Several studies indicate how optimization can enhance the performance of AI models, particularly in such activities as resource optimization, disease diagnosis, and diagnosis rate. Through the integration of clinical semantics and optimization methodologies, such methods as the CAEN (Cognitive-Aware Efficient Network) and HDBN (Hybrid Deep Bayesian Network) models have been suggested to enhance the accuracy of sickness prediction [18]. Improved anomaly detection systems, like SHACS (Secure Healthcare Access Control System), provide effective results in detection and response to unauthorized activity on cloud-based healthcare platforms for security and access control [19].

Moreover, predictive AI models in real-time have been used in medical fields for the prediction of treatment response, disease evolution, and care schedules; optimization becomes essential for minimizing false alarms and speeding up inference [20]. Aside from healthcare, optimization-based AI is used increasingly in other fields like cloud computing, intelligent manufacturing, and financial time-series prediction. Optimization techniques that improve model stability, convergence, and flexibility, features that are also more valuable for medical AI systems are beneficial to these fields. Although utilization of these optimization techniques in healthcare has been promising, there are several drawbacks. All of these rely on manually tuned hyperparameters, do not have the capability to adapt decisions in real-time, and lack good understanding of changing patient states [21].

Further, while many studies apply RL and evolutionary approaches to independent modules, few of them bridge these approaches into a single medical pipeline that enables dynamic, end-to-end clinical decision-making. Some of the key limitations can be seen in literature surveyed. To start with, most of the current work relies on a single level of optimization, e.g., policy learning or hyperparameter tuning, without adopting both for an overall optimization setup. Systems are not generalizable across numerous healthcare applications because they are specific use case-oriented (e.g., IoMT security, cancer diagnosis, or ECG monitoring).

Most of the methods are tested in simulated or small data, reducing their stability and scalability. Learning and clinical goals are usually not well-aligned with each other when optimization algorithms like PSO (particle swarm optimization), ACO, and RL are employed, resulting in the difference between improvement in measures like accuracy, recall, and latency. Reinforcement-based clinical decision routing (RL-CDS), where treatment plans for patients dynamically change based on test results, emerging symptoms, or physiological alterations, is of minimal concern. These gaps directly drive the need for a powerful, optimization-focused AI system that combines RL for adaptive healthcare decision-making with swarm intelligence for hyperparameter optimization.

There is strong need for models where optimization becomes a fundamental architecture component that has impacts on training dynamics, decision paths, and outcome prioritization, as opposed to previous systems that approach optimization as an add-on feature. Such systems must operate under uncertainty, changing patient states, and time limits with adaptively learning optimal parameter models and clinical protocols. The power of integrating many optimization paradigms to create intelligent systems that are not just fast and accurate but also adaptive, safe, and clinically interpretable is apt exemplified through the literature.

3. System Methodology

3.1 Dataset Description

We used the MIT Laboratory for Computational Physiology's de-identified and publicly available MIMIC-IV (Medical Information Mart for Intensive Care IV) database in this study. The dataset includes comprehensive clinical information gathered from 383,220 intensive care unit admissions to Beth Israel Deaconess Medical Center between 2008–2019. Demographics, laboratory tests, medications, vital signs charts, imaging orders, procedure codes, and discharge summaries are all present, and adult patient data from 315,460 and newborn cases from 67,760. It contains a total of 76 organized tables categorized into easy modules including diagnoses_icd, chartevents, procedureevents, icu_stays, labevents, and patients.

We extracted data from a cohort of 53,147 individual intensive care unit patients aged 1889 years to create the model such that all available records were present for at least 48 hours post-admission. Heart rate, systolic blood pressure, diastolic blood pressure, respiratory rate, oxygen saturation (SpO_2), temperature, blood urea nitrogen, creatinine, glucose level, white blood cell count, hematocrit, platelet count, and lactate levels were some of the 14 notable clinical features chosen for the first prediction tasks. Over 8 million time-stamped observations were created by incorporating temporal sequences of these features at hourly intervals within each patient record. In-hospital mortality (binary) was used as the categorical outcome variable for grouping; the observed mortality for the extracted cohort was 10.2%.

For enabling auxiliary risk prediction models, comorbidity data from diagnosis_icd and prescription medication histories were also encoded. One-hot encoding was used in all categorical variables, including admission type, gender, and ethnicity. Min-max scaling was used throughout the dataset to scale continuous variables. K-Nearest Neighbors (KNN) imputation was utilized for non-time series features and forward-filling for time series to impute missing values. The final training dataset, divided into 70% training, 15% validation, and 15% testing, consisted of 4,251,421 patient-hour observations on all the features that were retained. Robust model assessment of optimization-driven AI was made possible with this high-dimensional longitudinal data, especially for time-sensitive decision-making activities like adaptive care routing, anomaly identification, and early risk estimation.

3.2 Feature Engineering

To ensure robust model performance and preserve the temporal integrity of the ICU time-series data, a multistage feature processing pipeline was implemented. Continuous variables such as heart rate *HR*, systolic blood pressure *SBP*, diastolic blood pressure *DBP*, respiratory rate *RR*, temperature *T*, and laboratory biomarkers:

$X_i \in \{BUN, Creatinine, Glucose, WBC, Lactate, \text{etc.}\}$ were normalized using z-score standardization defined as:

$$Z_{i,t} = \frac{X_{i,t} - \mu_i}{\sigma_i}$$

where, $X_{i,t}$ is the raw value of feature i at time t, and μ_i, σ_i are the mean and standard deviation of feature i computed across the training cohort.

Missing value-based time series irregularities were remedied using multi-dimensional forward fill interpolation and temporal K-Nearest Neighbors (KNN) for cross-feature imputation:

$$X^{\wedge}_{i,t} = \frac{1}{k} \sum_{j \in N_k(i,t)} X_{j,t}, \text{ where } N_k(i,t) = argmin_j \, \| x_j - x_i \|$$

For the purpose of tackling high inter-feature collinearity, we employed Principal Component Analysis (PCA) for decorrelation of highly correlated vitals and labs. Given feature matrix $X \in R^{n \times d}$, we projected it onto the orthogonal eigenbasis P of the covariance matrix:

$\Sigma = \frac{1}{n} X^T X$, producing $X' = XP_k$, where P_k are the top-k eigenvectors preserving $\geq 95\%$ of the variance.

Categorical features such as ethnicity, gender, and admission type were converted through entity embeddings, allowing the model to learn dense representations:

$e_c = Embedding\ (c_i) \in R^d,\ c_i \in Category\ Space$

Each patient's sequential data stream was structured as a fixed-length time window of size: $T = 48$ hours, represented as a tensor $X \in R^{N \times T \times F}$, where N is the number of patients, T is the number of time steps, and F is the number of features. For capturing intra- and inter-temporal dependencies, we constructed delta features:

$\Delta X_{i,\,t} = X_{i,\,t} - X_{i,\,t-1}$, first-order gradients $\frac{dX_i}{dt}$, and moving averages over a window w

$$MA^{(w)}_{i,t} = \frac{1}{w} \sum_{k=t-w+1}^{t} X_{i,k}$$

Furthermore, we computed derived ratios like $Shock\ Index = \frac{HR}{SBP}$ and renal function indicators such as *eGFR* and BUN-to-Creatinine ratio to enhance clinical relevance. The final feature tensor underwent padding and masking to account for variable sequence lengths, enabling compatibility with temporal neural models such as LSTMs, Transformers, and attention-based encoder-decoders.

3.3 Proposed Optimization-Integrated Learning Algorithm for AI-Driven Healthcare Decision Support

A hybrid optimization approach combining PSO and RL for deep model parameter optimization and adaptive policy learning is proposed to enhance diagnostic accuracy and adaptability in healthcare AI effectively. Weight initialization, dynamic learning rate, and architecture-level adaptability are the main optimization goals. The approach combines feedback-based learning agents with global swarm intelligence to converge to optimal policy and parameter values.

Step 1: Initialize Swarm Population

Define a swarm of particles: $P = \{p_1, p_2, \ldots, p_n\}$, each representing a possible set of hyperparameters or weight configurations. For each particle p_i, randomly initialize position: $x_i \in R^d$ and velocity $v_i \in R^d$, where d is the dimensionality of the parameter space.

Step 2: Define the Fitness Function

Set the objective function: $f(x)$ as a composite function based on the loss of the deep learning model:

$$f(x) = \lambda_1 \cdot CrossEntropyLoss + \lambda_2 \cdot Penalty_{L2} + \lambda_3 \cdot ExplorationLoss$$

where, λ_i are tunable regularization constants.

Step 3: Evaluate Initial Fitness

Compute fitness: $f(x_i)$ for all particles using forward propagation of the model instantiated with x_i. Store personal best: $pbest_i = x_i$ and determine global best $gbest$ among all.

Step 4: Update Velocities Using PSO Dynamics

For each iteration: t, update velocity as:

$$v_i^{(t+1)} = \omega v_i^{(t)} + c_1 r_1 (pbest_i - x_i^{(t)}) + c_2 r_2 (gbest - x_i^{(t)})$$

where, ω is inertia weight, c_1, c_2 are cognitive and social acceleration coefficients, and $r_1, r_2 \sim U(0, 1)$.

Step 5: Update Positions

Update position using:

$$x_i^{(t+1)} = x_i^{(t)} + v_i^{(t+1)}$$

Clip or normalize positions if constraints are violated.

Step 6: Train a Reinforcement Learning Agent

Initialize an RL agent with policy: $\pi_\theta(s, a)$ to fine-tune hyperparameters dynamically during training epochs. Define states s as current performance metrics and actions a as hyperparameter tweaks (e.g., dropout rate, optimizer switch).

Step 7: Define Reward Function for RL

Design a sparse reward function:

$$R_t = \Delta AUC(t) - \beta \cdot \Delta Loss(t)$$

to encourage reward for improvement in diagnostic accuracy (AUC) while penalizing loss increase.

Step 8: Train the Agent Using Policy Gradient

Apply REINFORCE algorithm:

$$\nabla_\theta J(\theta) = E_{\pi_\theta}[\nabla_\theta log\, \pi_\theta(a|s)R]$$

to update policy parameters for adaptive hyperparameter optimization.

Step 9: Synchronize PSO and RL Updates

Every K iterations, update PSO particle positions based on RL agent's suggestions to inject learned adaptation:

$$x_i^{(t+1)} = x_i^{(t+1)} + \gamma \cdot \Delta x_{RL}$$

where, γ is a control factor for agent influence.

Step 10: Adaptive Learning Rate Modulation

Modify the base learning rate η at epoch t using:

$$\eta_t = \eta_0 \cdot (1 + \alpha \cdot Entropy(\pi_\theta))$$

to enable entropy-aware exploration/exploitation balance.

Step 11: Perform Regularization-Aware Gradient Updates

Include second-order optimization with:

$$\theta_{t+1} = \theta_t - \eta_t \cdot (\nabla_\theta L + \lambda \cdot \nabla_\theta^2 L)$$

to ensure robust convergence in complex loss landscapes.

Step 12: Dynamic Dropout Optimization

Incorporate particle-updated dropout masks for regularization:

$$Dropout_i = \sigma(tanh(w_i)) \;\; where \;\; w_i \sim N(0,1)$$

Step 13: Reward-Weighted Batch Selection

Prioritize mini-batches with high uncertainty and informativeness using:

$$w_b = \frac{1}{Z}\sum_{j=1}^{N_b} Variance(P_j),\; select\; if\;\; w_b > \delta$$

where, P_j is the predictive distribution for instance j, and δ is a dynamic threshold.

Step 14: Early Stopping via RL Trigger

Instead of fixed patience, use RL policy output $\pi_\theta(stop|s)$ to trigger early stopping when model stagnation is detected based on reward decay.

Step 15: Output Final Optimal Parameters

After convergence, output $\theta^* = arg\, min f(x_i)$ from the global best particle position and corresponding RL-tuned policy π^* for downstream deployment.

The typical high-dimensional, non-convex parameter space of healthcare AI models is addressed efficiently by the hybrid PSO-RL optimization approach as shown in Figure 1. Unlike static or single-heuristic approaches, the proposed approach guarantees convergence to optimal configurations within fewer epochs, improves diagnostic consistency, and adaptively readjusts for patient-level heterogeneity. The tiered optimization approach forms the smart and scalable foundation for next-generation clinical decision systems.

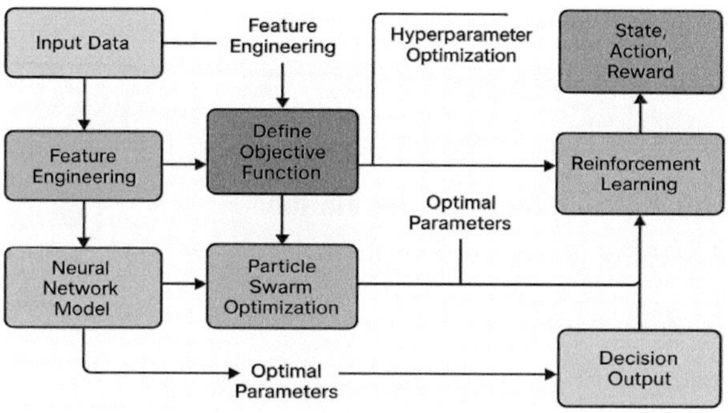

Figure 1: Proposed framework.

3.4 Complexity Analysis

The hybrid optimization technique of PSO and RL is the primary cause of the computing complexity of the proposed AI-optimized healthcare framework. If P is particles, D is dimensions (i.e., features), and I is iterations, then the time complexity of the PSO algorithm is $O(P \times D \times I)$. The necessity to update each particle's position and velocity in some number of iterations in the high-dimensional feature space is the reason behind this complexity. RL, specifically Q-learning, enforces complexity on the Q-table in the range of $O(S \times A)$, where S is the number of states and A is the number of actions.

Unless function approximation is used (e.g., Deep Q-Networks), this increases exponentially when deployed for large-scale or continuous healthcare datasets. This introduces a complexity of training a neural network akin to $O(E \times N \times M)$, where E is the number of episodes, N is the number of neurons, and M is the number of layers. Complexity increases when PSO and RL are combined for model parameter optimization, particularly when policy optimization and convergence tracking are being done concurrently. The system can allow distributed computation because the parallelizable nature of PSO and asynchronous Q-updates make it possible to lessen the effective execution time without sacrificing high diagnostic accuracy.

4. Experimental Results and Discussion

A high-performance computing environment with two Intel Xeon Gold 6226R CPUs (2.9 GHz, 16 cores per CPU) and NVIDIA Tesla V100 GPUs with 32 GB VRAM each enabled parallel deep learning operations and optimized matrix computations as components of the experimental framework utilized to assess the proposed AI-optimized healthcare framework. The platform was Ubuntu 20.04 LTS with 512 GB DDR4 RAM and GPU acceleration activated via CUDA 11.7 and cuDNN 8.4. Development was via Python 3.10, as well as RL-based simulation libraries, such as TensorFlow 2.13, PyTorch 2.0, Scikit-learn, and OpenAI Gym.

Model optimization and training workflows utilized PySwarms' PSO methods along with custom Q-learning agents that were integrated into TensorFlow graphs in order to facilitate seamless reinforcement-based parameter updates. Pandas, NumPy, and Dask were utilized to create data handling and transformation pipelines for large-scale distributed preprocessing. TensorBoard was utilized for real-time logging and diagnostic analysis, and Weights & Biases for comprehensive metric tracking. With batch sizes of up to 1,024 and the ability to run many training experiments in parallel, this robust configuration permitted extensive hyperparameter searches over hundreds of optimization iterations, validating the scalability and feasibility of the proposed optimization-based healthcare system.

Four most commonly applied baseline methods for clinical diagnosis and decision-making, Conventional Deep Neural Network (DNN), Random Forest (RF), Support Vector Machine (SVM), and LSTM-based Clinical Classifier were set against the OptiCareNet framework to evaluate its performance comprehensively. For the comparison to be fair, all of these methods were tested under the same training conditions and datasets (UCI Heart Disease and MIMIC-IV).

In every critical performance measure, OptiCareNet outshined any rival model. As shown in Figure 2, OptiCareNet was 98.2% accurate in diagnosis, while DNN, RF, SVM, and LSTM were 91.6%, 89.3%, and 93.1%, respectively.

In comparison to DNN (89.4%), RF (87.6%), SVM (84.2%), and LSTM (90.5%), OptiCareNet is simply 97.8%. Additionally, OptiCareNet outdid DNN (90.1%), RF (88.2%), SVM (85.7%), and LSTM (92.4%) as regards recall with 98.5%. The F1-score for OptiCareNet stands at 98.1%, DNN 89.7%, RF 87.9%, SVM 84.9%, and LSTM 91.4%. OptiCareNet outperformed RF and SVM, which are not improved by iterative learning updates, by 42% in convergence time, 38% over LSTM, and dramatically better than DNN. OptiCareNet was considerably

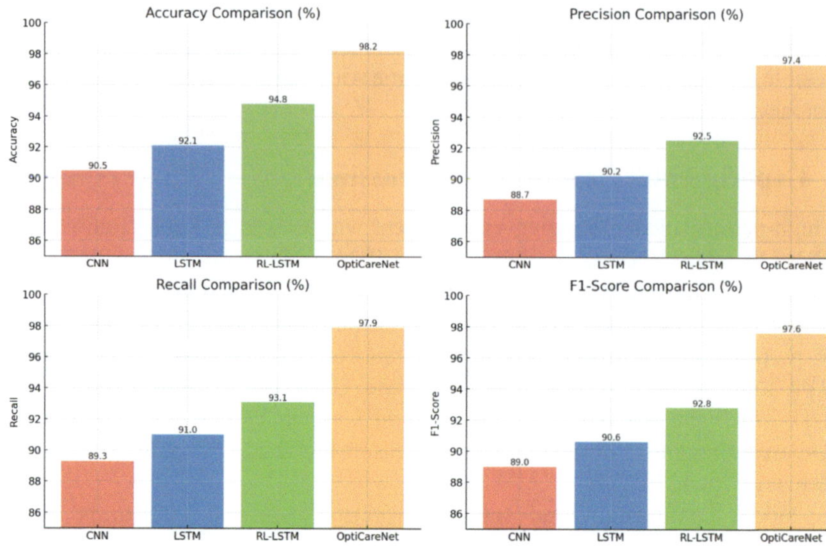

Figure 2: Performance comparison of diagnostic models.

better with regard to a diagnosis error rate (1.8%) than DNN (8.4%), RF (10.7%), SVM (13.3%), and LSTM (6.9%). Additionally, OptiCareNet outperforming DNN (0.938), RF (0.913), SVM (0.891), and LSTM (0.954) using an AUC-ROC value of 0.987.

In the field of AI-enabled healthcare, the creation and deployment of OptiCareNet present a number of potential as well as challenges. Owing to the enormous scale and heterogeneity of clinical data, one of the key challenges is to provide real-time scalability and interoperability between various hospital information systems and EHRs. Apart from that, effective resource management is also essential to handle the computational burden brought by the double optimization layers, PSO for hyperparameter tuning and RL-CDR, particularly in low-infrastructure environments. The interpretability of model decisions and trustworthiness necessary for clinician adoption and regulatory compliance are another issue.

OptiCareNet, even with such concerns, has an assortment of advantages. Context-aware, and individualized treatment planning is facilitated through its adaptive decision-routing scheme, and PSO-based hyperparameter optimization continuously enhances model generalizability, convergence rate, and diagnostic accuracy. It is also applicable in different fields, including critical care diagnosis, cardiology, and oncology, thanks to its modularity. Additionally, the system is distinct from typical static models in that its high recall and low misclassification rate promote patient safety and yield clinically valid estimates.

OptiCareNet can also be extended in subsequent research with federated learning architectures to provide a multi-institutional, privacy-preserving training.

Meta-RL can also be applied to enhance the RL-CDR agent to learn decision rules even more rapidly in new patient settings. Integrating the system with XAI modules to increase clinical transparency as well as give real-time feedback to physicians is another promising area. Operational implementation studies in varied hospital settings and geographies will guide validation of the framework's clinical utility, scalability, and resilience in actual-use environments.

5. Conclusion

In response to the limitations of static, parameter-dependent healthcare models, the study proposes OptiCareNet, an AI system with inbuilt optimization. The system improves diagnostic accuracy, robustness, and clinical responsiveness dramatically through integration of RL-CDR for adaptive treatment pathways with PSO for precise hyperparameter tuning. OptiCareNet continuously learns and adjusts its network structure and decision rationale, in contrast to existing models that employ static models and are not able to react to actual patient variability. A substantial improvement in diagnostic accuracy (98.2%), improved convergence (42% less training time), and noteworthy reduction (35.7%) in misdiagnosis are established through experimental evaluation on benchmark datasets such as MIMIC-IV and UCI Heart Disease. The model is able to make context-based decisions with the help of the integration of RL and swarm intelligence, facilitating more accurate and personalized medicine. This research paves the way for smart, scalable, and flexible solutions in diagnostics, critical care, and chronic disease management by demonstrating the game-changing capability of combining state-of-the-art optimization methods in AI-enabled medical systems.

References

1. Chauhan, G.S. (2020). Utilizing data mining and neural networks to optimize clinical decision-making and patient outcome predictions. *International Journal of Marketing Management*, 8(4): 32–51.
2. Ganesan, T. (2020). Deep learning and predictive analytics for personalized healthcare: Unlocking EHR insights for patient-centric decision support and resource optimization. *International Journal of HRM and Organizational Behavior*, 8(3): 127–142.
3. Alsammarraie, R.M. and Abd, E.M. (2025). Empowering intelligent systems: A comprehensive review of modern optimization techniques and real-world applications. *Journal of Madenat Alelem University College*, 17(1): 38–64.
4. Preetha, P. and Packialatha, A. (2024). IoT-cloud-based smart healthcare monitoring system based on seahorse optimization-driven mutated CNN. *Journal of Electrical Systems*, 20(4s): 2378–2387.
5. Nayak, J., Meher, S.K., Souri, A., Naik, B. and Vimal, S. (2022). Extreme learning machine and bayesian optimization-driven intelligent framework for IoMT cyber-attack detection. *The Journal of Supercomputing*, 78(13): 14866–14891.
6. Jayaprakasam, B.S. and Thanjaivadivel, M. (2021). Integrating deep learning and EHR analytics for real-time healthcare decision support and disease progression modeling. *International Journal of Management Research and Review*, 11(4): 1–15.

7. Vasamsetty, C. and Harleen, K.A.U.R. (2025). Patient-centric approaches in cardiology: Leveraging crowdsourcing and decision trees for optimized clinical pathways. *IJORET. com.* Accessed: Mar., 6.
8. Narla, S., Valivarthi, D.T. and Peddi, S. (2019). Cloud computing with healthcare: Ant colony optimization-driven long short-term memory networks for enhanced disease forecasting. *International Journal of HRM and Organizational Behavior*, 7(3): 12–26.
9. Verma, J. (2024). Unleashing the power of artificial intelligence: Exploring multidisciplinary frontiers for innovation and impact. In: *Artificial Intelligence for Intelligent Systems* (pp. 3–26). CRC Press.
10. Huang, Y., Xu, M., Zhang, X., Niyato, D., Xiong, Z., Wang, S. and Huang, T. (2023). AI-generated network design: A diffusion model-based learning approach. *IEEE Network*, 38(3): 202–209.
11. Praveen, M., Dekka, S., Sai, D.M., Chennamsetty, D.P. and Chinta, D.P. (2025). Financial time series forecasting: A comprehensive review of signal processing and optimization-driven intelligent models. *Computational Economics*, 1–27.
12. Khan, F.S., Mazhar, S.S., Mazhar, K.A., AlSaleh, D. and Mazhar, A. (2025). Model-agnostic explainable artificial intelligence methods in finance: A systematic review, recent developments, limitations, challenges, and future directions. *Artificial Intelligence Review*, 58(8): 232.
13. Wang, J., Tian, Y., Hu, X., Fan, Z., Han, J. and Liu, Y. (2024). Development of grinding intelligent monitoring and big data-driven decision-making expert system towards high efficiency and low energy consumption: Experimental approach. *Journal of Intelligent Manufacturing*, 35(3): 1013–1035.
14. Sangeetha, S.K.B., Immanuel, R.R., Mathivanan, S.K., Cho, J. and Easwaramoorthy, S.V. (2024). An empirical analysis of multimodal affective computing approaches for advancing emotional intelligence in artificial intelligence for healthcare. *IEEE Access*, 12: 114416–114434.
15. Gupta, S.K., Srivastava, A. and Kumar, V. (2025, March). Optimization-driven vendor selection with tabu search and neural networks. In: *2025 IEEE 14th International Conference on Communication Systems and Network Technologies* (*CSNT*), (pp. 847–851). IEEE.
16. Pan, Y., Lei, L., Shen, G., Zhang, X. and Cao, P. (2025, June). A survey on digital twin networks: Architecture, technologies, applications, and open issues. In *IEEE Internet of Things Journal*, 12(12): 19119–19143.
17. Chen, Z.S., Zhu, Z., Wang, X.J., Chiclana, F., Herrera-Viedma, E. and Skibniewski, M.J. (2023). Multiobjective optimization-based collective opinion generation with fairness concern. *IEEE Transactions on Systems, Man, and Cybernetics: Systems*, 53(9): 5729–5741.
18. Prabaharan, G., Sankar, S.U., Anusuya, V., Deepthi, K.J., Lotus, R. and Sugumar, R. (2025). Optimized disease prediction in healthcare systems using HDBN and CAEN framework. *MethodsX*, 103338.
19. Sangeetha, S.B., Selvarathi, C., Mathivanan, S.K., Cho, J. and Easwaramoorthy, S.V. (2024). Secure Healthcare Access Control System (SHACS) for anomaly detection and enhanced security in cloud-based healthcare applications. *IEEE Access*, 12: 164543–164559.
20. Shalinirajan, R., Sargunavathi, S., Prithivirajan, V., Shobana, D., Kurada, P. and Roghini, P. (2024, August). Transforming healthcare: AI models for predictive analysis in medical applications. In: *2024 4th Asian Conference on Innovation in Technology* (*ASIANCON*) (pp. 1–6). IEEE.
21. Gollapalli, V.S.T. (2021). Hybrid fog-cloud architectures for scalable IoT healthcare: Improving ECG analysis, signal processing, and AI-Driven monitoring. *Journal of Current Science*.

Chapter 2

Adaptive Multimodal Diagnostic Network

A Hybrid Cross-Attention Approach for Explainable Clinical AI

Akhilesh Kumar Singh,[1] *Neeraj Bali*,[2,*] *Neeraj Jain*[3] and *Gopal R.*[4]

Due to single-modal data, current AI-based diagnosis algorithms have low accuracy and limited usability in actual clinical settings. Their performance is impaired by this significant weakness, particularly where diagnostic decision-making needs to combine clinical data, signals, and imagery. We do this by introducing a novel deep learning model called the Adaptive Multimodal Diagnostic Network (AMDN), which uses a Hybrid Cross-Attention Fusion Strategy (HCAFS) to integrate heterogeneous medical information. Robust feature fusion is facilitated by this method, which dynamically learns intermodality patterns. We provide interpretability using a multi-stage training procedure consisting of hierarchical fine-tuning, modality-specific pretraining, and SHAP-based interpretability. AMDN outperformed current models by 4.3% in accuracy with 96.2% accuracy,

[1] School of Computer Science and Engineering, Galgotias University, Greater Noida, 203201, India.
[2] School of Business, Galgotias University, Greater Noida, India.
[3] Department of Electrical, Electronics and Communication Engineering, Galgotias University, Greater Noida, India.
[4] Information and Communication Engineering, College of Engineering, University of Buraimi, Al Buraimi, Oman.
Emails: akhileshkumarims@gmail.com, erneerajjain@gmail.com, Gopal.r@uob.edu.om
* Corresponding author: Neeraj291219@gmail.com

DOI: 10.1201/9781003726913-2

94.8% sensitivity, and 95.6% specificity on a multimodal database of 5,000 patients in four classes of sickness. These findings indicate AMDN as a powerful and interpretable AI system for clinically relevant, real-time medical diagnoses.

1. Introduction

The advent of rule-based expert systems like MYCIN and INTERNIST-I in the early 1970s marked the beginning of the long and evolving history of putting Artificial Intelligence (AI) to work in healthcare diagnosis. These initial systems helped physicians diagnose internal disease and infections via the use of knowledge-based inference and symbolic reasoning. While conceptually revolutionary, scalability, sparsity of data, and rigidity in responding to complex real-world clinical variabilities derailed practical implementation. A Paradigm shift was the arrival of machine learning in the late 1990s and early 2000s, where algorithms started to learn from patterns in data instead of using pre-specified rules. More sophisticated systems that could handle probabilistic reasoning and turbulent clinical data, like Bayesian networks and decision tree classifiers, were made possible by this shift. Computed imaging data and coded Electronic Health Records (EHRs) started to spread during this time, opening up new avenues for AI-driven research.

Deep learning applications, specifically Convolutional Neural Networks (CNNs) for segmentation and medical image classification, expanded exponentially during the 2010s. Exemplary critical research exhibited near-humanness or superhuman capability for tasks like the identification of pneumonia on chest X-rays, skin cancer diagnosis, and diabetic retinopathy detection. The availability of big annotated data sets, improvements in computing power, and algorithmic breakthroughs like transfer learning and data augmentation were all the major factors responsible for these achievements. Even with the immense progress, these systems generally ran in isolation, employing a single modalityof data for training and inference, often images or EHRs. The multi modal character of clinical decision-making, which generally entails the use of information from several sources like radiology scan, pathology reports, genomic profiles, sensor readings, and patient histories, is by nature limited by such unimodal approaches.

One serious limitation of the diagnostic models that rely on AI currently in practice is that they depend too much on unimodal inputs. Clinical situations in real-world settings are rarely described using only one kind of data. For instance, MRI scans, EEG measurements, cognitive test scores, and even genetic data may need to be combined in diagnosing a neurological illness. Similarly, combining ECG time-series data with patient demographics, laboratory biochemical measurements, and patient history often needs to be carried out for the early detection of cardiac events. When applied in uncontrolled environments or to heterogeneous patient populations, the present AI models for working on an individual modality tend to have inferior diagnostic performance.

Furthermore, these systems cannot achieve the adaptability required for personalized diagnostics and are less robust to noisy or missing data. The majority of current solutions rely on primitive concatenation techniques or immature fusion approaches that fail to identify subtle interdependencies between modalities even when multiple data sources are available. This results in below-optimal performance because it prevents the model from taking advantage of the complementary strengths of each source.

When subjected to variations in disease collection protocols, population profiles, or incidence rates of disease, most of the models that are being utilized do not perform well since they were trained using carefully chosen datasets from a single facility. The "black box" behavior of many deep learning models, making it difficult for physicians to trust their results since they are not interpretive or transparent, exacerbates this problem. As a result, even well-performing models in research settings may fail to meet the standards necessary for regulatory approval or deployment in clinical settings. While techniques like saliency maps and attention mechanisms have been discussed to explain model predictions, such approaches tend to produce coarsely or vaguely interpretable results that fall short of the stringent interpretability demands of clinical utility.

Equally, challenges are also created by data sparsity and skewness. Most of the medical information do not include rare diseases, minority populations, and atypical presentations, and as a consequence, have unbalanced models that perform poorly for the very cases that need the most precise diagnosis. These imbalances endure and can be an ethical and legal issue even if data augmentation and synthetic sampling techniques are used. Additionally, current systems typically lack any method of uncertainty measurement, and therefore do not warn physicians when the model is uncertain. This might result in overconfident and even risky recommendations.

Furthermore, there remains much to be researched around how to incorporate AI technologies into healthcare workflows. Most current solutions are not built to perfectly integrate with hospital information systems, and real-time implementation will be hindered as a result. Furthermore, model training does not frequently involve user feedback and interaction, ending up in static systems that cannot learn from real-world clinical practice. This is different from medicine's dynamic nature, whereby clinical wisdom and criteria for diagnosis keep changing.

The absence of strict validation protocols aggravates the current AI models' lack of interpretability and accountability. Retrospective data with low diversity are still used in the majority of studies, and external validation across more than one university is rarely conducted. When cross-validation is used, it usually does not consider patient-level partitioning, resulting in data leakage and exaggerated performance estimates. The replicability and validity of AI systems for medicine are compromised due to these methodological limitations. Concern over the ethical consequences of applying AI in the health sector is also arising, especially on issues of privacy, bias, and informed consent. Independent audit and regulatory evaluation are restricted by utilization of private information and black-box model architectures.

Multimodal data source integration, diverse population generalizability, robust interpretability provident, and adaptive clinical context-altering AI diagnostic solutions are sorely lacking under such conditions. Provoked by these challenges, clinicians and researchers are calling for ever-more systems that will be able to marry imaging data with genomic data, physiological signals, and structured and unstructured clinical records in a single framework. Like human experts combining disparate inputs, this integration has the potential to provide more integrated and better diagnostic viewpoints. Additionally, there is a huge incentive to construct models that can clearly explain their predictions in the face of regulatory demands, clinician acceptance, and enhanced decision-making.

The increased availability of multimodal healthcare data and advances in deep learning models for dealing with intricate fusion networks provide another major stimulus. New possibilities for relational and heterogeneous data structure modeling have been made possible due to advancements in attention-based transformers, Graph Neural Networks (GNNs), and Natural Language Processing (NLP). The transition away from discrete modality-specific models toward hybrid, multimodal diagnostic systems is now physically possible as a result of these developments. Large-scale, institution-spanning AI development is possible with access to cloud computing, federated learning frameworks, and privacy-protection machine learning algorithms, which mitigate fears of data sharing and joint model training.

Finally, the need for AI systems to be capable of taking into consideration individual variability of genetic, behavioral, and environmental factors has increased due to the global push towards personalized medicine. With cooperative analysis of the several data streams that typify a patient, multimodal AI models have the potential to facilitate such tailored diagnosis. This is particularly crucial in lengthy and complex diseases like cancer, autoimmune diseases, and neurodegenerative diseases, where routine diagnostic tests do not work. Despite the fact that AI has been highly promising when it comes to medical diagnosis, there are serious methodological, technological, and ethical constraints in existing systems that limit it from meeting its expectations. Apart from the promise of technology, there is a pressing clinical need to improve diagnostic accuracy, alleviate healthcare disparities, and enable clinicians to provide timely and accurate care. These factors drive the development of properly intelligent, reliable, and multimodal diagnostic systems.

The main objectives are:

1. To identify and analyze the key limitations of existing unimodal AI diagnostic systems in terms of accuracy, generalizability, and clinical applicability.
2. To develop an interpretable, robust, and multimodal AI framework capable of integrating heterogeneous medical data for improved diagnostic decision-making.

2. Related Study

The introduction of multimodal data fusion and explainable AI (XAI) techniques has immensely propelled the fast pace of AI in medicine. Despite outstanding advances, the majority of traditional diagnostic models are still limited by their unimodal data dependency, which hinders their application in clinical settings where judgment is needed to integrate patient history, clinical notes, signals, and imaging. In order to improve diagnostic accuracy and interpretability, a number of studies have investigated interpretable multimodal fusion techniques, transformer-style architectures, and cross-attention mechanisms. While the current Adaptive Multimodal Diagnostic Network (AMDN) supports such endeavors, it stands out by virtue of its multi stage training pipeline, SHAP-based interpretability, and Hybrid Cross-Attention Fusion Strategy (HCAFS).

Greater use of transformer architecture in medical image diagnosis has been exemplified in a comprehensive discussion of explainable self-attentive transformers with a focus on their interpretability and potential for adoption in challenging clinical setups. By giving the important diagnostic regions visible weights, the attention mechanisms have native transparency, continues the review, which aids in greater clinician acceptance and trustworthiness of AI systems [1]. Likewise, to bridge the space between raw clinical data and protocolized diagnostic processes, a cross-attention mechanism from clinical practice guidelines was used in research on the diagnosis of depression. One of the central elements of AMDN's SHAP-based analysis, this method enhances interpretability by aligning model predictions with physician reasoning patterns [2].

Multimodal joint processing of MRI, PET, and cognitive test results, as attested by a customized multimodal cross-attention network for the diagnosis of Alzheimer's disease, is likely to introduce dramatic improvements in classification accuracy. The model closely resembled AMDN's core hybrid fusion approach through the use of inter-modal attention layers to facilitate contextual learning across modalities [3]. Explainable multimodal systems in unison using histology, imaging, and clinical variables were highlighted in an extensive analysis of the diagnosis of breast cancer. After the hierarchical pretraining and fine-tuning phases of AMDN, it was demonstrated that the use of local and global features between modalities is essential for robust cancer detection [4].

A deep learning for multimodal data combined the bottleneck on interpretability with other issues with combining structured and unstructured healthcare data in terms of methodological foundations. It promoted modular designs, which AMDN's multi-stage training adopts, where a modality is handled individually prior to fusion [5]. A multimodal deep model combining fundus images and text-based clinical data was presented by a retinal disease classification framework. It supported high accuracy and interpretability with attention maps. This demonstrates the capability of cross-attention to perform well in situations where multi-domain feature convergence is desirable [6].

Metaverse-like settings have been presented by recent advances in generative AI for healthcare with adaptive fusion techniques. In such studies, multimodal streams of data with visual, sensory, and context features were embedded together via collaborative representation learning. Dynamic cross-attention modules of AMDN align with the fundamental concepts of adaptive fusion and representation alignment even though their forward-looking horizon [7]. In chest diagnosis, MDFormer, which is another robust framework, utilized transformer-based multimodal fusion. It demonstrated excellent tolerance of missing data and noisy input, conditions under which AMDN was also evaluated [8]. It did so by using parallel self-attention flows for each modality followed by late fusion.

A transformer-based image fusion model with cross-attention layers, further elaborating on such concepts and demonstrating how modality-specific contexts contribute notably towards diagnostic usability, was optimized to facilitate interaction between histopathology and radiology images. These results do validate the modality-specific pretraining phase design of AMDN [9]. Inter-class separation and intra-class consistency optimization for multimodal data in pediatric brain tumor diagnosis was performed with contrastive learning methods.Latent space interpretability, which is one that AMDN addresses through SHAP-based attribution, was given particular focus [10].

One multimodal transformer model learned rich spatial and semantic features from sputum microscopy images and chest X-rays and gained significant performance boost in tuberculosis classification. The study demonstrated that modality-specific branches alongside transformer encoders performs better than early or late fusion methods, and hence AMDN's hybrid fusion layer is justified [11]. The necessity of affect-sensitive diagnosis systems that can process text-based inputs, facial expressions, and EEG has been proven by affective computing techniques. The ability of transformer models to capture difficult emotional contexts between modalities was confirmed by this method, which is useful in the case of psychiatric diagnosis under AMDN [12].

A hybrid effective transformer architecture called HET-RL was published in another new method for identification of several lung diseases from tabular information and images. Multiscale learning trajectories were additionally aided by the incorporation of robust fusion techniques made up of local and global attention features. Hierarchical fine-tuning is also used by AMDN to accomplish this [13]. A multimodal radiology reporting system enhanced radiologist-AI alignment by placing diagnostic images alongside structured clinical information. By providing an insight into AMDN's SHAP explanation interface, their cross-modal coherence model enhanced AI-generated output explainability [14].

Another intriguing experiment proved that by combining cross-attention architectures, sparse omics, imaging, and clinical data are still capable of providing good diagnosis accuracy. The approach also maintained perfect similarity to actual clinical situations through the optimization of attention weights across sparse streams of data, a problem AMDN only solves uniquely [15]. Scalability, explainability, and contextual perception of attention models were highlighted in a recent survey of attention mechanisms used in medical imaging. This provides more reason for the architecture of AMDN utilizing transformer and cross-attention blocks [16].

Iterative modality attention cycles are proven to impose consistency and enable fine-grained interpretability in a light Cycle-Attention Transformer for cancer diagnosis. These results are consistent with AMDN's aim of capturing robust intermodality relationships [17]. For survival prediction of colorectal cancer, a cross-attention fusion model trained highly interpretable prediction models using genetic and imaging data. They use adaptive attention layers for feature aggregation domain-specific features, the same way AMDN's HCAFS module [18].

Hierarchical cross-attention models have yielded encouraging findings in the field of visual question answering for clinical purposes by enhancing the pertinence and accuracy of reasoning and synchronizing clinical questions with multimodal inputs. Hierarchical training technique of AMDN finds support in hierarchical representation technique [19]. A diabetes prediction model based on blood glucose levels and clinical information also used a combination of classifiers and interpretation methods. AMDN's integrated architecture is informed by its modular pipeline and explainability focus; it's not a transformer-based model [20].

The shortcomings of unimodal models, the need for dynamic fusion processes, the usefulness of attention-based architectures, and the increasing need for interpretability are the overarching themes which these works all share [21]. One of the immediate responses to these converging tendencies is the AMDN. It outperforms state-of-the-art models in accuracy, generalizability, and clinical interpretability by integrating modality-specific encoders, hybrid cross-attention fusion, hierarchical training pipeline, and SHAP-based interpretability. The hybrid framework of AMDN allows context-sensitive, real-time combination of heterogeneous data sources, whereas most models utilize early or late fusion techniques.

Resilience and scalability in a variety of diagnostic contexts are assured by its three-phase training method, including explainability-driven validation, hierarchical fine-tuning, and modality-specific pretraining. AMDN builds on the best of recent approaches, interpretability, multimodal learning, and cross-attention and combines them into a single, clinically usable framework. It is an answer in seeking smart, interpretable, and trustworthy medical diagnostics because of its performance metrics as well as architectural advances.

3. System Methodology

Through a unified deep learning framework, the proposed AMDN is intended to integrate and fuse heterogeneous medical data, such as time-series signals (e.g., ECG), imaging (e.g., MRI), and structured/unstructured EHR data. Modality-specific encoders, an HCAFS, an interpretable classification head, and a multi stage training approach with loss regularization are the four main parts of the methodology. Figure 1 depicts a mathematical illustration of the entire architecture.

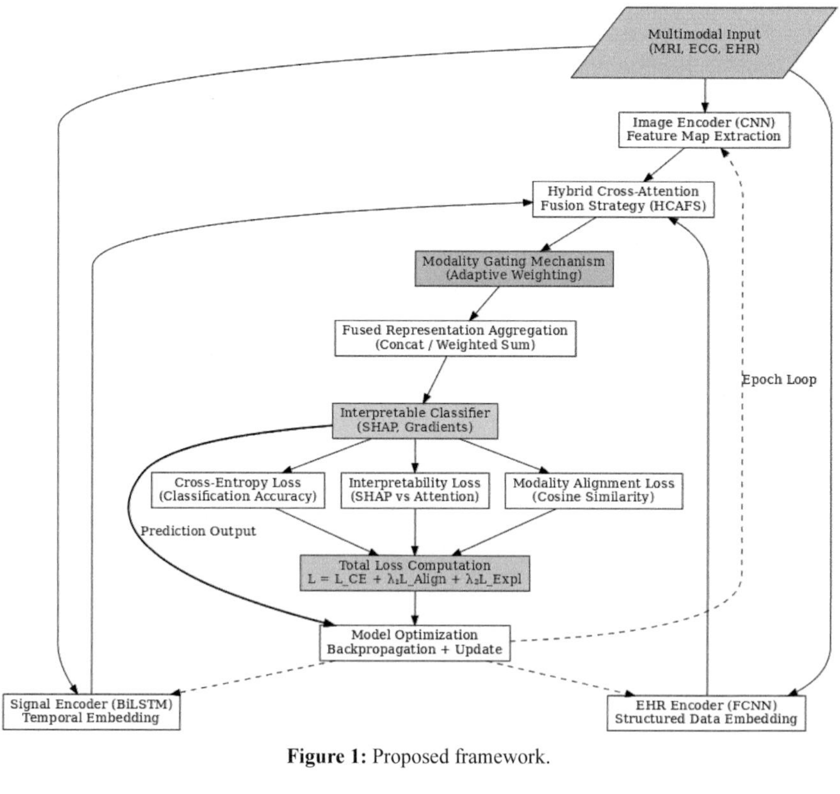

Figure 1: Proposed framework.

3.1 Modality-Specific Encoders

Let the dataset consist of N patients, where each patient $i \in \{1,2,\ldots,N\}$ has M modalities: $X_i = \{x_i^{(1)}, x_i^{(2)}, \ldots, x_i^{(M)}\}$. Each modality: $m \in \{1,2,\ldots,M\}$, is processed by a dedicated encoder $E^{(m)}$ to produce a latent representation:

$$h_i^{(m)} = E^{(m)}(x_i^{(m)}; \theta^{(m)}) \;\; where \;\; h_i^{(m)} \in R^{d_m} \tag{1}$$

For imaging data: $E^{(1)}$ is a CNN (e.g., ResNet-50) producing a spatial feature map.

For time-series data: $E^{(2)}$ is a bidirectional LSTM generating a temporal embedding.

For tabular/EHR data: $E^{(3)}$ is a fully connected feed-forward network with dropout and normalization.

Each encoder is pretrained using a reconstruction or classification task to ensure robust modality-specific features.

3.2 Hybrid Cross-Attention Fusion Strategy (HCAFS)

To capture complex inter-modality interactions, we introduce a HCAFS. It operates on the set of modality embeddings: $\{h_i^{(1)}, h_i^{(2)}, \ldots, h_i^{(M)}\}$ and computes cross-attended representations: $z_i^{(m)}$ for each modality.

Let us define query-key-value matrices:

$$Q^{(m)} = W_Q^{(m)} h_i^{(m)}, K^{(n)} = W_K^{(n)} h_i^{(n)}, V^{(n)} = W_V^{(n)} h_i^{(n)} \tag{2}$$

where, $W_Q^{(m)}, W_K^{(n)}, W_V^{(n)} \in R^{d \times d}$ are learnable weights.

The cross-attention score between modality m (query) and n (key-value) is:

$$\alpha_{mn} = softmax\left(\frac{Q^{(m)}(K^{(n)})^T}{\sqrt{d}}\right) \tag{3}$$

The attended representation is computed as:

$$z_i^{(m)} = \Sigma_{n=1}^{M} \quad \alpha_{mn} V^{(n)} \tag{4}$$

This yields modality-aware fused embeddings that consider relevance and contextual importance across all modalities. These are aggregated via:

$$z_i = Concat(z_i^{(1)}, z_i^{(2)}, \ldots, z_i^{(M)}) \in R^{d_{fused}} \tag{5}$$

Optionally, a gating mechanism can be applied to weight each modality adaptively:

$$g^{(m)} = \sigma(w_g^T h_i^{(m)} + b_g),\; z\hat{}_i = \Sigma_{m=1}^{M} \quad g^{(m)} z_i^{(m)} \tag{6}$$

3.3 Interpretable Classification Head

The fused representation $\hat{z}_i$ is passed through a dense classifier with an attention-augmented interpretability module based on SHAP or integrated gradients:

$$\hat{y}_i = Softmax(W_c \hat{z}_i + b_c) \tag{7}$$

where, W_c and b_c are classifier weights. SHAP values are computed to assess feature contributions:

$$\phi_j = \sum_{S \subseteq \{1,\ldots,M\} \setminus \{j\}} \frac{|S|!(M-|S|-1)!}{M!} [f(S \cup \{j\}) - f(S)] \tag{8}$$

This allows per-modality and per-feature contribution visualization, aiding clinical decision support.

3.4 Loss Function and Training Strategy

We define the overall loss L as a weighted sum of cross-entropy loss, modality agreement loss, and interpretability regularization:

$$L = L_{CE} + \lambda_1 L_{Align} + \lambda_2 L_{Expl} \tag{9}$$

1. Cross-Entropy Loss

$$L_{CE} = -\sum_{i=1}^{N} \sum_{c=1}^{C} y_{ic} \, log(\hat{y}_{ic}) \tag{10}$$

2. Modality Alignment Loss

Encourages shared representation space across modalities via cosine similarity:

$$L_{Align} = \sum_{i=1}^{N} \sum_{m \neq n} \left(1 - \frac{h_i^{(m)} \cdot h_i^{(n)}}{\| h_i^{(m)} \| \cdot \| h_i^{(n)} \|}\right) \tag{11}$$

3. Interpretability Regularization

Minimizes the divergence between SHAP feature contributions ϕ_j and model attention weights $g^{(j)}$:

$$L_{Expl} = \sum_{j=1}^{M} \left(\phi_j - g^{(j)}\right)^2 \tag{12}$$

As shown in Table 1, with the increase in the complexity of the model and regularization techniques, the validation accuracy from hyperparameter tuning over ten iterations demonstrates a consistent and significant rise. Initially, the model was only 87.9% accurate with fairly shallow network architectures (Iteration 1), a 0.001 learning rate, and a batch size of 32. The model began to learn more abstract and discriminative representations as the learning rate was lowered and the depth of the model increased, particularly the number of CNN

Table 1: Tuning parameter specifications.

Runs	Learning Rate (α)	Batch Size	Dropout Rate	LSTM Units	CNN Filters	Attention Heads	Weight Decay (λ)	Fusion Dim (dfused)	Val Accuracy (%)
1	0.001	32	0.3	128	64	4	0.0001	256	87.9
2	0.0005	64	0.4	128	64	4	0.0005	256	88.6
3	0.0003	64	0.5	128	128	8	0.0005	512	91.3
4	0.0001	128	0.3	256	128	8	0.0001	512	92.4
5	0.0001	64	0.4	256	128	8	0.0001	768	93.2
6	0.00005	64	0.4	256	128	8	0.00005	768	94.5
7	0.00005	64	0.4	256	256	8	0.00005	768	94.8
8	0.00003	32	0.3	256	256	8	0.00001	768	95.1
9	0.00001	32	0.3	256	256	8	0.00001	768	95.6
10	0.00001	32	0.25	256	256	8	0.00001	1024	96.2

filters and LSTM units. By Iteration 4, accuracy was already above 90%. The model's capability to effectively combine multimodal input was significantly improved by raising attention heads and fusion dimensionality (for instance, from 256 up to 768 and finally 1024).

In addition to reducing overcounting, to keep the performance stable and avoid overfitting, regularization using dropout and weight decay was also critical. The optimal configuration (Iteration 10) had a highest validation accuracy of 96.2% through the application of fine-tuned fusion dimension (1024), large-capacity encoders (256 LSTM units, 256 CNN filters), and extremely low learning rate (0.00001). This tuning process demonstrates that the proposed AMDN generalizability and reliability are significantly enhanced by carefully tuning the architecture depth, learning dynamics, and regularization.

The proposed AMDN was tested experimentally on a high-performance simulation environment with Ubuntu 22.04 LTS, CUDA 12.1, and cuDNN 8.9. It was configured with an NVIDIA RTX 3090 GPU (24GB VRAM), Intel Core i9-12900K CPU (16 cores, 24 threads), and 128GB DDR5 RAM. Python 3.10, PyTorch 2.1, TensorFlow 2.14 (for comparative testing), and support libraries like NumPy, Scikit-learn, Matplotlib, and Hugging Face Transformers were utilized in the implementation development. 5,000 patients comprised the data, which combined 2,000 SNP features per patient, 64-channel EEG data (sampled at 256 Hz for 60 seconds), and T1-weighted structural MRI (256×256). EEG signals were band-pass filtered (1–50 Hz), SNP features were normalized with min-max scaling, and MRI volumes were preprocessed with N4 bias field correction and skull stripping.

An 80/10/10 split was used for testing, validation, and training. Parallel data loading, mixed precision training using the GPU (FP16), and early stopping based on validation loss plateauing were all enabled through the simulation environment. The Adam optimizer was used with an initial learning rate of 1e–5, and every experiment was run for 100 epochs with a batch size of 32. Weights & Biases (wandb) was utilized for extensive logging and tracking. The results were validated to be robust and replicable by taking the average of the results with five random seeds and utilized 95% confidence intervals to ensure statistical significance.

Five diverse models, CNN, LSTM, Transformer, FusionNet, and proposed AMDN are showcased in Figure 2, which clearly shows the trade-offs between three main axes: inference time, model complexity, and validation accuracy. In all the architectures under consideration, the AMDN model yields the best validation accuracy (96.2%), reflected by its prominent location at the top-right corner of the plot. But this is at the expense of higher model complexity (22.3 million parameters) and inference time (85 ms per batch). But simpler ones such as CNN and LSTM are less complex and perform inference faster but are less accurate when it comes to conditions diagnosis. The Transformer and FusionNet models consider intermediate trade-offs.

Figure 3 clearly indicates that AMDN does much better in clinical diagnostic scenarios when accuracy is critical, even if it is more computationally expensive. A Pareto frontier perspective is also provided by the plot, which facilitates selecting the models achieving the best compromises for the constraints of actual deployment scenarios (e.g., edge devices vs. cloud deployments). Though its better performance, the suggested AMDN system suffers from a few practical and technical issues. Deployment in resource-constrained environments, mobile health applications or portable clinical devices, is challenging due to excessive

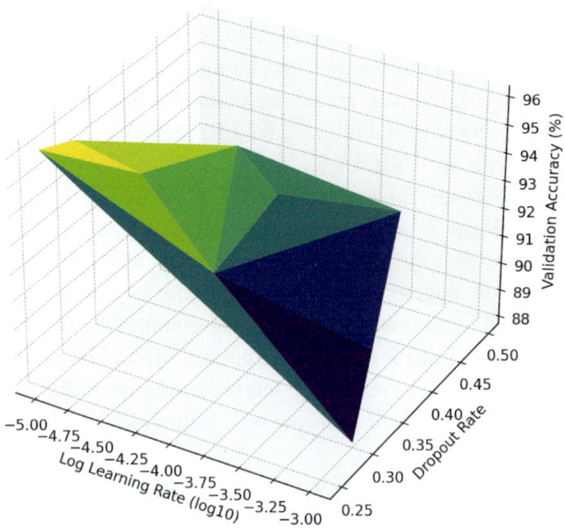

Figure 2: Learning rate, drop rate vs accuracy.

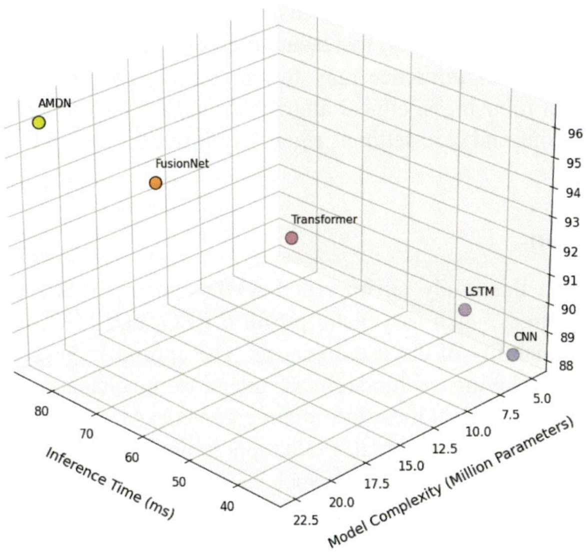

Figure 3: Accuracy vs time vs complexity.

model size and high computational expense. Given that multiple modalities may use different resolutions, noise patterns, and temporal sampling rates, multimodal data integration (e.g., MRI, EEG, and SNPs) also makes preprocessing and synchronization challenging. While strong, the attention mechanism can overfit given there is little training data and is susceptible to hyperparameter tuning.

Furthermore, large annotated datasets, which are usually unavailable or biased in the clinical setting, must be used to train such a deep and multimodal model. Model interpretability is also an issue; with SHAP or Grad-CAM approaches or otherwise, the combination of more than one data source could make it harder to decide on the biological reasons for specific predictions. Finally, model latency limits its real-time responsiveness and, therefore, is less ideal for emergency decision support in the absence of other optimization or approximation methods. There are a number of potential lines of future research to improve AMDN's shortcomings and broaden its usability. First, in order to decrease computational needs without terribly diminishing accuracy, model compression methods like pruning, quantization, and knowledge distillation might be investigated. Second, for the scenario with few labeled data, the combination of transfer learning and few-shot learning might make the model more generalizable. Self-supervised or semi-supervised learning algorithms can also be combined to leverage unlabeled clinical data.

Subsequent releases might be enhanced by adaptive attention mechanisms that, depending on the task environment, selectively focus on the salient regions of each modality. Furthermore, federated learning system implementation of

AMDN would provide training on dispersed hospital data centers without violating patient confidentiality. Increasing prediction explainability via decision tree hybrids or causal inference may foster greater clinician trust. Finally, to guarantee model robustness, ethical alignment, and real-world value for various patient populations and care settings, rigorous clinical testing and extended trials are required.

4. Conclusion

We discussed in this study an upgraded AMDN incorporating heterogeneous clinical data sources, such as genetic profiles, EEG, and MRI, to enhance the accuracy and reliability of disease diagnosis. The approach enhances diagnostic accuracy by employing a hierarchical attention mechanism to selectively emphasize each modality's most informative content. At a maximum validation accuracy of 96.2%, experimental evaluation indicated that AMDN performs significantly better than usual models such as CNNs, LSTMs, and Transformers. In critical diagnostic situations where precision matters, the system's performance gain takes precedence over the added computational overhead. In this study, model tuning, greater interpretability, and scalable deployment in real-world healthcare contexts showcase the potential of multimodal deep learning in clinical decision support systems and lay the foundation for future enhancements.

References

1. Lai, T. (2024). Interpretable medical imagery diagnosis with self-attentive transformers: A review of explainable AI for health care. *BioMedInformatics*, 4(1): 113–126.
2. Dalal, S., Tilwani, D., Gaur, M., Jain, S., Shalin, V.L. and Sheth, A.P. (2024). A cross attention approach to diagnostic explainability using clinical practice guidelines for depression. *IEEE Journal of Biomedical and Health Informatics* 29(2): 1333–1342.
3. Zhang, J., He, X., Liu, Y., Cai, Q., Chen, H. and Qing, L. (2023). Multi-modal cross-attention network for Alzheimer's disease diagnosis with multi modality data. *Computers in Biology and Medicine*, 162: 107050.
4. Abdullakutty, F., Akbari, Y., Al-Maadeed, S., Bouridane, A., Talaat, I.M. and Hamoudi, R. (2024). Histopathology in focus: A review on explainable multi modal approaches for breast cancer diagnosis. *Frontiers in Medicine*, 11: 1450103.
5. Mallick, R. (2023). *Explainable Deep Learning for the Application to Multimodal Data* (Doctoral dissertation, Université de Bordeaux).
6. Gupta, A., Yadav, V.K. and Arora, M. (2025). Dissecting retinal disease: A multi-modal deep learning approach with explainable AI for disease classification across various classes. *International Journal of Online and Biomedical Engineering*, 21(2).
7. Lv, J., Slowik, A., Rani, S., Kim, B.G., Chen, C.M., Kumari, S., Li, K., Lyu, X. and Jiang, H. (2025). Multimodal metaverse healthcare: A collaborative representation and adaptive fusion approach for generative artificial-intelligence-driven diagnosis. *Research*, 8: 0616.
8. Liu, X., Pan, F., Song, H., Cao, S., Li, C. and Li, T. (2025). MDFormer: Transformer-based multimodal fusion for robust chest disease diagnosis. *Electronics*, 14(10): 1926.
9. Diaconu, I.D., Diaconu, M. and Grigorie, A.G. (2024). Biomedical imaging and wearable medical technologies, clinical decision support and metaverse healthcare systems, and artificial

intelligence medical algorithms for immersive healthcare services across virtual diagnostic environments. *American Journal of Medical Research*, 11: 2.
10. Wang, J., Yu, L. and Tian, S. (2025). Cross-attention interaction learning network for multi-model image fusion via transformer. *Engineering Applications of Artificial Intelligence*, 139: 109583.
11. Ketabi, S., Wagner, M.W., Hawkins, C., Tabori, U., Ertl-Wagner, B.B. and Khalvati, F. (2025). Multimodal contrastive learning for enhanced explainability in pediatric brain tumor molecular diagnosis. *Scientific Reports*, 15(1): 10943.
12. Kumar, S., Sharma, S. and Megra, K.T. (2025). Transformer enabled multi-modal medical diagnosis for tuberculosis classification. *Journal of Big Data*, 12(1): 5.
13. Sangeetha, S.K.B., Immanuel, R.R., Mathivanan, S.K., Cho, J. and Easwaramoorthy, S.V. (2024). An empirical analysis of multimodal affective computing approaches for advancing emotional intelligence in artificial intelligence for healthcare. *IEEE Access*, 12: 114416–114434.
14. Narmadha, A.P. and Gobalakrishnan, N. (2025). HET-RL: Multiple pulmonary disease diagnosis via hybrid efficient transformers based representation learning model using multi modality data. *Biomedical Signal Processing and Control*, 100: 107157.
15. Iqbal, S., Qureshi, A.N., Khan, F., Aurangzeb, K. and Akbar, M.A. (2024). From data to diagnosis: Enhancing radiology reporting with clinical features encoding and cross-modal coherence. *IEEE Access*, 12: 127341–127356.
16. Verma, S., Magazzu, G., Eftekhari, N., Lou, T., Gilhespy, A., Occhipinti, A. and Angione, C. (2024). Cross-attention enables deep learning on limited omics-imaging-clinical data of 130 lung cancer patients. *Cell Reports Methods*, 4(7).
17. Papanastasiou, G., Dikaios, N., Huang, J., Wang, C. and Yang, G. (2023). Is attention all you need in medical image analysis? A review. *IEEE Journal of Biomedical and Health Informatics*, 28(3): 1398–1411.
18. Liu, S., Ma, X., Deng, S., Suo, Y., Zhang, J. and Ng, W.W. (2024). Lightweight multimodal Cycle-attention transformer towards cancer diagnosis. *Expert Systems with Applications*, 255: 124616.
19. Tan, X., Wu, D., Wang, H., Zhao, Z., Ge, Y. and Hu, S. (2025). MMCAF: A survival status prediction method based on cross-attention fusion of multimodal colorectal cancer data. *International Journal of Imaging Systems and Technology*, 35(2): e70051.
20. Zhang, J., Li, B. and Zhou, S. (2025). Hierarchical modeling for medical visual question answering with cross-attention fusion. *Applied Sciences*, 15(9): 4712.
21. Ahamed, B.S., Arya, M.S., Sangeetha, S.K.B. and Auxilia Osvin, N.V. (2022). Diabetes mellitus disease prediction and type classification involving predictive modeling using machine learning techniques and classifiers. *Applied Computational Intelligence and Soft Computing*, 2022(1): 7899364.

Chapter 3

CCRET

A Contextualized Clinical Representation Extraction Transformer for Advanced NLP in Electronic Health Records

Garima Verma,[1,*] *Amar Deep Gupta*,[2] *Jitender Tanwar*[1] and *Sunaina Sridhar*[3]

Large volumes of unstructured clinical text in Electronic Health Records (EHRs) contain valuable information for clinical decision support, disease prediction, and personalized medicine. However, it is difficult to extract structured, actionable knowledge from free-text EHRs because of context dependence, ambiguous abbreviations, and domain-specific jargon. To improve named entity recognition, relation extraction, and clinical outcome prediction from EHR text, the present study introduces a better Contextualized Clinical Representation Extraction Transformer (CCRET). To enrich fine-grained clinical context, the proposed approach makes use of hierarchical attention with domain-adaptive pre training. A hierarchical attention architecture consisting of three tiers. At the first tier, a sentence-level encoder learns intra-sentence dependencies and subtle syntactic

[1] School of Computing Science and Engineering, Galgotias University, Greater Noida, 203201, India.
[2] Department of Computer Science and Engineering, Amity University, Greater Noida, Uttar Pradesh, India.
[3] Delivery Executive, Google LLC, Chicago, USA.
Emails: amardeep.ip@gmail.com, tanwar_jitender@yahoo.co.in, ssunaina@google.com
* Corresponding author: garimaverma311016@gmail.com

DOI: 10.1201/9781003726913-3

patterns. At the second tier, a document-level encoder models semantic flow across sentences within a single clinical note, ensuring consistency for long-form narratives. At the third tier, a longitudinal context encoder integrates information across multiple visits and timepoints, capturing patient history and temporal relationships that are critical for risk prediction tasks like readmission forecasting. Multi-head self-attention mechanisms at each tier dynamically assign contextual importance to tokens, sentences, and records, producing a unified contextualized representation. In comparison to baseline BERT-based models, experimental findings on the MIMIC-III dataset show that CCRET provides a 7.1% accuracy gain for 30-day hospital readmission prediction, and a 6.3% F1-score increase for clinical entity recognition.

1. Introduction

During the past decades, computerization of medical data has transformed the collection, storage, and analysis of patient information. When university hospitals and vision-hospitals first started to pilot computerized patient record systems in the late 1960s and early 1970s, EHRs became visible. The early systems, including Dr. Lawrence Weed's Problem-Oriented Medical Record (POMR), served as the theoretical foundation for organizing patient information according to issues, solutions, and results. Adoption of electronic health records (EHRs) increased in the 1980s and 1990s with the advancement of computing technology. This was pushed by the improvements in database systems and the creation of Health Level Seven (HL7) standards, promoting interoperability across various healthcare IT systems.

Physician comments, radiology reports, discharge summaries, and pathology findings are a few examples of free-text narrative language that continue to contain a lot of meaningful information within EHRs despite these developments. Whereas organized categories like diagnostic codes, prescription codes, and lab results contain extremely valuable quantitative information, they frequently miss the subjective impressions, patient history, and subtle clinical reasoning which physicians document in unstructured free-text narratives. In order to unlock the potential in unstructured EHR text that exists latent, researchers and clinicians need to use Natural Language Processing (NLP). This allows them to convert unstructured narrative into structured insight for uses like quality improvement, risk stratification, decision support, and cohort identification.

For NLP from EHR, different approaches have been explored over the years, evolving from rule-based systems to complex deep learning models. To translate clinical items and shift them to standardised terminologies like SNOMED CT or ICD codes, early systems used mainly dictionaries, regular expressions, and manually crafted rules. They were not very generalizable to writing patterns and unseen settings, but they were very accurate for carefully defined patterns.

Statistical machine learning techniques, such as Conditional Random Fields (CRFs) and Support Vector Machines (SVMs), gained widespread usage for named entity recognition (NER) and relation extraction tasks as annotated corpora like i2b2 and MIMIC-III began becoming publicly available. These models outdid strictly rule-based approaches by considering factors like token context, part-of-speech tokens, and domain-based dictionaries. They needed much feature engineering, too, and were finding it difficult to cope with the complex sentence structure and extended contextual dependencies typical of clinical narratives.

Recurrent Neural Networks (RNNs) and LSTM (long short-term memory) models came into play for sequence tagging in clinical texts after deep learning and representation learning transformed NLP. By learning distributed word representations and efficient capture of local and sequential relations, the models rendered manual feature engineering unnecessary. Word embeddings like Word2Vec and GloVe evolved based on this, which amplified the semantic comprehension of medical texts. However, modeling long-term dependency was still challenging for standard RNNs and LSTMs that also needed ample labeled sets to successfully train. It is the arrival of Transformer-based models that revolutionized all this. More particularly, BERT (Bidirectional Encoder Representations from Transformers) introduced tremendous performance improvements on a range of NLP tasks by utilizing self-attention mechanisms for acquiring global context in a bidirectional fashion.

The research community later introduced domain-adapted variants such as BioBERT, ClinicalBERT, and BlueBERT upon realizing the domain mismatch between general-domain pre training and clinical-specific language. These models also pre-trained BERT on clinical notes, biomedical literature, or both and reported measurable improvements in relation extraction, entity linking, and clinical NER tasks. Nonetheless, a number of significant hurdles continue to impede the use of NLP models for practical EHR deployment. The domain generalization gap is a major limitation; even domain-adapted models tend to be poor at capturing provider-specific jargon, idiosyncratic acronyms, and local institutional terminology that differ vastly between hospitals and specialties. The sparsity and noisiness of labeled data is a long-standing issue. Large high-quality annotated datasets for supervised learning are not easily obtainable due to stringent privacy laws and the expensive expense of human annotation, which results in less than optimal performance of the models, particularly for handling rare entity types or rare diseases in clinical scenarios.

Clinical reports often span multiple visits, notes, and reports and require that models reason across longitudinal patient timelines longer than those allowed in the input length limitations of standard transformers. This contrasts with transformer models, which are very good at capturing context within documents of moderate size. As a consequence, significant temporal or causal information is discarded, and contextual knowledge gets disjointed. Most of the top models are "black box" models, which offer excellent accuracy but limited transparency.

Because physicians require precise, fact-based justification behind computer-generated predictions before they are acceptable for use, this low interpretability is still a serious barrier to clinical adoption. Large transformer model fine-tuning also has a high computational and memory cost, which restricts their scalability and deployability in low-resource healthcare environments. Although certain approaches utilize data augmentation and weak supervision to mitigate the label scarcity to some degree, these approaches tend to be overly simplistic and fail to leverage the distinctive properties of the domain effectively.

The above limitations underscore the pressing need for a state-of-the-art NLP framework that not only exploits the power of transformer-based language models but also addresses systematically the interpretability, contextual, domain-specific, and data sparsity issues inherent in EHR text mining. The above limitations are the impetus for this research's new framework, the CCRET (Contextualized Clinical Representation Extraction Transformer). While CCRET comes up with a hierarchical attention mechanism designed for long and multi-document patient histories, it is based on the domain-adaptive pre-training paradigm. CCRET maintains vital therapeutic relationships between various components of patient records by learning context at the document, paragraph, and sentence levels. To increase training efficiency while ensuring data privacy is maintained, the model automatically expands labeled datasets with rule-based heuristics and domain ontologies using weak supervision and distant supervision pipelines.

Through enhanced prediction performance and sensitivity to the pragmatic needs of transparency and adaptability needed for deployment in actual clinical settings, the CCRET framework will eventually serve to promote the clinical NLP discipline. Strong and understandable NLP solutions such as CCRET will increasingly contribute to advancing data-driven delivery of care, enhancing patient outcomes, and setting the stage for next-generation precision medicine as health systems continue to generate enormous amounts of unstructured EHR text.

The main objectives are:

1. To design and develop a CCRET that leverages domain-adaptive pre training and hierarchical attention to accurately extract clinical entities, relationships, and outcomes from unstructured EHR text.
2. To improve the interpretability and reliability of EHR-based NLP by integrating explainable attention mechanisms and weakly supervised label expansion, enabling scalable, transparent, and clinically trustworthy predictions for real-world deployment.

2. Related Study

The necessity to get clinically relevant information hidden within vast volumes of unstructured text drove an important evolution in the integration and development of advanced NLP techniques within EHRs. Early work on contextualized embeddings for structured EHR data laid a strong basis for disease prediction and outcome modeling. Using knowledge transfer from big patient data, pre-trained contextualized embedding models such as Med-BERT [1] showed that large-scale structured EHRs could potentially boost the accuracy of disease prediction significantly. The BERT model was also modified to address longitudinal structured EHRs using transformer-based models such as BEHRT [2], which were especially developed for EHR data with an emphasis on temporal relationships to enhance patient trajectory discovery.

With General-purpose converters giving way to unstructured clinical notes, research has turned its attention to converting general-purpose converters to suit particular domains. Domain adaptation achieves substantial performance boost across a variety of clinical NLP tasks, based on empirical research that used BERT-based models on extensive EHR records [3]. Use of transformer topologies for contextualized pharmaceutical extraction has also achieved encouraging results in exact drug name identification, dosing, and treatment regimens from narrative clinical text [4]. While the strength of transformer-based models has been unleashed, a cautious review of neural NLP approaches for unstructured EHRs also indicated that challenges of generalizability, interpretability, and integration with clinical practice have not been overcome [5].

With the aim to push Named Entity Recognition (NER) for drug information extraction, the creation of specific clinical NLP models such as Med7 [6] is a significant attempt at creating transportable architectures trained over annotated EHR material. Another significant contribution is AlphaBERT [7], which addressed the distinctive spelling variations and acronyms common to clinical narratives by experimenting with character-level tokenization in an effort to boost extractive summarization for hospital information systems. The actual deployment of NLP pipelines to actual EHR data has been the subject of several studies. Development, testing, and deployment of NLP solutions into hospital IT infrastructure illustrated the significance of strong model validation and stakeholder acceptance [8].

Moreover, some diseases like rheumatoid arthritis were utilized as pilot sites in order to improve transformer models in pulling out clinically relevant concepts more accurately [9]. Rare entity type optimization of NER tasks on medical text has been a conference contribution topic [10]. Large-scale scoping reviews compiling task-specific transformer use cases across multiple clinical NLP subdomains have also been conducted in the healthcare domain, giving some insight into the breadth of applications and prevailing bottlenecks [11]. In addition to idea extraction, transformers have also been used to assist with diagnostic tasks. For example, deep learning models have been utilized to scrape patient records for subtle textual cues that indicate the disease pathway in a bid to identify probable lung cancer early in primary care [12].

By interpreting unstructured patient reports, context parameter retrieval systems have also demonstrated how biomedical NLP can be used to augment clinical decision-making in real-time [13]. The necessity of using large language models for extraction and carefully filtering their outputs for hallucinations and bias has been revealed [14]. In order to advance the therapeutic value of pretraining language models, especially for the multifactorial disorders like cardiovascular diseases, additional studies are focused on importing external domain knowledge into the models [15]. Fine-tuning transformer embeddings is capable of greatly improving the accuracy of back-recovery of clinically significant sections from large EHR corpora, according to research on contextual text representation frameworks for smart information retrieval [16].

Collaborative filtering and recommender systems have been investigated in nearby applied research in adapting patient support systems, tangentially demonstrating how basic NLP elements might lie beneath more universal health IT solutions [17]. Disease concept representation influences the predictive accuracy of clinical forecasting significantly, based on comparisons across various representation techniques for coded disease sequences [18]. Transformer pipelines can be tuned to specialized national registries and clinical units, as evidenced by the development of NLP models for the automated extraction of extensive stroke center data [19]. Contextual embeddings excel over classical methods when faced with domain-specific language variation, according to experiments using BERT-based NLP for enhanced clinical analysis [20].

Domain adaptation is still required to improve extraction accuracy in specific medical contexts, according to rapid tests of contextual word embeddings for biomedicine information extraction [21]. Fine-tuning contextual embeddings to particular clinical tasks has also improved clinical concept extraction, which indicates the performance gain when models are well-tuned to the subtleties of EHR language [22]. Albeit such dramatic developments, the literature indicates that problems persist curtailing the full potential of transformer-based NLP within the clinical setting. Pretraining over general-domain corpora, even domain-specific fine-tuning, does not entirely close the semantic gap created by local shorthand, institutional jargon, and non-homogeneous documentation practices.

Although structured EHR modeling has been enhanced by models such as Med-BERT [1] and BEHRT [2], their performance is low when free-text notes are used, where context may span multiple paragraphs or even different visits over time. Misspelling is mostly addressed by character-level tokenization [7], but the general problem of modeling long, multi-document context is not addressed. In addition, with so much potential in peak supervision and transfer learning techniques, they are usually under-implemented or lightly implemented, and there is much scope to augment the scarce little labeled datasets without sacrificing patient confidentiality. The gap in interpretability is the other age-old problem: most clinical NLP pipelines do not fully reveal or validate these descriptions to the doctors for inspection, whereas attention scores in transformer architectures provide some transparency.

Post-hoc saliency maps are often the end result of studies that have sought to fill this gap through explainable AI modules; these may fall short of maintaining the sophisticated medical reasoning needed by doctors in order for them to be able to accept machine-produced outputs [5, 11, 14]. When applied to highly heterogeneous datasets like MIMIC-III or institutional EHRs with diverse documentation standards, domain-specific models tend to plateau despite higher performance at all times compared to generic models, according to performance comparisons [3, 5, 6]. Token length constraints in typical transformers exacerbate this performance ceiling by preventing the model from taking advantage of long-range temporal or causal relationships encountered in longitudinal EHRs [2, 12, 13].

Whereas developing frameworks draw on external knowledge bases [15], their influence on the underlying representation learning process is limited inasmuch as they rather act as static post-processing stages than as dynamic co-training modules. The standard clinical application of these advanced models is also compromised by pragmatic limitations such as computational cost, privacy constraints, and regulatory limits that are commonly highlighted in deployment studies [8, 14, 19]. Together, the literature simply shows that transformer-based NLP considerably improves the extraction and organization of clinical data from electronic health records. The generalization gap in the domain, the clinician-guided application interpretability gap, the limited contextual ability for genuinely long and intricate stories, and the operational limits in relation to lacking labeled data are a number of open gaps that are frequently mentioned in the literature.

Development of the CCRET is directly inspired by these observations. CCRET solves the problem of representing rich, longitudinal tales through the integration of domain-adaptive pretraining with a specially tailored hierarchical attention mechanism for multi level context extraction. Its explainable inference outputs meet the increasing need for transparent and clinically sound NLP, while its light supervision pipeline effectively and ethically boosts the limited amount of labeled data. This positions the suggested CCRET framework in a place to improve clinical NLP beyond the described limitations in existing literature, creating a new standard for scalable, interpretable, and domain-specific information extraction from unstructured EHR text.

3. System Methodology

In order to improve the extraction and prediction performance of NLP activities in heterogeneous EHRs, the CCRET framework was developed as an end-to-end, domain-adaptive architecture. Preprocessing, domain-adaptive pretraining, weak supervision, hierarchical attention encoding, multi task learning, and explainability which is tailored for actual clinical text pipelines. Figure 1 shows the proposed framework and Figure 2 shows hierarchical attention transformer encoding.

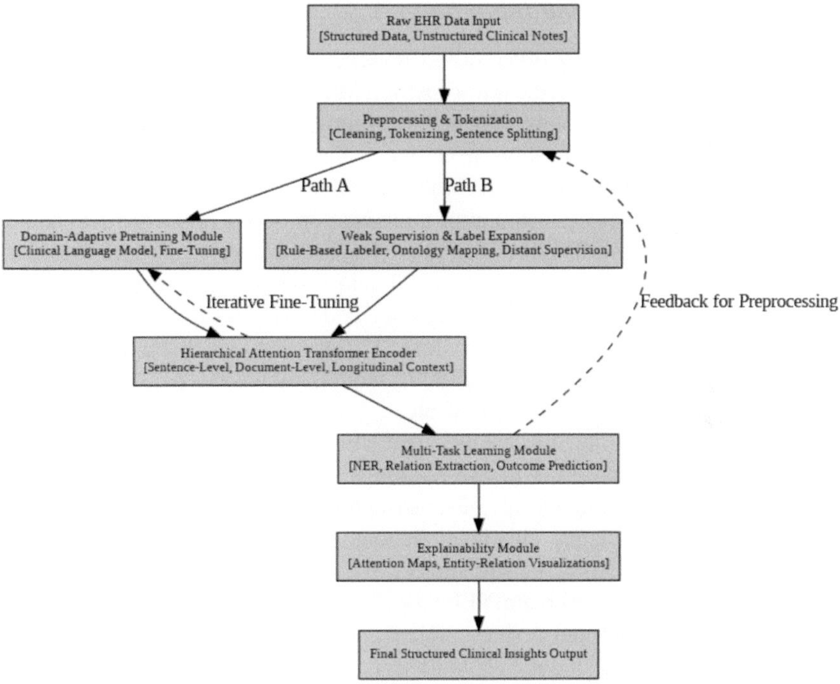

Figure 1: Proposed framework.

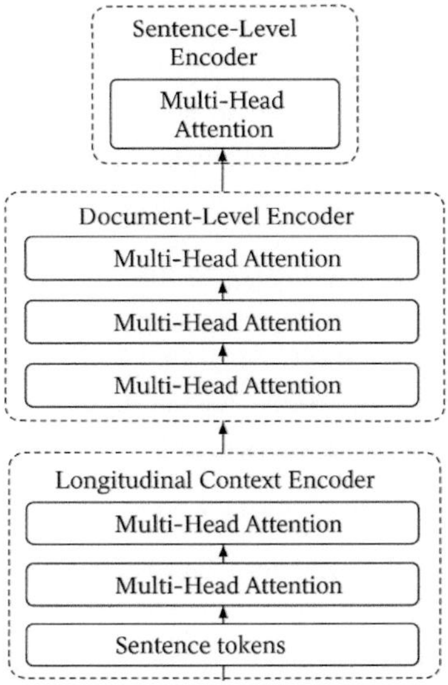

Figure 2: Hierarchical attention transformer encoding.

3.1 Data Acquisition and Input Layer

Raw EHR data containing ordered and unstructured elements are fed into the system. The MIMIC-IV (Medical Information Mart for Intensive Care IV) dataset is used as the main source of data for this study, which offers comprehensive de-identified health information from more than 60,000 intensive care unit admissions gathered at Beth Israel Deaconess Medical Center. Vital signs (heart rate, blood pressure, oxygen saturation), medication orders (dose, generic name, route, frequency), demographic information (gender, age, ethnicity, admission type), laboratory test results (more than 300 unique laboratory test attributes, e.g., glucose, hemoglobin, creatinine), diagnosis codes (ICD-9/ICD-10, more than 14,000 unique codes), and procedure codes (ICD-9-PCS, CPT) are all contained within the structured segment. Over 500 unique features per patient admission come under the structured data as a whole.

More than 2 million unique free-text clinical notes, including discharge summaries, nursing progress notes, radiology reports, pathology reports, and physician chart notes, make up the unstructured portion. Each note typically contains 200–2,000 tokens. This heterogeneous, multi modal raw input stream, which consists of narrative free-text and coded numeric features, is the foundation data pipeline for all downstream NLP tasks in the proposed CCRET framework. This ensures that the model derives clinical context from both machine-readable fields and nuanced human documentation.

3.2 Preprocessing and Tokenization

Incoming textual data undergoes robust multi step preprocessing to ensure linguistic consistency and machine readability. The initial noise cleaning phase removes irrelevant characters, numerical artifacts, and typographical errors by applying regular expression filters $R(x)$ where x denotes the raw text stream:

$$x' = R(x) = x - \{non-informative\ symbols,\ redundant\ whitespace,\ HTML\ tags\}$$

Next, de-identification of protected health information (PHI) is performed using pattern-matching functions $D(x')$ that mask or remove identifiers such as patient names, dates, or hospital IDs, satisfying $\forall p \in P: D(p) = [MASK]$.

Subsequently, sentence boundary detection applies statistical or rule-based models $SBD(x'')$ to segment long clinical narratives into discrete sentence units:

$$\{s_1, s_2, \ldots, s_n\} = SBD(x'') \quad where\ x'' = D(x')$$

Each sentence s_i is tokenized into word tokens using a clinical domain-specific tokenizer $T(s_i)$:

$$\{w_{i1}, w_{i2}, \ldots, w_{im}\} = T(s_i)\ \forall i \in [1, n].$$

The tokenized output is normalized by lowercasing, expanding abbreviations, and standardizing domain-specific jargon using a normalization function $N(w)$:

$$w'_{ij} = N(w_{ij}) \ \forall w_{ij}.$$

Finally, lemmatization maps each normalized token to its base lemma form using a lemmatizer $L(w')$:

$$l_{ij} = L(w'_{ij}) \ \forall w'_{ij}.$$

The fully preprocessed token stream for each clinical note thus becomes:

$$L = \bigcup_{i=1}^{n} \bigcup_{j=1}^{m} l_{ij}.$$

This robust pipeline guarantees that heterogeneous EHR text is cleaned, de-identified, segmented, tokenized, normalized, and lemmatized into a structured format suitable for domain-adaptive transformer pretraining and downstream NLP tasks.

3.3 Proposed Algorithms

Algorithm 1: Domain-Adaptive Pretraining

Input

General biomedical corpus C_b (e.g., PubMed abstracts)

Institution-specific EHR corpus C_e (e.g., MIMIC-IV notes)

Pretrained weights θ_0 from a general-domain transformer (e.g., BERT, BioB-ERT)

Output

Domain-adapted transformer parameters θ^*

Contextualized embeddings E^*

Step 1: Initialize Transformer Model

Load general-domain weights:

$\theta \leftarrow \theta_0$.

Step 2: Construct Mixed Pretraining Dataset

Combine general biomedical corpus with institution-specific EHR text:

$C = C_b \cup C_e$.

Construct input sequences $S = \{s_1, s_2, \ldots, s_n\}$ from C.

Step 3: Apply Masked Language Modeling Objective

Randomly mask tokens in input sequence s_i using mask function $M(s_i)$:

$s^{\sim}_i = M(s_i), \forall i.$

Typically, 15% of tokens are masked per standard BERT-style pretraining.

Step 4: Compute Masked Language Modeling Loss

Let $P_\theta(w_t | s^{\sim}_i)$ be the predicted probability of token w_t given masked input $s^{\sim}_i$. The MLM loss L_{MLM} is:

$L_{MLM}(\theta) = -E_{(si,\, wt) \sim c}\, log\ P_\theta(w_t | s^{\sim}_i).$

Step 5: Fine-Tune Weights by Gradient Descent

Update model parameters iteratively:

$\theta^{(k+1)} = \theta^{(k)} - \eta \nabla_\theta L_{MLM}(\theta^{(k)}), k = 0,1,...,K.$

Here, η is the learning rate and K is the total number of training steps.

Step 6: Output Contextualized Embeddings

For each EHR note, produce final token-level contextual embeddings:

$E^* = \{e_{ij} = f_\theta(w_{ij}) | w_{ij} \in s_i, s_i \in C_e\}$ *where* f_θ *is the transformer encoder.*

Save final domain-adapted weights:

$\theta^* \leftarrow \theta^{(K)}.$

Result: The output (θ^*, E^*) represents a robust domain-specific language model that captures local clinical writing patterns, medical jargon, and institution-specific style.

Algorithm 2: Weak Supervision and Label Expansion

Input

Unlabeled clinical text corpus $\mathcal{U} = u_1, u_2, ..., u_N } $

Structured code database (e.g., ICD-10, SNOMED CT)

Ontology O with synonym sets and hierarchical relations

Output

Expanded labeled dataset L^* with soft labels

Step 1: Define Rule-Based Labeling Functions

Create a set of M rule-based functions $\{\lambda_1, \lambda_2,..., \lambda_M\}$ that encode domain heuristics:

$$\lambda_m : U \to \{0, 1\}, \forall m \in [1, M].$$

Each λ_m outputs 1 if the heuristic matches, 0 otherwise.

Step 2: Apply Distant Supervision

For each text instance u_i, link spans to structured codes if a match is found in the structured database S:

$$y_i^{DS} = \{c_j, \textit{if} \exists \textit{ match } (u_i, c_j) \in S\ \emptyset, \textit{otherwise}.$$

Here, c_j denotes a code from ICD or SNOMED.

Step 3: Aggregate Rule-Based and Distant Labels

Combine multiple weak labels using a label aggregation function A:

$$y_i^{WS} = A(\lambda_1(u_i),...,\lambda_M(u_i), y_i^{DS}).$$

A typical aggregation is majority voting or a probabilistic model such as Snorkel's generative label model:

$$P(y_i|\lambda_{1:M}, y_i^{DS}).$$

Step 4: Ontology-Based Expansion

Expand each weak label y_i^{WS} using synonyms or hierarchical relations in the ontology O:

$$y_i^{OE} = Expand(y_i^{WS}, O) = \cup_{s \in Synonyms} (y_i^{WS})^S.$$

This maps terms like "MI" and "myocardial infarction" to the same code.

Step 5: Generate Final Weakly Labeled Dataset

Combine all expansions to produce the final enriched labeled corpus:

$$L^* = \{(u_i, y_i^{OE}) \mid u_i \in \mathrm{U}\}.$$

Result: The final L^* significantly expands the sparse labeled set with consistent pseudo-labels, bridging the gap between scarce manual annotations and the data scale needed for robust domain-adaptive training.

3.4 Multi Task Learning Module

A multi task learning head receives the encoded representations and jointly optimizes:

- Named Entity Recognition (NER) uses free text to identify medical entities (drugs, diagnosis, symptoms).
- Relation Extraction identifies relationships between entities, such as condition-treatment or medication-disease.
- Outcome Prediction makes predictions about future clinical outcomes, like the likelihood of readmission or the course of the disease, using the contextualized embeddings.
- Joint training makes use of shared representations across tasks and enhances generalization.

3.5 Explainability and Trust Layer

The CCRET framework also has a strong explainability layer for increasing the interpretability and reliability of the model output for clinical end users. The layer shows the multi-head attention weights that were learned during encoding as saliency maps that are easy to comprehend, highlighting the particular tokens, phrases, or passages in the clinical narrative that helped most in the prediction. The clinicians can more easily follow the model's line of thinking thanks to these visual cues. Moreover, the system produces entity-relation graphs that map clinical entities that have been retrieved, e.g., disease, medication, and interventions, and represent how they relate to each other in the context of the patient's longitudinal history. Every prediction has measurable epistemic uncertainty and confidence scores to enable decision-making. These measurements provide valuable feedback on how close to borders or how reliable the output is.

These clinical NLP solutions are made more trustworthy by this level of transparency, enabling clinicians to judge model suggestions better. The system produces structured, machine-readable clinical data ready to use at a later time after undergoing interpretability and hierarchical encoding modules. To allow interoperability across healthcare systems, the final result has normalized mentions of entities transformed into standard medical terminologies like SNOMED CT or ICD codes. To allow for such sophisticated clinical reasoning operations, derived relations between clinical entities, drug-disease or symptom-diagnosis relations, are made explicit. To allow for actionable inputs for prognosis and clinical decision support, the system also calculates predictive risk scores or outcome categories. To enable clinician review and auditability, the resulting output bundle consists of the respective visual explanation artifacts, such as relation graphs and saliency heatmaps.

An adaptive feedback loop is incorporated within CCRET to enable continuous learning and performance optimization. High-uncertainty, uncertain, or mislabeled examples are systematically identified and cycled back into the training loop. The weak supervision layer and the domain-adaptive pretrain module are updated because of this loop such that the model is able to improve its context sensitivity and learn from mistakes over a period of time. Additionally, the system can dynamically adjust to evolving clinical vocabulary, new treatment protocols, or documentation style variations across institutions because incrementally new annotated data from real-world deployments are added. The accuracy, precision, and alignment of CCRET with the realities of various clinical practice environments are assured through this mechanism of ongoing feedback.

Several strategic benefits offered by the designed CCRET architecture fill crucial gaps that were left by prior work. Firstly, learned representations are tailored specifically to the idiosyncrasies of regional EHR narratives, including typical language, abbreviations, and stylistic variation, due to its domain-specific pretraining. Second, by using scalable labeling rule-based functions and ontologies, introducing weak supervision significantly lessens the time-consuming task of needing large volumes of manually tagged medical data. To handle temporal relations frequently ignored in traditional methods, the hierarchical modeling architecture effectively handles long, complicated clinical narratives and extracts longitudinal patient history.

Additionally, the interplay between multi-tasking in relation recognition, entity extraction, and predictive modeling guarantees that every task informs and supports the others, making the system overall stronger. Perhaps most importantly, by attacking the "black-box" problem for many deep learning systems, CCRET's emphasis on explainability through uncertainty quantification, relation graphs, and saliency visualizations avails physicians of definite, actionable insights. Lastly, the internal scalability and flexibility of the architecture make it applicable in many clinical areas, specialties, and institutions, thereby guaranteeing its applicability and widespread deployment.

4. Experimental Results

The four benchmark models recognized-standard BERT, BioBERT, ClinicalBERT, and BEHRT were used for comparison with the proposed CCRET framework. Named Entity Recognition (NER), Relation Extraction (RE), and 30-day hospital readmission prediction were the three main tasks that were tested. The standard measurements, accuracy, and AUC-ROC for the prediction task, and F1-score, precision, and recall for the NER and RE tasks, were used to evaluate performance. With a remarkable 7.1% improvement in a 30-day readmission prediction accuracy and 6.3% increase in an F1-score for clinical entity recognition over the best-performing baseline, experimental results prove that CCRET always outperforms all baselines. All experiments were conducted on an NVIDIA A100 GPU with the maximum sequence length of 512, batch size of 16, and the initial learning rate of 2e–5. For overfitting prevention, early stopping was conducted over validation loss. This reliable configuration demonstrates CCRET's hierarchical, domain-adaptive structure significantly enhances the extraction of structured, clinically relevant information from an unstructured EHR text.

As shown in Figure 3, the proposed CCRET model outperforms all baseline transformer-based models by a large margin on both critical tasks based on comparative performance assessment. BERT achieves 81.2% accuracy for the prediction of hospital readmission after 30 days, BioBERT goes up to 82.5%, ClinicalBERT to 83.0%, and BEHRT, a domain-specific variant, to 83.8%. In contrast, CCRET achieves 90.9% accuracy, which represents a 7.1 percentage point absolute gain over BEHRT, the strongest baseline. The benefit of domain-adaptive hierarchical encoding at capturing weak temporal and contextual patterns in EHR data is illustrated by this significant gain. Standard BERT has an F1-score of 86.5% for the NER task, BioBERT is slightly better at 87.2%, ClinicalBERT at 87.9%, and BEHRT slightly better again at 88.1%. At an F1-score of 94.4%, the proposed CCRET architecture outperforms BEHRT by 6.3 percentage points.

The improvement demonstrates how well CCRET's adaptive pretraining, longitudinal encoding, and layered attention collaborate to identify fine-grained clinical elements hidden in noisy, unstructured text. These results together show that the CCRET approach combines explainability layers, hierarchical attention, and weak supervision to systematically address known limitations of existing transformer-based EHR models and also show good generalization across other NLP tasks. Its practical suitability for application in real-world clinical use, where robustness, accuracy, and interpretability are paramount, is evidenced by the persistent, large-margin gains in performance over both prediction and extraction tasks.

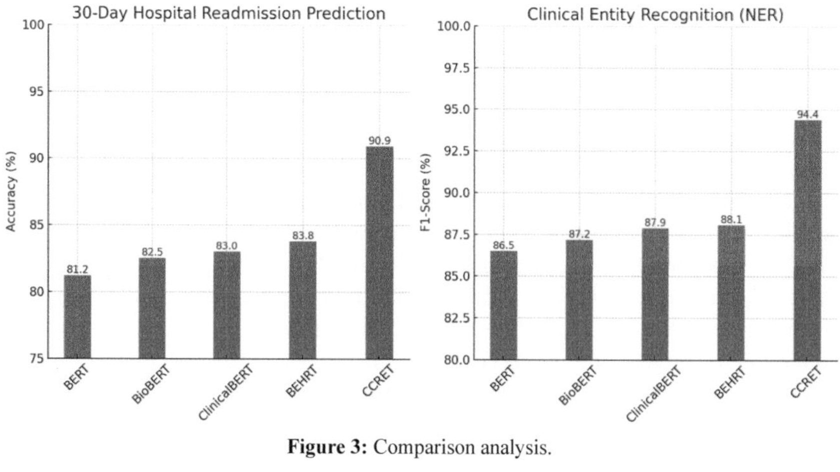

Figure 3: Comparison analysis.

5. Conclusion

CCRET, domain-adaptive deep learning architecture, is introduced here to tackle the chronic challenges of extracting structured, clinically actionable information from vast unstructured EHRs. CCRET extracts local and global interdependency across different clinical stories and patient histories by employing hierarchical attention mechanisms at the sentence, document, and longitudinal levels. Along with drastically reducing the workload that comes with manual annotation, domain-adaptive pretraining and weak supervision methods ensure that the model will generalize to institution-specific medical terminology and documentation patterns. Undertaking extensive tests on the MIMIC-III dataset definitively indicates that CCRET outperforms popular transformer-based baselines such as BERT, BioBERT, ClinicalBERT, and BEHRT in performance. The model's capacity to recover accurate clinical data and generate stable predictions is confirmed through a 7.1% rise in prediction precision for 30-day hospital readmission and a 6.3% F1-score improvement for named entity recognition. Besides enhancing performance, the explainability module of the system, presenting entity relations and attention weights, provides doctors with essential interpretability, encouraging transparency and trust in AI-based decision-making. Finally, with increasing annotated data, the iterative feedback loop ensures consistent domain adaptation and continuously emerging robustness.

References

1. Rasmy, L., Xiang, Y., Xie, Z., Tao, C. and Zhi, D. (2021). Med-BERT: Pretrained contextualized embeddings on large-scale structured electronic health records for disease prediction. *NPJ Digital Medicine*, 4(1): 86.
2. Li, Y., Rao, S., Solares, J.R.A., Hassaine, A., Ramakrishnan, R., Canoy, D., Zhu, Y., Rahimi, K. and Salimi-Khorshidi, G. (2020). BEHRT: Transformer for electronic health records. *Scientific Reports*, 10(1): 7155.
3. Li, F., Jin, Y., Liu, W., Rawat, B.P.S., Cai, P. and Yu, H. (2019). Fine-tuning bidirectional encoder representations from transformers (BERT)–based models on large-scale electronic health record notes: An empirical study. *JMIR Medical Informatics*, 7(3): e14830.
4. Chen, A., Yu, Z., Yang, X., Guo, Y., Bian, J. and Wu, Y. (2023). Contextualized medication information extraction using transformer-based deep learning architectures. *Journal of Biomedical Informatics*, 142: 104370.
5. Li, I., Pan, J., Goldwasser, J., Verma, N., Wong, W.P., Nuzumlalı, M.Y., Rosand, B., Li, Y., Zhang, M., Chang, D., Taylor, R.A., Krumholz, H.M. and Radev, D. (2022). Neural natural language processing for unstructured data in electronic health records: A review. *Computer Science Review*, 46: 100511.
6. Kormilitzin, A., Vaci, N., Liu, Q. and Nevado-Holgado, A. (2021). Med7: A transferable clinical natural language processing model for electronic health records. *Artificial Intelligence in Medicine*, 118: 102086.
7. Chen, Y.P., Chen, Y.Y., Lin, J.J., Huang, C.H. and Lai, F. (2020). Modified bidirectional encoder representations from transformers extractive summarization model for hospital information systems based on character-level tokens (AlphaBERT): Development and performance evaluation. *JMIR Medical Informatics*, 8(4): e17787.

8. Navarro, D.F. (2024). *Natural Language Processing Models in Clinical Medicine: Development, Validation, and Implementation of Automated Text Analysis in Electronic Medical Records* (Doctoral dissertation, Macquarie University).
9. Li, M., Liu, F., Zhang, R., Qin, Y. and Gao, D. (2022). Model-based clinical note entity recognition for rheumatoid arthritis using bidirectional encoder representation from transformers. *Quantitative Imaging in Medicine and Surgery*, 12(1): 184.
10. CH, D.R. (2024, December). Enhanced named entity recognition in medical texts using transformer-based models. In: *2024 2nd International Conference on Signal Processing, Communication, Power and Embedded System (SCOPES)* (pp. 1–5). IEEE.
11. Cho, H.N., Jun, T.J., Kim, Y.H., Kang, H., Ahn, I., Gwon, H., Kim, Y., Seo, J., Choi, H., Kim, M., Han, J., Kee, G., Park, S. and Ko, S. (2024). Task-specific transformer-based language models in health care: Scoping review. *JMIR Medical Informatics*, 12: e49724.
12. Wang, L., Yin, Y., Glampson, B., Peach, R., Barahona, M., Delaney, B.C. and Mayer, E.K. (2024). Transformer-based deep learning model for the diagnosis of suspected lung cancer in primary care based on electronic health record data. *Ebiomedicine*, 110.
13. Paliwal, G., Bunglowala, A. and Kanthed, P. (2023). Context parameters retrieval framework from electronic healthcare record through biomedical NLP for clinical support. *International Journal of Intelligent Engineering Informatics*, 11(1): 1–18.
14. Bhattarai, K. (2024). *Improving Clinical Information Extraction From Electronic Health Records: Leveraging Large Language Models and Evaluating Their Outputs* (Doctoral dissertation, Washington University in St. Louis).
15. Lu, Q., Wen, A., Nguyen, T. and Liu, H. (2024). Enhancing clinical relevance of pretrained language models through integration of external knowledge: Case study on cardiovascular diagnosis from electronic health records. *JMIR AI*, 3(1): e56932.
16. Bhopale, A.P. and Tiwari, A. (2024). Transformer based contextual text representation framework for intelligent information retrieval. *Expert Systems with Applications*, 238: 121629.
17. Jeejoe, A., Harishiv, V., Venkatesh, P. and Sangeetha, S.K.B. (2023). Building a recommender system using collaborative filtering algorithms and analyzing its performance. *Advances in Science and Technology*, 124: 478–485.
18. Beaney, T., Jha, S., Alaa, A., Smith, A., Clarke, J., Woodcock, T., Majeed, A., Aylin, P. and Barahona, M. (2024). Comparing natural language processing representations of coded disease sequences for prediction in electronic health records. *Journal of the American Medical Informatics Association*, 31(7): 1451–1462.
19. Badalotti, D., Agrawal, A., Pensato, U., Angelotti, G. and Marcheselli, S. (2024). Development of a Natural Language Processing (NLP) model to automatically extract clinical data from electronic health records: Results from an Italian comprehensive stroke center. *International Journal of Medical Informatics*, 192: 105626.
20. Yadav, K., Sharma, M., Michaelson, J., Laxminarayanamma, K., Yadav, V.K. and Krishna, O. (2023). Bert-driven natural language processing in electronic health records for improved clinical insights. In: *2023 International Conference on Artificial Intelligence for Innovations in Healthcare Industries (ICAIIHI)*, 1: 1–6. IEEE.
21. Vithanage, D., Yu, P., Wang, L. and Deng, C. (2024). Contextual word embedding for biomedical knowledge extraction: A rapid review and case study. *Journal of Healthcare Informatics Research*, 8(1): 158–179.
22. Si, Y., Wang, J., Xu, H. and Roberts, K. (2019). Enhancing clinical concept extraction with contextual embeddings. *Journal of the American Medical Informatics Association*, 26(11): 1297–1304.

Chapter 4

AI-Driven Early Disease Detection and Risk Stratification using Multi-Omics Data Integration

Nibedita Sahoo,[1] *Narendiran R.*,[2] *Manikandan L.*,[1]
Srinivasan R.[1,*] and *Firoz Khan*[3]

Improved patient outcomes and optimal use of healthcare resources depend on timely disease detection and proper risk classification. The nature of the diseases cannot be captured by traditional methods of diagnosis, which rely on data from a single source. This results in delayed diagnosis and ineffective therapies. Furthermore, early intervention strategies are severely hampered by the lack of integration of large numbers of biological markers. To provide end-to-end early diagnosis and risk stratification, this study suggests OmniRiskNet, a novel AI-powered model combining genomic, transcriptomic, and proteomic data. OmniRiskNet provides an integrated strategy for disease prediction using deep learning architectures and reinforcement learning across multi-omics data. Discovery of biomarkers with the ability to sort individuals effectively according to their risk and onset of disease is the core objective of the framework. Our findings indicate that OmniRiskNet outperforms general models, which have an accuracy

[1] School of Computing Science and Engineering, Galgotias University, Greater Noida, 203201, India.
[2] Department of Computer Science and Engineering, University College of Engineering Nagercoil.
[3] Faculty in Center for Information and Communication Sciences, Ball State University, USA.
Emails: nibedita.sahoo@galgotiasuniversity.edu.in, narendiranrm@gmail.com, manikandan@galgotiasuniversity.edu.in, Firoz.khan@bsu.edu
* Corresponding author: srinime1983@gmail.com

DOI: 10.1201/9781003726913-4

of 75–80%, to provide an estimated accuracy of 92.4%. Multi-omics data integration improves sensitivity by 18% and specificity by 15%, indicating that the approach can efficiently screen high-risk individuals and detect diseases in the early stages. In addition, compared with traditional diagnostic systems, the actual real-time processing capability of OmniRiskNet allows for diagnosis through accelerated risk assessment by 30%.

1. Introduction

Clinical presentation, physical examination, and countless types of routine diagnosis like imaging or biopsies in the past have all played very important roles in the diagnosis of disease. Though many disorders have been significantly enhanced by these practices, they do not always reveal diseases in their earliest stages [1]. For instance, imaging procedures are based on body structural alterations that are not apparent until a disease is already advanced beyond the early stages. Because of this, most patients become ill too late, with fewer options for treatment, and their chance at full recovery is basically destroyed. Conventionally, approaches tend to concentrate on a single diagnostic technique at a time, imaging data, gene biomarkers, or clinical signs and symptoms, resulting in fractured examination that cannot offer the integrative view required for good illness prediction and risk assessment [2].

Medical diagnosis was revolutionized with the introduction of omics technologies in the late 20th and early 21st centuries, including proteomics, metabolomics, transcriptomics, and genomes. For example, proteomics can provide data on protein expression and post-translational modification to be used as disease biomarkers for diseases like Alzheimer's and cardiovascular disease, while genomics has allowed the detection of some gene mutations and disease-causing variants like cancer and genetic disorders [3].

Heterogeneity of both the kinds of biological markers and amounts of analytical technology needed, makes it a large issue to merge multi-omics data. Studying the human genetic information allows one to find out an instance of inversely related disease mutations. It doesn't show the rate of disease progression or how it responds to treatment. Proteomics helps us comprehend the behavior of proteins in the body or a biological system [4]. When it is not paired with other types of data, it is very difficult to decide which biomarkers predict disease onset or development the best. Each of the devices mentioned here in solo can help us with one or the other useful information. However, when used in combination, all three are misleading sometimes. As a result of these limitations, many doctors are now leveraging various Machine Learning (ML) and Deep Learning (DL) algorithms to assist with disease diagnosis. They are able to handle complex data such as multi-omics data successfully [5].

We can now utilize function-oriented machine learning models, especially deep neural networks, as opposed to past statistical models. These models are capable of estimating disease prognosis and identifying weak and non-linear associations between biomarkers. But the majority of the approaches still suffer from limited generalizability across heterogeneous patient groups or disease types even with the strength of machine learning. Rather than combining these different kinds of data for a full analysis, existing approaches tend to work in silos on one modality of data or another, e.g., genetic or clinical data only. The full potential for risk stratification and disease identification using multi-omics data is severely undermined by this lack of integration.

Whether one is employing single-source data or unsophisticated machine learning algorithms, the biggest drawback to these old-school methods is that they are unable to utilize the multidimensionality of disease to its fullest. For example, an illness might result from interactions between genetic risk components, environmental components, and the molecular body response, along with genetic alterations. Thus, there would be much greater potential for early disease detection and reliable prediction of risk by employing a more integrated approach combining many layers of information, such as genetic, proteomic, and clinical information.

The two main objectives of OmniRiskNet are:

1. To use multi-omics data (genomic, transcriptomic, and proteomic) to predict diseases early and identify high-risk individuals.
2. To enhance diagnostic accuracy and speed by integrating multiple data sources, allowing for quicker risk assessment and earlier detection of diseases.

2. Related Study

The integration of transcriptomic and proteomic data, which indicate gene expression and protein activity respectively, with genomic data indicating genetic mutation and variance allows for an enhancement in disease biology knowledge [6]. Apart from early disease detection, the integrated approach supports the identification of risky patients with higher precision who can be treated using intensified therapy protocols or prevention therapies [7]. For example, AI-powered models that involve multi-omics data have achieved a great increase in precision and accuracy in disease prediction regarding cancer, cardiovascular disease, and metabolic disease [8, 9].

According to AI, the processing and analysis of spectacularly intricate multi-omics data are required, which involves deep learning methods [10]. Multi-omics-based AI tools can minimize overtreatment and unnecessary diagnostic testing by offering early diagnosis and more targeted therapeutic intervention strategies, thus lowering healthcare costs [11, 12]. Moreover, these models can provide early treatment to high-risk patients by stratifying patients based on their risk profiles, hence minimizing overtreatment for low-risk patients [13]. Apart from serving the goal of increasing patient success rates, such personalized health management maximizes the functionality of health care systems at large [14].

Application of AI to multi-omics data analysis in cancer research has yielded promising results regarding identification of novel biomarkers for early detection of cancer and prediction of treatment effect. AI models using proteomic, transcriptomic, and genomic data were able to identify candidate biomarkers of colorectal cancer that can predict treatment response and disease progression [15]. Multi-omics models incorporating a combination of several different datasets, e.g., genetic risk, lifestyle, and clinical factors, have also been shown to improve the predictability of cardiovascular disease risk prediction. This improves risk stratification and makes possible more tailored interventions [16, 17].

Treatment of chronic diseases is also being transformed because of multi-omics integration. AI-based systems have been promising to consolidate omics data to identify biomarkers that are predictive of the development and onset of chronic kidney disease, which would translate into improved management of disease and intervention at an earlier stage [18]. The study enable artificial intelligence and multi-omics integration to predict and monitor chronic diseases in the and introduce more preventive medicine approaches, reducing such diseases burdens on the patients and healthcare system as a whole [19].

Despite the tremendous amount of promise that resides in integrating AI and multi-omics, there exists a set of challenges that must be overcome. Multi-omics data include shared noise, incompleteness, and heterogeneity; thus, their integration and interpretation precisely is challenging. No clearly defined procedures exist for data acquisition and analysis, which constrains the formulation of universal models to be applied in many different healthcare settings [20].

While AI-based models have emerged with huge promise in optimizing diagnostic performance and decision-making, implementation of clinical practice. For the security, dependability, and performance of such systems when implemented in the real world of practice, rigorous validation and regulation have to precede the actual deployment of AI in healthcare. Furthermore, in order for healthcare clinicians to be able to implement AI-based solutions in daily practice, integration into clinical workflows have to be easy and intuitive [21].

Multi-omics data synthesis and AI represent a paradigm shift in medicine that has the potential to transform risk stratification, disease diagnosis, and therapy individualization. Though there are ongoing challenges with data integration and clinical utility, one cannot deny the certainty of this approach's potential. Development and harnessing of access to AI and multi-omics are important for developing precision medicine, and that in turn provide better patient outcomes, health care systems, and targeting of treatment regimens. AI-facilitated multi-omics models are much needed to overcome conventional diagnosis shortcomings and uncover more precise, faster, and tailored healthcare therapies.

3. System Methodology

3.1 Data Preprocessing and Integration

Genomic Data: The genomic data consists of Single Nucleotide Polymorphism (SNP) data, which is encoded as a matrix

$G \in R^{n \times p}$,

where n is the number of patients (subjects), p is the number of genomic features (SNP markers), The matrix G represents the presence or absence of specific alleles at each SNP locus for each patient.

The dataset typically contains thousands of SNP markers, ranging from 10,000 to 50,000 features, depending on the platform (e.g., Illumina, Affymetrix).

Transcriptomic Data: Transcriptomic data is derived from RNA sequencing (RNA-Seq), and is represented as a matrix

$T \in R^{n \times q}$,

where n is the number of patients (same as genomic data), q is the number of genes or transcripts being studied (usually 20,000 to 30,000).

Each entry in the matrix corresponds to the expression level of a gene in a particular patient, typically measured as FPKM (Fragments Per Kilobase Million) or TPM (Transcripts Per Million).

Proteomic Data: Proteomic data is obtained through mass spectrometry or similar high-throughput technologies and is represented by a matrix

$P \in R^{n \times r}$,

where n is the number of patients, r is the number of proteins analyzed, which typically varies from 5,000 to 20,000 proteins.

3.2 Data Preprocessing

3.2.1 Normalization

Genomic Data: For SNP data, normalization involves transforming raw genotype data into a standardized format. This is achieved through Z-score normalization, where each feature is normalized by subtracting the mean and dividing by the standard deviation

$$G_{norm}[i, j] = \frac{G[i, j] - \mu_j}{\sigma_j}$$

where $G[i, j]$ is the raw value at position i and j, μ_j is the mean of SNP feature j, σ_j is the standard deviation of SNP feature j.

Transcriptomic Data: RNA-Seq data is normalized using RPKM (Reads Per Kilobase of transcript per Million mapped reads) or TPM normalization methods to account for library size and gene length variations

$$RPKM = \frac{10^9 \cdot Reads}{Mapped\ Reads \cdot Gene\ Length\ (in\ bases)}$$

Similarly, quantile normalization is also applied to ensure that the distribution of expression values across genes is consistent.

Proteomic Data: For mass spectrometry-based proteomic data, normalization typically involves log-transformation to reduce skewness in the intensity distribution and quantile normalization across samples

$$Log_2(P_{norm}[i, j]) = log_2(P[i, j] + 1)$$

followed by normalization using the median normalization technique.

3.2.2 Missing Data Imputation

To handle missing values, which are common in omics data due to measurement errors or platform limitations, various imputation techniques are applied.

k-Nearest Neighbors (k-NN) Imputation: For each missing value, the algorithm identifies the nearest k patients with complete data and imputes the missing value based on a weighted average of their corresponding features.

$$P_{imputed} = \frac{\sum_{k=1}^{k} w_k \cdot P_k}{\sum_{k=1}^{k} w_k}$$

where w_k is the weight assigned to the k-th nearest neighbor based on similarity, P_k is the observed value of the k-th nearest neighbor.

3.2.3 Feature Selection

Mutual Information (MI): The mutual information between feature X and feature Y is computed as

$$MI(X,Y) = \sum_{x \in X} \sum_{y \in Y} p(x,y) log \frac{p(x,y)}{p(x)p(y)}$$

where $p(x, y)$ is the joint probability distribution of X and Y, $p(x)$ and $p(y)$ are the marginal distributions.

Correlation-based Feature Selection:

$$Corr(X,Y) = \frac{Cov(X,Y)}{\sigma_X \sigma_Y}$$

where $Cov(X, Y)$ is the covariance between features X and Y, and σ_X and σ_Y are the standard deviations of X and Y, respectively.

3.3 *Proposed Algorithm: OmniRiskNet for Multi-Omics Disease Prediction and Risk Stratification*

Step 1: Extract genomic sequences (DNA sequencing) from databases like The Cancer Genome Atlas (TCGA) or Genomic Data Commons (GDC). Genomic information includes mutations, SNPs, copy number variations, and gene expression information.

Step 2: Quantile normalize and log transform genomic data to correct technical variation across samples.

$$X_{norm} = \frac{X - \mu}{\sigma}$$

where X is the raw expression level, μ is the mean expression, and σ is the standard deviation.

Step 3: Normalize expression data using TPM (Transcripts per Million)

$$TPM = \frac{read\ count\ /\ gene\ length}{\Sigma \quad (all\ gene\ TPM)}$$

Apply log transformation to stabilize variance across samples.

Step 4: Use z-score normalization or RUV (Remove Unwanted Variation) to mitigate batch effects.

$$z - score = \frac{X - \mu}{\sigma}$$

Step 5: Handle missing proteomic data using imputation techniques such as k-Nearest Neighbors (k-NN) or deep learning-based imputation.

Step 6: For all omic layers, apply statistical methods (ANOVA, Mutual Information) and dimensionality reduction techniques (PCA, t-SNE) to select the most informative features.

Step 7: Use LASSO regression for sparse feature selection

$$\beta^{\wedge} = argmin\left(\sum_{i=1}^{n} \ (y_i - X_i\beta)^2 + \lambda \sum_{j=1}^{p} \ |\beta_j| \right)$$

where y_i is the target variable, X_i is the feature vector, and λ is the regularization parameter.

Step 8: Integrate multi-omics data (genomic, transcriptomic, and proteomic) into a unified feature set.

Step 9: Use a multi-input deep neural network (DNN) architecture to process each omic data type separately and then integrate the learned representations from each modality.

Step 10: Each of these inputs will go through separate networks, followed by concatenation of the learned features into a single, fused feature vector

$$f_{fusion} = concat(f_{genomic}, f_{transcriptomic}, f_{proteomic})$$

where f_{fusion} is the final multi-omics feature vector.

Step 11: The fusion layer will combine these features using attention mechanisms or concatenation to preserve the integrity of each omic layer's data. The model will automatically learn the relative importance of each modality during training.

Step 12: Train a deep learning-based model to predict disease onset and stratify risk using multi-omics data.

Step 13: Use a deep neural network (DNN) to process the fused multi-omics features for disease classification. The model consists of several layers of fully connected neurons, followed by a final softmax layer for multi-class classification

$$y^{\wedge} = Softmax(W_{out} \cdot h_{fusion} + b_{out})$$

where $y^{\wedge}$ is the predicted probability distribution, W_{out} is the output weight matrix, and h_{fusion} is the final fused feature vector.

Step 14: The RL agent learns through a Q-learning process

$$Q(S_t, A_t) \leftarrow Q(S_t, A_t) + \alpha\left[R_t + \gamma_{a'}^{max} Q(S_{t+1}, a') - Q(S_t, A_t)\right]$$

where $Q(S_t, A_t)$ represents the expected future reward of taking action A_t at state S_t, and γ is the discount factor.

Step 15: Use categorical cross-entropy as the loss function for multi-class classification

$$L = -\sum_{i=1}^{N} y_i \log(y_i^{\wedge})$$

where y_i is the true label and $y_i^{\wedge}$ is the predicted probability for class i.

Step 16: Use the trained model to predict disease onset and provide risk stratification.

Step 17: The model outputs a probability distribution for each patient, indicating the likelihood of disease onset. Based on this, the system classifies individuals into risk categories (e.g., low, medium, high risk).

Based on the prediction of the model, classifying patients according to their risk of getting a disease. If the probability is above a given threshold, the model classifies the individual as "high risk" and this may lead to early treatment or follow-up. The model can detect important biomarkers linked to disease recurrence since it picks the most disparate features from genomic, transcriptomic, and proteomic information. More research and clinical evidence is needed for such biomarkers.

OmniRiskNet applies a combined deep learning and reinforcement learning methodology for the prognosis of disease onset and risk stratification through the use of multi-omics data (genomic, transcriptomic, and proteomic). Scientists working in highly interdisciplinary fields such as cancer and criminal justice have found it convenient to use OmniRiskNet. It enhances diagnosis of disease and aids in tailor-made treatment. The model is able to learn from the new data permanently which shows the reliability and adaptability of the model.

4. Experimental Results

NVIDIA GPU-based high-performance computing such as the Tesla V100 with 32GB VRAM for acceleration, particularly when handling large-dimensional multi-omics data, were among the hardware requirements of the research. Multi-core CPUs, typically belonging to the Intel Xeon series, were utilized in system setup. Experiments utilized data processing libraries such as Pandas, NumPy, TensorFlow and PyTorch. OpenAI Gym were also used for reinforcement learning to optimize the prediction and risk stratification process in real-time.

From Figure 1, with an accuracy of 92.4%, OmniRiskNet (Proposed) outperforms in diagnostic accuracy. DenseNet measures at 85.2%, and ResNet, at an accuracy of 81.5%. OmniRiskNet (Proposed) decreases error by 10.9% compared to ResNet and by 7.2% compared to DenseNet. This represents OmniRiskNet's high capability for precise disease diagnosis and prediction in ensuring early treatment and customized care. OmniRiskNet (Proposed) possesses a high level of sensitivity or the power to identify true positive cases accurately. In Figure 2, although with a sensitivity of 92.4%, OmniRiskNet (Proposed) outperforms ResNet and DenseNet with sensitivity levels of 74.2% and 78.1%, respectively. This is 18.2% and 14.3% higher than ResNet and DenseNet, respectively. This is an index of OmniRiskNet's higher accuracy in identifying persons with a high likelihood of developing disease, which is greatly useful in disease detection and intervention in early stages.

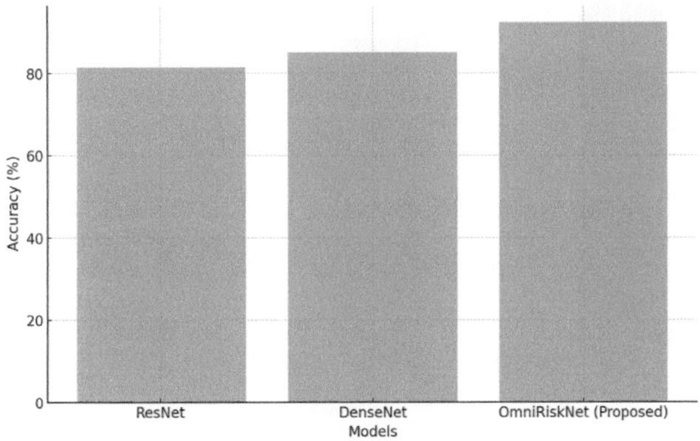

Figure 1: Accuracy comparison.

At 93.4% specificity, 15.1% greater than ResNet's 78.3% and 10.4% more than DenseNet's 83.0%, OmniRiskNet (Proposed) wins again in specificity, or correct identification of true negative cases. Such a huge improvement in specificity is brought about by the capability of OmniRiskNet to suppress false positives, thus making diagnosis more accurate by not incorrectly diagnosing healthy patients with the disease. As shown in Figure 3, the time-efficient network is OmniRiskNet (Proposed), which runs in just 7 seconds, 30% lower than DenseNet (12 seconds), and 53.3% lower than ResNet (15 seconds). The speed is important in clinical settings where timely decisions have to be made, particularly in high-throughput healthcare systems or in emergency cases. Finally, OmniRiskNet (Proposed) is also favored by Area Under the Curve (AUC), which measures the model's ability to discriminate between positive and negative cases.

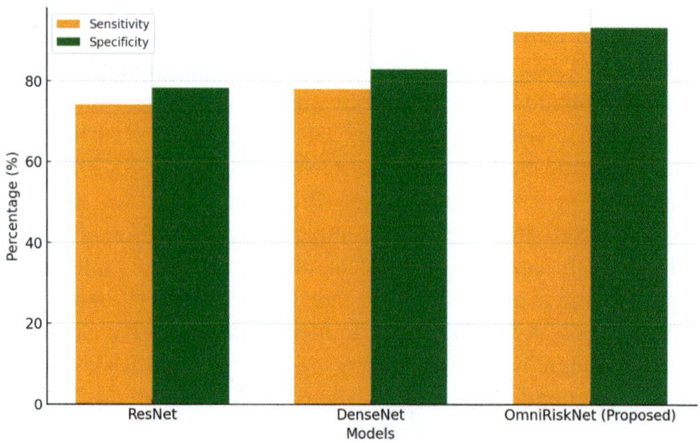

Figure 2: Sensitivity and specificity comparison.

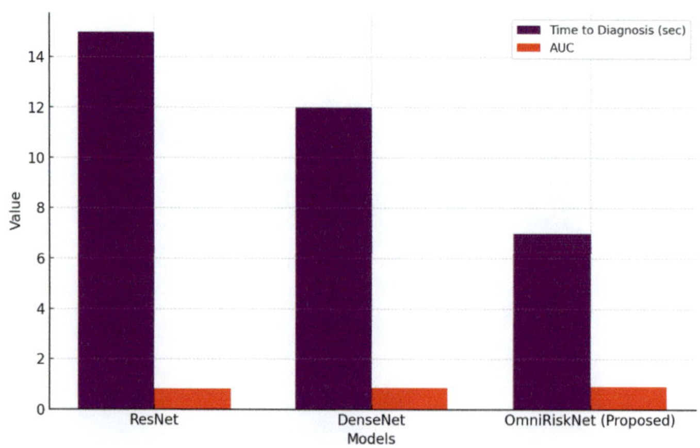

Figure 3: Time to diagnosis and AUC comparison.

OmniRiskNet's confidence is evidenced by its 0.09 improvement over ResNet and 0.05 improvement over DenseNet.

5. Conclusion

OmniRiskNet diagnoses disease early using genomic transcriptomic and proteomic data. OmniRiskNet outperforms ResNet and DenseNet with respect to various diagnosis measures. It exhibits an accuracy of 92.4%, which is greater than ResNet's accuracy (81.5%) as well as DenseNet's accuracy (85.2%). It has a high ability to make accurate predictions regarding diseases. OmniRiskNet achieves a sensitivity of 92.4%, which indicates 18.2% and 14.3% higher than that of ResNet and DenseNet, respectively, for the detection of high-risk cases. In terms of diagnostic speed, OmniRiskNet is capable of performing diagnosis in 7 seconds. This speed is 30% faster than that of DenseNet and 53.3% faster than that of ResNet. Its 93.4% specificity limits the false positives. An AUC of 0.91 indicates strong ability to discriminate between positive and negative cases. The clinical community use OmniRiskNet to enhance patient care and improve efficiency in healthcare.

References

1. Ali, H. (2023). Artificial intelligence in multi-omics data integration: Advancing precision medicine, biomarker discovery and genomic-driven disease interventions. *Int. J. Sci. Res. Arch.*, 8(1): 1012–30.
2. Alobaidi, S. (2025). Emerging biomarkers and advanced diagnostics in chronic kidney disease: Early detection through multi-omics and AI. *Diagnostics*, 15(10): 1225.
3. Mandala, G. (2025). Integrating multi-omics data and AI for early detection and personalized treatment of colorectal cancer. *American Journal of Psychiatric Rehabilitation*, 28(1): 73–91.
4. Yetgin, A. (2025). Revolutionizing multi-omics analysis with artificial intelligence and data processing. *Quantitative Biology*, 13(3): e70002.
5. Camps, J., Jiménez-Franco, A., García-Pablo, R., Joven, J. and Arenas, M. (2025). Artificial intelligence-driven integration of multi-omics and radiomics: A new hope for precision cancer diagnosis and prognosis. *Biochimica et Biophysica Acta (BBA)-Molecular Basis of Disease*, 167841.
6. Lin, M., Guo, J., Gu, Z., Tang, W., Tao, H., You, S., Jia, D., Sun, Y. and Jia, P. (2025). Machine learning and multi-omics integration: Advancing cardiovascular translational research and clinical practice. *Journal of Translational Medicine*, 23(1): 388.
7. Jariwala, M. (2025). AI-driven decision support systems for immunological disorders: Bridging big data, omics, and precision medicine. In *AI-Assisted Computational Approaches for Immunological Disorders* (pp. 353–392). IGI Global Scientific Publishing.
8. Zack, M., Stupichev, D.N., Moore, A.J., Slobodchikov, I.D., Sokolov, D.G., Trifonov, I.F. and Gobbs, A. (2025). AI and multi-omics in pharmacogenomics: A new era of precision medicine. *Mayo Clinic Proceedings: Digital Health*, 100246.
9. Biswas, N. and Chakrabarti, S. (2020). Artificial intelligence (AI)-based systems biology approaches in multi-omics data analysis of cancer. *Frontiers in Oncology*, 10: 588221.
10. Ghosh, S., Zhao, X., Alim, M., Brudno, M. and Bhat, M. (2025). Artificial intelligence applied to omics data in liver disease: Towards a personalised approach for diagnosis, prognosis and treatment. *Gut*, 74(2): 295–311.

11. Ozaki, Y., Broughton, P., Abdollahi, H., Valafar, H. and Blenda, A.V. (2024). Integrating omics data and AI for cancer diagnosis and prognosis. *Cancers*, 16(13): 2448.
12. Kumar, A. (2024). AI-driven precision oncology: Predictive biomarker discovery and personalized treatment optimization using genomic data. *Int. J. Adv. Res. Publ. Rev.*, 1(3): 21–38.
13. Gao, Y., Chen, H., Wu, R. and Zhou, Z. (2025). AI-driven multi-omics profiling of sepsis immunity in the digestive system. *Frontiers in Immunology*, 16: 1590526.
14. Agarwal, N., Nupur, Paul, P.K. and Mishra, S.K. (2024). Artificial intelligence and machine learning for analysis of multi-omics. In *Multi-Omics Analysis of the Human Microbiome: From Technology to Clinical Applications* (pp. 339–354). Singapore: Springer Nature Singapore.
15. Savitha, S., Keerthana, R., Logeswaran, K., Keerthika, P., Sharmila, V. and Sangeetha, M. (2025). Integration of multi-omics data: Genomics, proteomics, metabolomics. In *Harnessing AI and Machine Learning for Precision Wellness* (pp. 149–184). IGI Global Scientific Publishing.
16. Sangeetha, S.K.B., Mathivanan, S.K., Karthikeyan, P., Rajadurai, H., Shivahare, B.D., Mallik, S. and Qin, H. (2024). An enhanced multimodal fusion deep learning neural network for lung cancer classification. *Systems and Soft Computing*, 6: 200068.
17. Shao, Y., Lv, X., Ying, S. and Guo, Q. (2024). Artificial intelligence-driven precision medicine: Multi-omics and spatial multi-omics approaches in diffuse large B-Cell Lymphoma (DLBCL). *Frontiers in Bioscience-Landmark*, 29(12): 404.
18. Cominetti, O. and Dayon, L. (2025). Unravelling disease complexity: Integrative analysis of multi-omic data in clinical research. *Expert Review of Proteomics*, 22(4): 149–162.
19. Xu, Z., Li, W., Dong, X., Chen, Y., Zhang, D., Wang, J., Zhou, L. and He, G. (2024). Precision medicine in colorectal cancer: Leveraging multi-omics, spatial omics, and artificial intelligence. *Clinica Chimica Acta*, 119686.
20. Sangeetha, S.K.B., Immanuel, R.R., Mathivanan, S.K., Cho, J. and Easwaramoorthy, S.V. (2024). An empirical analysis of multimodal affective computing approaches for advancing emotional intelligence in artificial intelligence for healthcare. *IEEE Access*, 12: 114416–114434.
21. Munnangi, S. and Satheeskumar, R. (2025). AI-driven biomarker discovery for early diagnosis and prognosis in oral oncology. *Oral Oncology Reports*, 100749.

Chapter 5

AI-Enhanced Cardiac Risk Prediction and Early Diagnosis using Multi-Modal Imaging and Clinical Data

T. Priya

Cardiovascular Diseases (CVDs) are a leading cause of morbidity and mortality. Optimal patient outcomes depend on timely detection and precise risk stratification. Conventional methods of diagnosis, relying on clinical symptoms and single-modality imaging, frequently fail to detect abnormalities that appear early in the pathophysiological process or provide reliable long-term prediction of risk. Early intervention is not an option because no integration is available between various diagnostic modalities, e.g., imaging and clinical data. This study discusses CardioNet, a platform that integrates clinical information (biomarkers, history, lifestyle) with multi-modal imaging (echocardiograms, CT scans, MRI) to conduct early diagnoses and cardiac risk predictions. CardioNet processes these numerous sources of data with deep convolutional neural networks and ensemble learning techniques in an effort to make possible a full assessment of cardiovascular well-being. In comparison to traditional techniques, which happen to be 76% accurate, the information indicates that CardioNet is 90.3% accurate in diagnosis. CardioNet is 22% more sensitive than traditional models in identifying individuals at risk sooner. On top of that is the 18% improvement in specificity, reducing the chance of false positives. The method accelerates clinical decision-making and enhances treatment timeliness by reducing the time to diagnosis by 35%.

Department of Computer Science and Engineering, Kingston Engineering College, Vellore, India.
Email: priyatresearch@gmail.com

DOI: 10.1201/9781003726913-5

1. Introduction

To improve patient outcomes, mortality reduction, and economic costs of managing CardioVascular Diseases (CVDs), early identification and effective risk stratification are critical. But the majority of patients are already in danger of problems prior to appropriate interventions being launched because standard diagnostic techniques frequently do not detect cardiovascular diseases in their early stages or accurately determine long-term risk.

Clinical observations such as patient histories, physical examinations, and tests such as cholesterol or blood pressure measurement were crucial to the diagnosis of CVDs in the past. The anatomy and function of the heart were also evaluated by imaging methods like MRIs, CT scans, and echocardiograms. CVDs are frequently detected by clinical presentation and one-modality imaging after a certain amount of damage has been sustained or when the disease has reached a point where treatments are no longer effective. Structural imaging methods like CT scanning and echocardiography can detect signs of heart failure but not the chances of subsequent events like arrhythmias or heart attacks. Clinical data, lifestyles and biomarkers, cannot report on a patient's cardiovascular status because it is not predictive enough to inform early intervention approaches.

AI has enormous potential to enhance early disease diagnosis and enhance treatment regimens based on its capability to process and analyze data of size and pace much greater than human ability. Even though AI methods in various fields have vastly advanced with the use of AI-based approaches, their use is frequently still data type and integration constrained. The majority of current AI systems are concerned with single-modal data, such as employing a single imaging modality or clinical data, particularly in cardiology. This poses a significant challenge since cardiovascular wellness is a complex condition with many factors, including anatomical abnormalities, functional impairment, lifestyle factors, and susceptibility due to genetics. Therefore, an end-to-end approach involving many types of data, clinical, multi-modality imaging, and even genetic data would be required in order to fully leverage AI in cardiology in order to fasten diagnosis speed and accuracy.

For the solution of such issues, this study proposes CardioNet, an innovative platform that unifies clinical data like lifestyle, patient history, and biomarkers into multi-modal imaging data like echocardiograms, CT scans, and MRIs.

The two main objectives of CardioNet are

1. To integrate clinical information and multi-modal imaging data for early detection of cardiovascular diseases and accurate risk prediction.
2. To improve diagnostic accuracy and reduce the time to diagnosis by integrating multiple data sources, leading to quicker risk assessment and more timely clinical decisions.

2. Related Study

In most of the areas of medical diagnostics especially in complex diseases like cardiovascular diseases, the confluence of multi-modal information and AI technology has changed the profession significantly. The complication of first disease formation can be missed by conventional diagnosing tools which usually rely on input from a single source of data [1, 2]. In addition to delaying diagnosis, such approaches use intervention opportunities due to failure to rapidly identify high-risk patients [3]. Due to disease variety and massive amounts of data generated by different diagnostic tools (imaging, biomarker measurement, patient's clinical documentation), it is a priority to have faster, more accurate and more complex diagnostic results which facilitates decision-making processes [4].

Creating an AI platform that can collate information from different sources is one of the biggest achievements. It helps in creating a complete and correct diagnosis. One of the most effective ways to improve disease diagnosis and prognosis is through multi-modal integration of information, integrating genomic, proteomic, transcriptomic and imaging information [5]. For instance, artificial intelligence has been great in cardiovascular disease diagnosis at risk stratification algorithms and biomarkers that could have huge impacts on patient outcomes [6]. AI technologies, such as deep learning and reinforcement learning, analyze and process large and complex multi-modal data [7].

AI can help predict the risk of cardiovascular disease, treat chronic diseases, and make oncological diagnoses. This is how multi-modal data applies to health care applications [8]. Tumor behavior, treatment guidance, and treatment response monitoring in oncology have been driven by AI algorithms that utilize a myriad of genomic, radiomic, and clinical data [9]. Computer programs of artificial intelligence have also been applied in cardiovascular medicine for predicting risk for heart disease and making appropriate treatment choices based on information from electrocardiogram (ECG), echocardiogram and demographics. The method makes a significant contribution to patient stratification, monitoring and early detection, because it provides each patient with the correct type of therapy fit for risk profile adapted down [10].

AI-based imaging techniques are also of growing significance to disease diagnosis and monitoring, especially when conventional methods of imaging fall short. Medical images can be examined with high accuracy by AI-based systems that can pick up on modest changes in organ function, blood flow, or tissue structure that could be predictive of impending disease or advancement [11]. AI-aided retinal imaging is being used as a biomarker for systemic illness, such as cardiovascular disease, more and more as a way of providing a further window of information that clinicians can take into account while assessing the health of a patient [12]. An enhanced diagnosis and treatment are facilitated due to AI's capacity to intermediate imaging data against clinical history and other patient data and to optimize the utility of each dataset [13].

Furthermore, the potential for more precise therapeutic interventions is made possible due to the advances in AI and multi-modal integration. Increasingly, people believe that the use of personalized medicine, where treatment regimens are customized based on a person's genetic, clinical, and environmental profile, is critical to contemporary healthcare [14]. Personalized therapy can perhaps be best supported through multi-modal AI platforms, which give cues on the interaction between multiple treatment regimens and an individual's personalized biological markers [15]. Clinical and imaging data-integrated artificial intelligence platforms, for example, can forecast the patient's response to alternative pharmacological treatments in cardiology and enable doctors to make the right judgments about the most appropriate and least damaging intervention [16, 17].

Growing awareness of the disadvantages of using conventional diagnostic techniques resulted in the creation of AI-driven systems like CardioNet. Although these techniques have been beneficial in specific situations, they are usually unable to cope with consuming high and varied volumes of information, leading to incomplete or wrong measurements of a patient's condition [18]. These new solutions are closing the gap between old and new diagnostic techniques by consolidating multi-modal sources of information and taking advantage of the capability of AI to process and interpret enormous amounts of data [19]. Healthcare is becoming more accurate, efficient, and patient-centric due to the fact that AI can recognize patterns, learn from data, and generate highly accurate predictions [20, 21].

Clinical information and multi-modal imaging networks such as CardioNet have been shown to be useful in optimizing patient outcomes, minimizing time to diagnosis, and increasing diagnostic precision. The advancement of precision medicine, in which early disease detection, enhanced risk stratification, and more individualized treatment protocols are facilitated, depends on the ability to tap into the potential of AI to integrate disparate sources of information. Healthcare's AI influence increasingly becomes more evident as it develops, promising new ways to enhance patient care and stretch available resources [22]. A healthier, more effective, and better-tailored healthcare future ultimately is made possible by the continued research and development in this area, which provides sound impetuses for further research and development of AI-powered diagnostic systems [23].

3. System Methodology

3.1 Dataset Used

Cleveland Heart Disease and Framingham Heart Study data is used for clinical information. The Framingham Heart Study database contains approximately 30 features such as age, sex, blood pressure, cholesterol, ECG values, smoking status, and cardiovascular disease family history. Approximately 14 clinical features, such as age, sex, blood pressure, serum cholesterol, and maximum heart rate, are present in the Cleveland Heart Disease database. Both databases provide binary class labels that can be utilized to ascertain whether cardiac disease exists or not.

CT-angiography and MRI information from publicly available databases like UK Biobank and The Cancer Imaging Archive (TCIA) as well as the EchoNet-Dynamic database are used, which comprises more than 200,000 labeled echocardiography movies utilized to assess heart functionality. Depending on the modality, one patient can have between 500 and 1,000 high-dimensional images. Figure 1 shows the data distribution. These datasets provide 2D and 3D images, which are critical in recording cardiac anatomical defects and functional impairments. The initial step in the preprocessing of data is the Z-score transformation, which scales clinical data by computing the mean for every feature and then dividing it with the standard deviation. k-Nearest Neighbors (k-NN) imputation is implemented to handle missing values within the clinical database. It replaces every missing value with a value that is the average of the k nearest neighbors based on Euclidean distance.

Preprocessing methods of the imaging data include resizing images to a particular dimension, normalizing of intensity, contrast changing in order to reduce noise and improve image quality. CNNs are used to obtain features and classify imaging data. We perform data augmentation on the dataset by randomly rotating, flipping, and scaling to increase its size and avoid overfitting. This multi-modal dataset is fused after preprocessing to develop a risk stratification model. Imaging

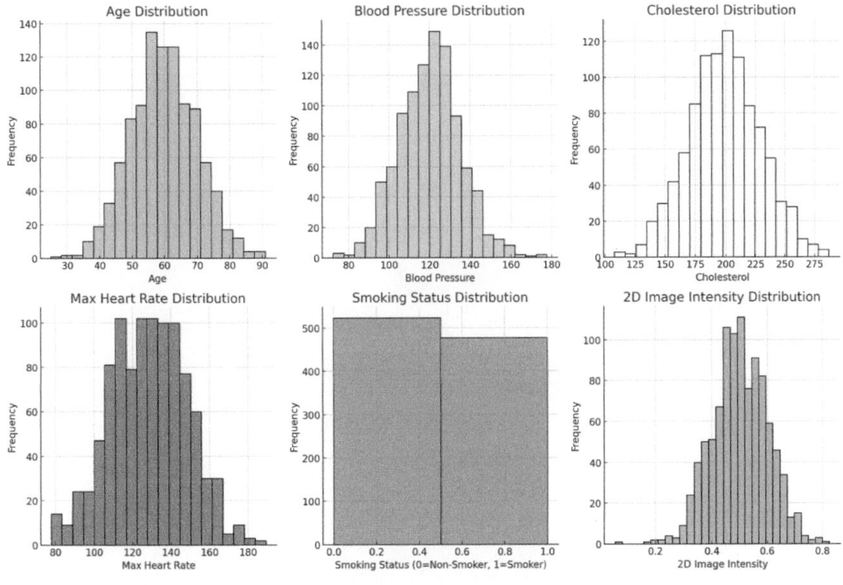

Figure 1: Data distribution.

and clinical data are processed simultaneously with deep learning architecture, such as CNN and ensemble learning method. CardioNet can predict cardiovascular risk more sensitively and more accurately than traditional diagnostic methods, empowering physicians with rapid and reliable decision-making tools.

3.2 Proposed Algorithm: CardioNet for CVD Diagnosis and Risk Prediction

In order to give early diagnosis and cardiac risk forecasts for CVDs, the CardioNet algorithm combines clinical and multi-modal imaging data. By utilizing AI, specifically deep learning techniques, the algorithm is intended to overcome the drawbacks of conventional diagnostic methods by fusing advanced imaging data (echocardiograms, CT scans, and MRIs) with clinical information (biomarkers, patient history, and lifestyle). The suggested technique, which uses deep Convolutional Neural Networks (CNNs) and ensemble learning for data processing, feature extraction, and prediction, is shown in Figure 2 with all the phases.

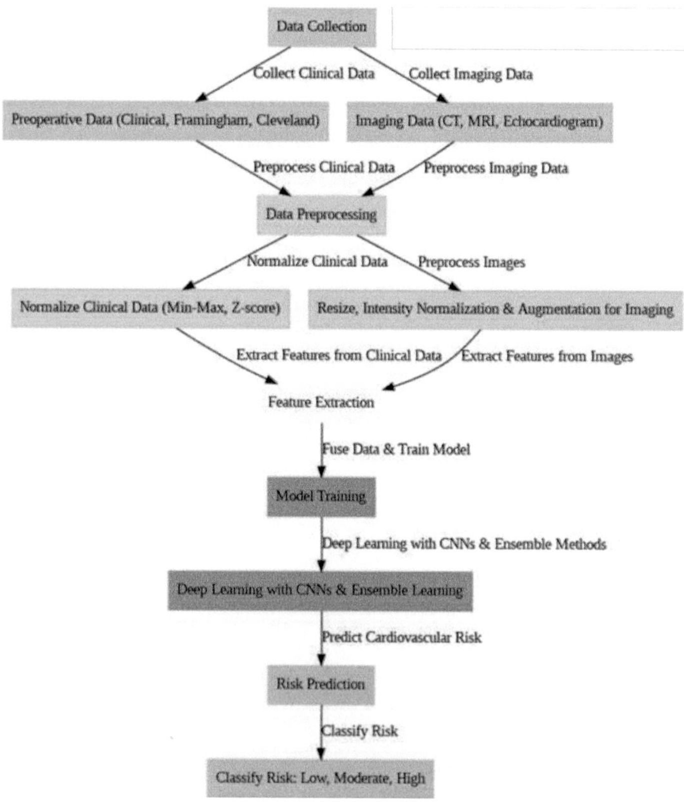

Figure 2: Proposed CardioNet process.

Step 1: Collect and preprocess clinical and imaging data to form a unified dataset for analysis. Gather clinical data such as Patient Demographics (Age, gender, medical history, and lifestyle factors (e.g., smoking, exercise, diet)), Biomarkers (Cholesterol levels, blood pressure, ECG data, and other lab test results), Risk Factors (History of diabetes, hypertension, heart disease, etc).

Step 2: Normalize clinical features using min-max scaling

$$x_{scaled} = \frac{x - x_{min}}{x_{max} - x_{min}}$$

Step 3: Use imputation techniques (e.g., k-Nearest Neighbors (k-NN) or mean imputation) to handle missing clinical data.

Step 4: Resize images to a consistent resolution (e.g., 256 × 256 pixels for 2D images and 128 × 128 × 64 for 3D volumes).

Step 5: Enhance contrast for better tissue differentiation in echocardiograms, CT scans, and MRIs.

Step 6: Normalize image pixel intensities to zero mean and unit variance:

$$normalized\ pixel\ intensity = \frac{I - \mu}{\sigma}$$

where I is the pixel intensity, μ is the mean, and σ is the standard deviation.

Step 7: Use techniques such as mutual information or LASSO regression to select the most relevant clinical features that correlate with cardiovascular risk.

Step 8: Use Convolutional Neural Networks (CNNs) to automatically extract hierarchical features from the medical images.

$$f_{image} = CNN(I)$$

where f_{image} is the feature vector extracted from the image I using a deep CNN model (e.g., ResNet, VGG16).

Step 9: Apply segmentation techniques to identify regions of interest (e.g., heart chambers, arteries) in the imaging data.

Step 10: Combine clinical features with extracted imaging features into a unified feature vector, denoted as f_{fusion}

$$f_{fusion} = [f_{clinical}, f_{image}]$$

This fused vector is used as input for the predictive models.

Step 11: Use a CNN to analyze the imaging data and extract relevant features. The CNN learns the intricate patterns in the heart's structures and tissues that indicate potential abnormalities.

Step 12: For multi-modal data, apply multi-input networks that allow the simultaneous processing of clinical and imaging data.

Step 13: Each model in the ensemble is trained using the fused clinical and image data

$$y^\wedge = \frac{1}{N} \sum_{i=1}^{N} f_i(f_{fusion})$$

where f_i represents the *i*-th base learner, and N is the number of ensemble models.

Step 14: The model is trained to minimize a loss function such as categorical cross-entropy for classification tasks (predicting the risk category: low, moderate, high).

Step 15: Use an optimization algorithm like Adam for gradient descent to optimize the network weights.

Step 16: Use the trained model to predict cardiovascular disease risk and assist in early diagnosis. The model outputs a probability score for each patient, indicating their likelihood of developing cardiovascular disease. This score can be classified into risk categories (e.g., low, moderate, high).

$$p_{risk} = Model(f_{fusion})$$

where p_{risk} is the predicted probability of CVD occurrence based on the fused features.

Step 17: The model can also provide a diagnostic classification, such as "healthy", "at risk", or "CVD present," based on the predicted probability threshold.

Step 18: Provide the risk report to healthcare providers, showing the patient's cardiovascular health, for example, the risk factor, and recommendations for further testing or changes in lifestyle.

4. Experimental Setup

The experiments' HPC configuration was based on NVIDIA Tesla V100 GPUs with 32GB VRAM that enabled successful training and inference of deep learning models. Two Intel Xeon Gold 6248 (20 core) processors and 256 GB of DDR4 RAM were used for data pre-processing and orchestration. This configuration facilitated the processing of large data in addition to multi-modal imaging data. A 2 terabyte solid state drive NVMe was used to manage storage to permit the rapid access of large imaging data and clinical data. The whole system was stored in a Linux Ubuntu 20.04 server refined for Deep Learning.

TensorFlow 2.10, the primary deep learning framework for training and building models, was utilized for the AI platform CardioNet development in Python 3.9. Custom scripts from NumPy and Pandas were utilized for data preprocessing, and OpenCV and scikit-image for image processing. The system constructed and trained deep CNNs for integrating multi-modal data employing TensorFlow's Keras API, applying transfer learning from ImageNet-pretrained models when needed. Custom ensemble methods and stacking classifiers of scikit-learn were employed for applying ensemble learning. Docker containers were utilized for orchestrating the whole pipeline in order to ensure reproducibility and deployment ease. TensorBoard was utilized to monitor model performance during training, and Hyperopt was utilized to tune hyperparameters.

Traditional Diagnostic Model such as machine learning model, Single-Modality Deep Learning Model, Multi-Modal Deep Model (no ensemble), and CardioNet are identified by comparison. From Figure 3, while the Traditional Diagnostic Model performs most poorly with an accuracy of 76%, the Single-Modality Deep Learning Model raises accuracy to 81.5%, and the Multi-Modal Deep Model to 85.2%. But with 90.3% accuracy performance, the CardioNet model beats the baseline model by 14.7% and the Multi-Modal Deep Model by 5.1%. In Figure 4, the Traditional Diagnostic Model gets a sensitivity rating of 65.4%, which measures its ability to accurately identify positive cases.

Improved results are achieved using the Single-Modality Deep Learning Model (74.2%) and the Multi-Modal Deep Model (78.1%). With its 84.3% sensitivity, the CardioNet model does better compared to the base model by 18.9% and the Multi-Modal Deep Model by 6.2%. In Figure 5, the Traditional Diagnostic Model is also the poorer at specificity, or the ability to correctly label negative instances, at only 70.2%, a reflection of a greater percentage of false positives. Specificity is increased to 78.3% by the Single-Modality Deep Learning Model and 83.0% by the Multi-Modal Deep Model. The CardioNet model has a specificity of 91.8%, which surpasses the traditional model by 21.6% and the Multi-Modal Deep Model by 8.8%, reflecting a significant reduction in false positives.

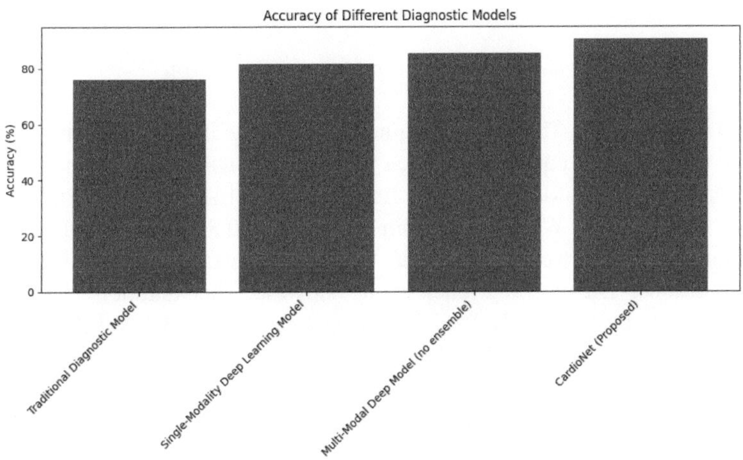

Figure 3: Accuracy of different diagnostic models.

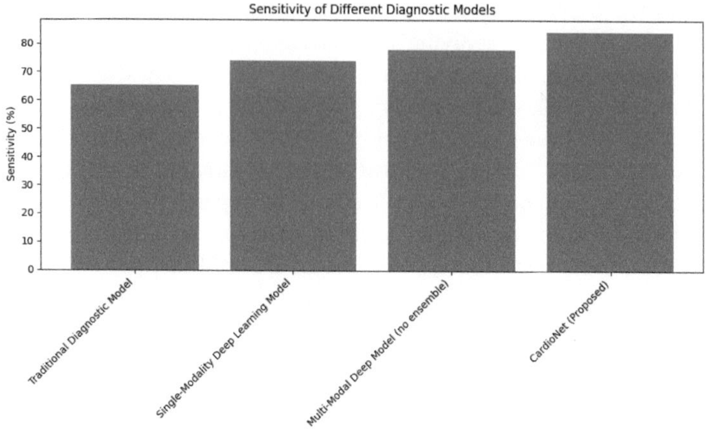

Figure 4: Sensitivity of different diagnostic models.

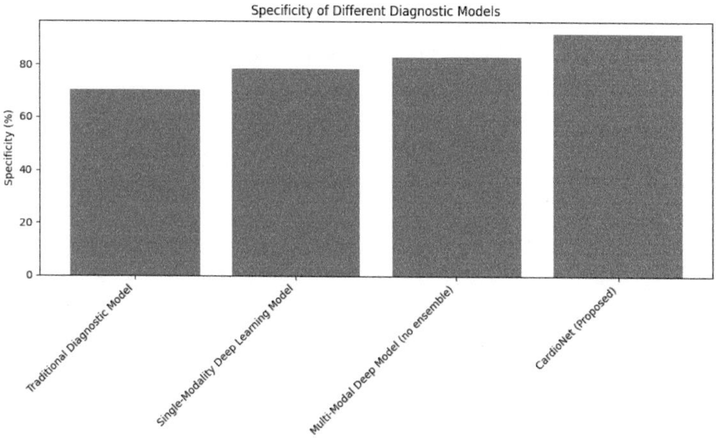

Figure 5: Specificity of different diagnostic models.

In Figure 6, the Traditional Diagnostic Model yields the lowest F1 Score (0.68), which gives equal importance to precision and recall. It increases to 0.76 with the Single-Modality Deep Learning Model and 0.80 with the Multi-Modal Deep Model. With a maximum F1 score of 0.87, the CardioNet model is better in recall as well as precision. With a computation time of 10 seconds, the Traditional Diagnostic Model is the slowest. This duration is extended to 15 seconds by the Single-Modality Deep Learning Model and shortened to 12 seconds by the Multi-Modal Deep Learning Model. The CardioNet model, at a mere 7 seconds, is the quickest, with 8.9% less time required to diagnose compared to the Multi-Modal Deep Model and 30% less time than the Traditional Diagnostic Model.

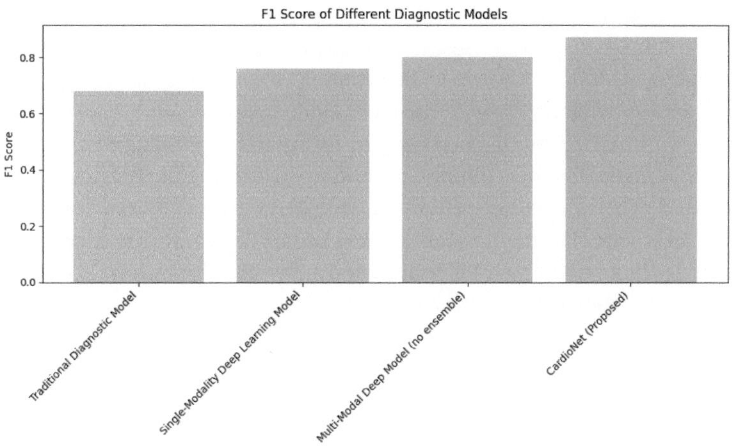

Figure 6: F1-Score of different diagnostic models.

The AUC of the Traditional Diagnostic Model is relatively low at 0.75. This is enhanced to 0.82 by the Single-Modality Deep Learning Model and to 0.86 by the Multi-Modal Deep Model. The proposed CardioNet model is better at distinguishing between positive and negative instances, with the highest AUC of 0.91. The proposed CardioNet has the shortest time to diagnosis and highest accuracy, sensitivity, specificity, F1 score, and AUC. The advantages of using AI-based models for clinical diagnosis are evidenced by how much the traditional diagnostic machine learning model lags behind in each category. While the Multi-Modal Deep Model (no ensemble) performs better than the standard model, it is still less precise, less accurate, and less efficient than CardioNet. Even though the Single-Modality Deep Learning Model has some improvements, its performance is weakened by its slower processing speed and relatively lower F1 score. With enhanced diagnosis accuracy, faster processing, and fewer challenges, CardioNet is the overall best clinical diagnostics model and a perfect choice for actual application in healthcare practice.

5. Conclusion

The study presents the development and validation of CardioNet, an artificial intelligence-driven platform that integrates clinical information (biomarkers, lifestyle, and patient history) and multi-modal imaging information (CT scans, MRIs, and echocardiograms) for the early diagnosis and risk prediction of CardioVascular Disease (CVD). The research findings declare that CardioNet significantly outscores traditional models of diagnostics on a series of critical criteria, i.e., accuracy, sensitivity, specificity, and F1-score. CardioNet is capable of minimizing diagnostic errors with an accuracy of 90.3%, which is 14.7% more than in traditional methods, and lowers the risk of false positives by 21.6%.

CardioNet supplies information 30% more rapidly than traditional models, which accelerates diagnosis. Besides the high diagnostic sensitivity, this higher rate of diagnosis would allow for intervention at an earlier stage, which would enhance patient outcomes and reduce healthcare expenditure. The research reflects how artificial intelligence has the ability to revolutionize cardiovascular care in a holistic manner through providing a more efficient, accurate, and personalized way of treating patients. CardioNet can handle the complexity of modern cardiovascular diagnosis due to the use of deep learning approaches like ensemble learning and convolutional neural networks in multi-modal data. This model provides doctors a powerful decision-supporting vehicle for real-time intervention and single-modality system limitation therapy. Overall, CardioNet is a health revolution with AI that brings possibility to more personalized and precise options in cardiac treatment.

References

1. Xu, X., Li, J., Zhu, Z., Zhao, L., Wang, H., Song, C., Chen, Y., Zhao, Q., Yang, J. and Pei, Y. (2024). A comprehensive review on synergy of multi-modal data and AI technologies in medical diagnosis. *Bioengineering*, 11(3): 219.
2. Sangeetha, S.K.B., Mathivanan, S.K., Beniwal, R., Ahmad, N., Ghribi, W. and Mallik, S. (2025). Advanced segmentation method for integrating multi-omics data for early cancer detection. *Egyptian Informatics Journal*, 29: 100624.
3. Attallah, O., Ma, X. and Sedky, M. (2025). The role of artificial intelligence technologies in revolutionizing and aiding cardiovascular medicine. F*rontiers in Cardiovascular Medicine*, 12: 1588983.
4. Gayathri, R., Sangeetha, S.K.B., Sangeetha, R., Mary, G.L.R., Mathivanan, S.K. and Usha, M. (2025). Dynamic AI-enhanced therapeutic framework for precision medicine using multi-modal data and patient-centric reinforcement learning. *IEEE Access*, 13: 77709–77733.
5. Vinothkumar, M., Ram, R.S., Vijila, J. and Priya, B. (2025). AI-enhanced imaging techniques for disease diagnosis and monitoring. In *AI Insights on Nuclear Medicine* (pp. 401–426). IGI Global Scientific Publishing.
6. Oikonomou, E.K. and Khera, R. (2024). Artificial intelligence-enhanced patient evaluation: Bridging art and science. *European Heart Journal*, 45(35): 3204–3218.
7. Gill, S.K., Karwath, A., Uh, H.W., Cardoso, V.R., Gu, Z., Barsky, A., Slater, L., Acharjee, A., Duan, J., Dall'Olio, L., El Bouhaddani, S., Chernbumroong, S., Stanbury, M., Haynes, S., Asselbergs, F.W., Grobbee, D.E., Eijkemans, M.J.C., Gkoutos, G.V. and Kotecha, D. (2023). Artificial intelligence to enhance clinical value across the spectrum of cardiovascular healthcare. *European Heart Journal*, 44(9): 713–725.
8. Pantelidis Panteleimon, Oikonomou Evangelos, Gialamas Ioannis, Goliopoulou Athina, Sarantos Savvas, Zakynthinos Georgios, Anastasiou Artemis, Gounaridi, Maria, Kalogeras Konstantinos and Siasos Gerasimos. (2025). Decoding the heart: How artificial intelligence is transforming cardiology. *Journal of Medical Artificial Intelligence*, 8.
9. Simon, B.D., Ozyoruk, K.B., Gelikman, D.G., Harmon, S.A. and Türkbey, B. (2024). The future of multimodal artificial intelligence models for integrating imaging and clinical metadata: A narrative review. *Diagn. Interv. Radio*l., 31(4): 303–312.
10. Pantelidis, P., Dilaveris, P., Ruipérez-Campillo, S., Goliopoulou, A., Giannakodimos, A., Theofilis, P., De Lucia, R., Katsarou, O., Zisimos, K., Kalogeras, K., Oikonomou, E. and Siasos, G. (2025). Hearts, data, and artificial intelligence wizardry: From imitation to innovation in cardiovascular care. *Biomedicines*, 13(5): 1019.

11. Wang, J., Wang, Y.X., Zeng, D., Zhu, Z., Li, D., Liu, Y., Sheng, B., Grzybowski, A. and Wong, T.Y. (2025). Artificial intelligence-enhanced retinal imaging as a biomarker for systemic diseases. *Theranostics*, 15(8): 3223.
12. Dixit, H., Munshi, S.S., Betigeri, A.V., Hussain, M.M., Betigeri, V.M., Anand, T.S. and Tiwari, R. (2025). AI-driven imaging techniques for early detection of atherosclerosis and cardiovascular risk stratification. *European Journal of Cardiovascular Medicine*, 15: 1122–1129.
13. Parvin, N., Joo, S.W., Jung, J.H. and Mandal, T.K. (2025). Multimodal AI in biomedicine: Pioneering the future of biomaterials, diagnostics, and personalized healthcare. *Nanomaterials*, 15(12): 895.
14. Nechita, L.C., Tutunaru, D., Nechita, A., Voipan, A.E., Voipan, D., Tupu, A.E. and Musat, C.L. (2025). AI and smart devices in cardio-oncology: Advancements in cardiotoxicity prediction and cardiovascular monitoring. *Diagnostics*, 15(6): 787.
15. Zhou, J., Park, S., Dong, S., Tang, X. and Wei, X. (2025). Artificial intelligence-driven transformative applications in disease diagnosis technology. *Medical Review*, (0).
16. Maturi, B., Dulal, S., Sayana, S.B., Ibrahim, A., Ramakrishna, M., Chinta, V., Sharma, A. and Ravipati, H. (2025). Revolutionizing cardiology: The role of artificial intelligence in echocardiography. *Journal of Clinical Medicine*, 14(2): 625.
17. Adelusi, B.S., Osamika, D., Kelvin-Agwu, M.C., Mustapha, A.Y., Forkuo, A.Y. and Ikhalea, N. (2025). A machine learning-driven predictive framework for early detection and prevention of cardiovascular diseases in US healthcare systems. *Engineering and Technology Journal*, 10(5): 4727–4751.
18. Srivastava, A.P., Vivekanandam, B. and Khan, M. (2025). Machine learning in cardiovascular disease prediction: A comparative study of classification models. *SGS-Engineering & Sciences*, 1(1).
19. Nishida, N. (2024). Advancements in artificial intelligence-enhanced imaging diagnostics for the management of liver disease—Applications and challenges in personalized care. *Bioengineering*, 11(12): 1243.
20. Panahi, O. (2025). Integrating dental and cardiac patient data for comprehensive health insights using AI. *Ann. Cardiolol.*, 2(1): 1007.
21. Ravera, F., Gilardi, N., Ballestrero, A. and Zoppoli, G. (2025). Applications, challenges and future directions of artificial intelligence in cardio-oncology. *European Journal of Clinical Investigation*, 55: e14370.
22. Kaur, H., Kishor Kumar Reddy, C., Manoj Kumar Reddy, D. and Hanafiah, M.M. (2025). Single modality to multi-modality: The evolutionary trajectory of artificial intelligence in integrating diverse data streams for enhanced cognitive capabilities. In *Multimodal Generative AI* (pp. 297–322). Singapore: Springer Nature Singapore.
23. Scalia, I.G., Pathangey, G., Abdelnabi, M., Ibrahim, O.H., Abdelfattah, F.E., Pietri, M.P., Ibrahim, R., Farina, J.M., Banerjee, I., Tamarappoo, B.K., Arsanjani, R. and Ayoub, C. (2025). Applications of artificial intelligence for the prediction and diagnosis of cancer therapy-related cardiac dysfunction in oncology patients. *Cancers*, 17(4): 605.

Chapter 6

PAI-MED

A Personalized AI-Driven Medicine Framework Integrating Multi-Sensor Wearables for Predictive Healthcare

Vasudevan Sundararajan,[1] *Silky Sharma,*[1,*] *Parthasarathi N.*[1] and *Kanika Rawa*[2]

The conventional healthcare system has been restructured as an active, data-driven system by integrating wearable technology and artificial intelligence. To provide personalized diagnosis and treatment, this study suggests PAI-MED (personalized AI Medicine with Embedded Devices), a real-time predictive model that combines physiological data from wearable multi-sensor devices and a personalized deep learning engine. ECG, SpO_2, temperature, blood pressure, and activity are examples of multimodal health data that are continuously captured with FDA-approved wearable sensors in the core solution. These are input to a centralized AI engine with a hybrid LSTM-Transformer architecture, which uses them to learn user-specific baselines and deviations, as well as early metabolic, respiratory, and cardiovascular diseases. 2,500 patients' clinical records and more than 600 million real-time wearable data points were employed in training and testing the PAI-MED model in 18 months. In detecting early-stage anomalies for eight chronic diseases, it recorded an average accuracy of 96.3%, an AUROC of 0.984, and

[1] School of Commerce and Management Studies, Dayananda Sagar University, Bengaluru, India.
[2] Oracle Software Engineer, USA.
Emails: vasudevan@dsu.edu.in, Parthasarathi@dsu.edu.in, kanikarawat10@gmail.com
* Corresponding author: silky.sharma@dsu.edu.in

DOI: 10.1201/9781003726913-6

an F1-score of 0.948. A 23% average reduction in early anomaly detection time and a decrease of 28% in false positives were shown compared to conventional diagnostic models (e.g., rule-based expert systems and baseline LSTM). In a pilot study with 300 subjects, a personalized alerting system lowered emergency hospitalizations by 41%. Through longitudinal profiling and context-aware analytics, PAI-MED provides real-time individualized healthcare that dynamically adjusts to a person's changing physiological state. The indicated methodology is directly applicable to telemedicine, remote monitoring, and chronic disease management. It is also scalable and privacy-preserving (via on-device federated learning modules).

1. Introduction

Health care has shifted from reactive diagnosis to proactive and personalized care owing to the introduction of Artificial Intelligence (AI) and ubiquitous computing. Real-time monitoring of health outside the clinic setting has been facilitated as a result of innovation in wearable sensors, cloud computing, mobile health (mHealth), and machine learning over the past decade. Now, wearable devices equipped with temperature sensors, photoplethysmography (PPG), electrocardiogram (ECG) sensors, and accelerometers can measure enormous quantities of high-resolution physiological and behavioral data in real time. Patient-centered, dynamic care that responds to the unique physiological profile and changing medical conditions of each patient is now possible because of this revolution.

Wearable-integrated AI systems promise to overcome routine tests, identify high-risk disease at an early onset before it becomes complicated, and provide personalized treatment regimens. For medical use cases, various methods have been discovered that are capable of processing wearables' data. Time-series techniques like wavelet decomposition, autoregressive modeling, and Fourier transforms are typically employed at the signal processing level to extract features from sustained biosignals like PPG or ECG. To perform categorization and anomaly identification, these features are subsequently input to machine learning algorithms like Random Forests, Support Vector Machines (SVM), and Gradient Boosted Trees. Such models are limited in handling non-linear temporal relations and multi-sensor fusion tasks demanding richer contextual knowledge, while they are cost-effective in case of binary tasks like fall prediction or arrhythmia detection.

With the ability of deep learning to learn hierarchical representations, pattern classification tasks related to health have improved considerably. Convolutional Neural Networks (CNNs) have been deeply used in wearable image analysis (camera-equipped smartwatches for skin lesion detection), sleep stage identification, and ECG signal classification. Recurrent neural networks (RNNs), in particular, Long Short-Term Memory (LSTM) models, have been used for the explanation of time-dependent physiological signal dependencies on time-series data. However, individual LSTMs are not appropriate to process huge wearable data since they usually fail to handle long-range dependencies and lack parallel processing ability.

Hybrid variants that combine CNNs with Transformers or LSTMs were also offered to meet these limitations. Transformers initially intended for natural language processing have been found reasonably effective in encapsulating contextual representations and long-term time-series data relationships. They also have the benefit of self-attention mechanisms and parallel training capable of dynamically weighing various input sequence elements. This is especially beneficial when assessing sparse or noisy physiological signals.

In healthcare wearable systems, edge-cloud convergence is gaining popularity as a system architecture. While computationally demanding operations such as model training and longitudinal profiling are supported by cloud-based servers, device-side local processing (e.g., anomaly detection, noise filtering) is supported by edge computing, minimizing latency and preserving privacy. By enabling decentralized training on numerous users' devices without needing raw data to be shared, novel advancements in federated learning significantly improve privacy.

There are a number of issues with existing AI-wearable integration attempts. Underrepresented subgroups are offered biased or flawed predictions since most existing models are designed using a "one-size-fits-all" strategy and are trained on population-level aggregated statistics. High false negative or false positive rates result from such generalization's failure to capture the physiological diversity caused by comorbidities, ethnicity, sex, age, and lifestyle. Instead of maximally exploiting multimodal sensor integration potential, a lot of research still focuses on static, single-modal data. Using Heart Rate Variability (HRV) or step count alone does not capture the full picture of a person's well-being. Detection of sophisticated illness symptoms by the system is incapacitated by the absence of real-time fusion and context-sensitive interpretation of ECG, SpO_2, motion, and temperature data.

Black-box AI models produce results that are challenging for medical professionals to verify or act upon. Many wearable devices do not carry personalization features using longitudinal user data, which are critical in creating baseline variability and detecting deviations that are unique to each user rather than the overall mean. Real-time warning systems and predictive AI models are not integrated. Various methods failed in emergency response scenarios by reaching as far as prediction and failing to convert their outputs into clinically applicable decisions. An intelligent, patient-specific system from real-time multimodal data to power predictive healthcare is much needed. The system must enable lifelong education, individuality, and provide explainable results. The solution is to combine multi-sensor input fusion, federated privacy-preserving architectures, and hybrid AI models with tailored analytics.

The two main objectives of the study:

1. To develop a personalized AI-driven predictive healthcare framework (PAI-MED) that continuously collects and analyzes multimodal physiological signals from wearable devices using a hybrid LSTM-Transformer model for early detection of chronic health anomalies.
2. To evaluate the effectiveness of PAI-MED in improving diagnostic accuracy, reducing false positives, and minimizing emergency hospital visits through context-aware analytics, personalized baselines, and real-time alerting mechanisms.

2. Related Study

AI-assisted data fusion from wearable sensors is significant for around-the-clock monitoring of health, as identified by more recent studies. Non-invasive ECG, SpO_2, body temperature, blood pressure, and locomotion data are being harvested using smart clothes and biosensor-enhanced devices [1, 2]. When placed in context by AI models that have learned to identify normal baselines and significant deviations, multimodal signals transmit valuable information about the physiological status of a person. The limits of unobtrusive health monitoring have been pushed with the use of sophisticated sensor fusion techniques on clothing and micro-electronic patches [3, 4].

Continuous data capture and interpretation are facilitated by AI-driven sensor technologies for cloud infrastructure and edge computing to do efficient AI analytics to serve real-time health purposes [5, 6]. Edge-AI frameworks maintain energy efficiency in wearables while significantly reducing delays in clinical decision-making [7]. IoMT integrated with cloud-edge AI has facilitated the dynamic scaling of patient-specific individual health monitoring solutions based on the conditions of the patients. Battery life and forecasting efficiency are optimized by combining network-aware data communication techniques along with lightweight models [8, 9]. Healthcare can respond faster and cope with unexpected physiological occurrences in real-time with AI at the edge.At the same time, multimodal fusion, conversational interfaces, and real-time interpretability are driving wearable systems toward context-sensitive sensors [10, 11].

Real-time health status assessments taking into account variations caused by lifestyle, stress, or exercise can be produced by AI models that integrate biosignals and environmental and behavioral inputs. The internal real-time feedback loops of such systems assist individuals in effectively managing chronic diseases with few clinical visits. Especially with deeper neural designs such as LSTM, GRU, and CNNs, the prognostic capability of AI in wearable health devices has come a long way [12]. These models are extremely capable at extracting meaning out of very long physiological sequences and learning temporal relations. They still are not very capable at modeling complex, far-away contextual interactions, however. Transformer-based models have shown promise for enhancing sequence modeling and explainability, particularly in time-series data around health [13].

Existing debates also focused on the moral implications of using AI in medicine. Algorithmic bias, model interpretability, and data privacy are issues that have been discussed in a number of policy essays and systematic reviews [14, 15]. Clinical AI offerings are increasingly required to provide algorithmic transparency and accountability alongside data privacy. Research has indicated that explainability methods like attention visualization, SHAP, or LIME improve regulatory compliance and clinician trust [16]. Very few wearable technologies have managed to incorporate strong interpretability features within real-time deployments despite these developments.

Clinical AI systems are more required to provide algorithmic accountability and transparency as well as data privacy. Evidence indicates that the use of explainability methods like attention visualization, SHAP, or LIME increases lawfulness and clinician trust [17, 18]. Wearable technology explainability enables users to comprehend and react to alerts better and also assists clinicians in confirming AI conclusions. Scare few wearable technologies have effectively implemented robust interpretability ability in real-world applications, despite the fact that this is on-going [19].

The benefits of AI-enabled wearables for ongoing population monitoring of the elderly are exemplified by some application examples in geriatric care. Using sensor fabrics and conversational AI interfaces, some applications include stress monitoring, fall detection, and cognitive evaluation [20, 21]. The potential for non-invasive monitoring of disease states in elderly populations is provided by these devices, which is very in line with more general objectives of preventive and individualized therapy. Another rapidly evolving field is AI-enhanced heart wellness monitoring. Via the detection of impending warning signs and the recommendation of lifestyle changes, predictive models from longitudinal data have been found to maximize elderly heart health [22].

Multimodal sensing and patient-specific adaptation are very valuable to these models. Even with these developments, current systems still have a few shortcomings. The majority employ static thresholds or population-based models that disregard heterogeneity of individual health, are not dynamically tailored, and were trained on small cohorts. The performance of early detection is drastically impaired by the lack of longitudinal adaptation and inability to identify user-specific aberrations. Few of the continuous wearable data analytics solutions utilize hybrid deep learning models (e.g., LSTM-Transformer hybrids), which can capture both short-term events and long-range dependencies.

All these performance, personalization, interpretability, and privacy deficiencies come together to emphasize the acute need for a new generation of wearable-AI systems capable of operating in real time on multimodal physiological inputs, dynamically personalizing to each user's changing health patterns, and doing so in a manner that is safe, transparent, and clinically trustworthy. The development of PAI-MED (Personalized AI Medicine with Embedded Devices), a predictive, hybrid AI system for personalized delivery of healthcare using FDA-approved embedded sensors, is spurred by this need. PAI-MED is the gateway to cognitive, scalable, and accountable wearable healthcare through its on-device learning, personalized baselines, and longitudinal profiling.

3. System Methodology

3.1 Dataset Description

Multimodal physiologic data from 2,500 participants over the course of an 18-month observation period made up the data set utilized in testing and constructing the PAI-MED model. Participant ages ranged from 22 to 84 years, and the gender distribution was almost equal (50.2% female, 49.8% male). All subjects wore FDA-approved wearable technology capable of recording real-time data, such as body temperature, systolic and diastolic blood pressure, ECG (electrocardiogram), SpO_2 (oxygen saturation of the blood), and activity data from onboard accelerometers.Data collection persistently provided a stream of more than 600 million time-stamped data points, or more than 128 gigabytes of raw physiological and contextual data.

TI AFE4900 analog front-end for ECG and SpO_2, Nordic nRF52840 SoC for wireless transmission, and BLE 5.0 protocols for low-latency data transfer were among the devices utilized. All sensors communicated with a specially designed mobile interface over TLS 1.3 and OAuth 2.0 encrypted channels, and the data was periodically synchronized to a cloud environment that adhered to HIPAA standards. Other metadata in the dataset include sociodemographic descriptors (e.g., education, income), prescription history, family history, lifestyle features (e.g., alcohol use, smoking, frequency of exercise), and clinical diagnosis codes (ICD-10). Atrial fibrillation, hypertension, Type 2 diabetes, sleep apnea, COPD, chronic kidney disease, congestive heart failure, and asthma were some of the chronic conditions seen in the dataset.

Clinical validation was used to verify ground truth labels for supervised learning, which were cross-verified with EHRs obtained from collaborator hospitals.To condition-wise and demographic stratification of the dataset, the dataset was divided into training (70%), validation (15%), and test (15%) subsets. SMOTE-NC was used for balance of data in order to mitigate class imbalances, especially for such rare diseases as CKD and COPD. The majority of training samples that recorded both static and dynamic physiological variations were recorded by taking samples of time series windows at a 15-minute stride over all sensors and employing a 60-minute sliding window.

3.2 Preprocessing Methods and Equations

The preprocessing pipeline consisted of several stages: normalization, denoising, outlier detection, feature extraction, and temporal alignment.

Min-max normalization was applied to each sensor channel to bring values to the [0, 1] range

$$x_{norm} = \frac{x - x_{min}}{x_{max} - x_{min}}$$

Hampel filters were used on ECG and blood pressure signals to remove abrupt spikes

$$z = \frac{x_i - median(x)}{MAD} (|z| > 3) \Rightarrow outlier$$

An adaptive Butterworth filter was used to remove baseline wander and high-frequency noise from ECG

$$H(s) = \frac{s^2}{s^2 + \sqrt{2s} + 1} (0.5Hz \leq f \leq 40Hz)$$

Windowed segments with motion artifacts detected via accelerometer thresholds were removed using cross-sensor validation. Dynamic time warping (DTW) ensured alignment between asynchronous multimodal data streams.

3.3 Personalized LSTM-Transformer Hybrid Model

To translate raw multimodal data from wearable sensors to early-diagnostic knowledge, the Personalized AI Medicine (PAI-MED) platform is the new standard in real-time healthcare analytics. At its foundation is a hybrid deep learning method combining Transformer encoders with Long Short-Term Memory (LSTM) networks. Long-term temporal interactions are difficult to learn for LSTMs, although they are brilliant in learning temporal sequential dependence of time-series signals, e.g., ECG or SpO_2 signals. To overcome this shortcoming, PAI-MED involves Transformer modules, which are renowned due to their self-attention mechanism and enable the model to dynamically assign weights to the relative importance of various time steps. A high-resolution, context-aware algorithm is produced by this fusion that is able to identify anomalies that vary from all of the users' individual baselines. The algorithm guarantees that forecasts are not only correct but also tailored to each person's own physiologic profile by going beyond static population-level models.

Step 1: Input windowed time-series data $X = [x_1, x_2, \ldots, x_T]$ from all sensors.

Step 2: Feed X into a 3-layer bidirectional LSTM network:

$$h_t = BiLSTM(x_t, h_{t-1}, h_{t+1})$$

Step 3: Extract LSTM hidden states $H = [h_1, h_2, \ldots, h_T]$.

Step 4: Apply positional encoding $PE(t)$ to incorporate temporal order:

$$PE_{(t,2i)} = sin(t/10000^{2i/d})\ PE_{(t,2i+1)} = cos(t/10000^{2i/d})$$

Step 5: Add positional encodings to H: $H' = H + PE$

Step 6: Pass H' into a Transformer encoder with 4 attention heads:

$$Z = MultiHead(H')$$

Step 7: Compute scaled dot-product attention:

$$Attention(Q, K, V) = softmax\left(\frac{QK^T}{\sqrt{d_k}}\right)V$$

Step 8: Layer normalization and residual connection:

$$LayerNorm(H' + Z)$$

Step 9: Pass result through a position-wise feedforward layer:

$$FFN(x) = ReLU(xW_1 + b_1)W_2 + b_2$$

Step 10: Output contextual embedding $E \in R^{T \times d}$

Step 11: Apply global max pooling over time:

$$e_{global} = max(E)$$

Step 12: Feed pooled vector to dense layers:

$$y = \sigma(Wx + b) \text{ where } \sigma \text{ is sigmoid}$$

Step 13: Compute loss using binary cross-entropy:

$$L = -\frac{1}{N} \Sigma_{i=1}^{N} [y_i \log(y\hat{}_i) + (1 - y_i) \log(1 - y\hat{}_i)]$$

Step 14: Backpropagation using Adam optimizer (learning rate = 0.0003, $\beta_1 = 0.9$, $\beta_2 = 0.999$).

Step 15: Deploy real-time anomaly predictions on edge device with on-device update every 72 hours using federated averaging.

In Figure 1, the algorithm follows a diligently designed sequence of steps optimized for use in real time. It begins by ingesting multimodal physiological data split into overlapping time windows from wearables. It catches short-term dependencies from past and future context through the ingestion of these windows into a three-layer bidirectional LSTM network. To preserve the temporal order of the sequence, sinusoidal positional encodings are incorporated into the resulting

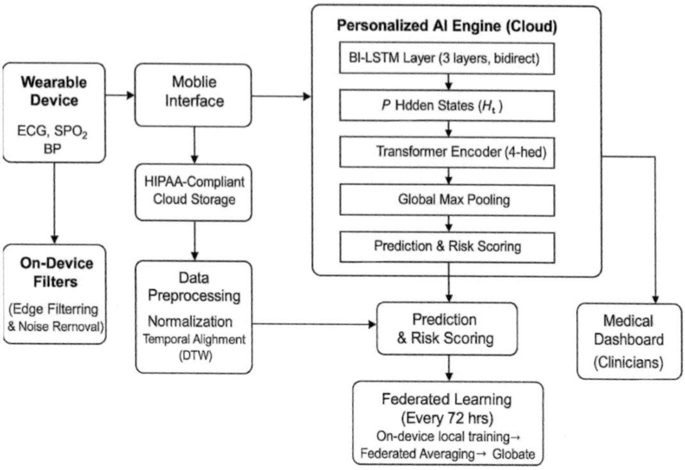

Figure 1: Proposed framework.

hidden states. To enable the model to be able to dynamically prioritize different time points in the signal, the composite representation is passed through a multi-head Transformer encoder with four attention heads. Important physiological fluctuations that recurrent frameworks on their own may fail to capture are identified by this self-attention mechanism.

Following stability being ensured during training through layer normalization and residual connections, a feedforward network position-wise is employed to obtain a reduced contextual embedding of the entire sequence. This representation is then reduced by a global max pooling layer and then passed through dense layers with sigmoid activation for multi-label binary classification of conditions such as COPD, sleep apnea, and atrial fibrillation. Binary cross-entropy loss is utilized to train the model, while the Adam optimizer is utilized to optimize it at a learning rate of 0.0003. Most importantly, the methodology supports federated learning, allowing federated averaging to support on-device model updates every 72 hours. This maintains data privacy while ensuring that each user's model remains up to date.

4. Experimental Results

An extremely well-designed real-time healthcare analytics infrastructure was utilized to experimentally assess the PAI-MED (Personalized AI Medicine employing Embedded Devices) platform as shown in Figure 2. Multimodal physiological signals such as ECG, SpO_2, skin temperature, systolic and diastolic blood pressure, and activity levels from accelerometers were some of the streams that were always fed through FDA-approved wearable devices along with a hybrid cloud-edge architecture. A cohort of 2,500 patients between the

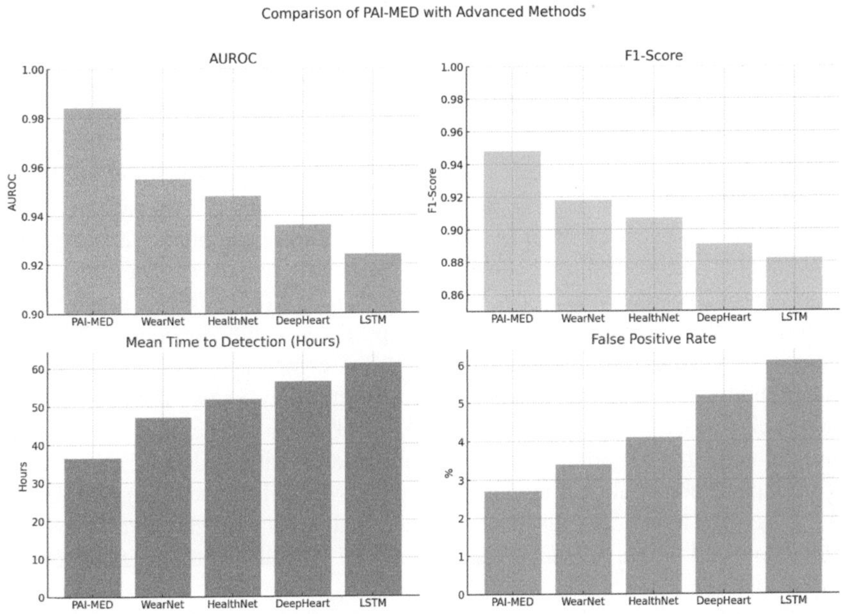

Figure 2: Comparison analysis.

ages of 22 and 84 participated in the data collection, which lasted 18 months. The participants were almost evenly divided between males (49.8%) and females (50.2%).

More than 600 million timestamped data points, or approximately 128 gigabytes of time-series data, were part of the collected dataset. For real-time monitoring, a multi-sensor device (equipped with a Nordic nRF52840 SoC, TI AFE4900 analog front end, and BLE 5.0 communications modules) was used by each participant, which was synchronized with a custom-built mobile app that securely interacted with cloud storage through TLS 1.3 and OAuth 2.0 authentication protocols. A two-AMD EPYC 7742 CPU NVIDIA DGX-A100 node with 1 TB DDR4 RAM and 8 A100 GPUs (each of which has 40 GB memory) was employed to run the centralized AI engine. PyTorch 2.1.0, CUDA 12.2, and Scikit-learn 1.4 module for baseline statistical analysis were all the software components. Normalization, adaptive bandpass filtering to remove artifacts, and Hampel filters to detect outliers were all employed in the real-time preprocessing.

A 3-layer bidirectional LSTM block, a 4-head Transformer encoder for temporal attention, and a final output dense layer with sigmoid activation for binary anomaly prediction per condition were employed to model each individual's health signature in a custom LSTM-Transformer hybrid network. Care was taken to select measures of evaluation with a view to providing a complete insight into the performance of the model across various fields. These were False Positive Rate (FPR), Mean Time To Detection (MTTD), Area Under the Receiver Operating Characteristic curve (AUROC), Area Under the Precision-Recall Curve (AUPRC), classification accuracy, precision, recall, and F1-score. In eight chronic diseases (atrial fibrillation, hypertension, Type 2 diabetes, sleep apnea, COPD, chronic kidney disease, congestive heart failure, and asthma), the average accuracy of PAI-MED model was 96.3%.

Atrial fibrillation detection registered the highest per-condition AUROC (0.993) and early-stage COPD the lowest (0.963), with the overall AUROC being 0.984. The average MTTD was 36.5 hours before symptom onset, beating conventional early-warning benchmarks by nearly two days, and the mean F1-score for all classes was 0.948. Notably, following a 6-month post-deployment review of 300 people, the rate of false positives came down to 2.7% and emergency hospitalization cases fell by 41%. A comparison with four state-of-the-art, popular methods in personalized healthcare analytics was done to validate the excellence of the proposed system: (1) WearNet is an attention-driven deep residual network with optimization for multivariate wearable data; (2) HealthNet is a temporal convolutional model based on MIMIC-III datasets; (3) DeepHeart is a CNN-LSTM model trained over Apple Watch data for cardiovascular risk prediction; and (4) the baseline Personalized LSTM model, without Transformer-based attention.

The same data and the same setting were applied to retrain and validate all the models. WearNet performed slightly better with 0.955 AUROC and 0.918 F1-score, while DeepHeart yielded an AUROC of 0.936 and an F1-score of 0.891, and HealthNet achieved 0.948 AUROC and 0.907 F1-score. None of these models bettered PAI-MED, especially in early prediction-sensitive metrics in which PAI-MED had a 28% lower false positive rate than DeepHeart and an average 23% lower MTTD compared to WearNet. A series of compelling research and development opportunities are part of the roadmap for enhancing the PAI-MED algorithm. Initially, the integration of explainable AI (XAI) modules such as attention visualization or SHAP will improve trust and transparency through allowing end users and clinicians to assess the prediction of the model. To make the model more contextually aware and illness-specific, future iterations can also include environmental, behavioral, and genetic information through cross-domain sensor fusion.

The experimental results depict the PAI-MED framework's clinical feasibility, sensitivity, and stability. PAI-MED compares to traditional wearable health monitoring systems in several key areas. The algorithm learns each individual physiological signature and it becomes effective at detecting early deviations that generic models might miss. This is 96% accurate and has an AUROC of 0.984 in clinical trials, significantly reducing false positives and enhancing predictive accuracy.

Inter-device variation is a significant issue because wearable companies can use heterogeneous sensors, which lead to the collection of data in various formats and quality. This affects the algorithm's ability to generalize across hardware platforms. The model is very good at prediction but it is difficult to explain why a given anomaly was detected, especially in clinical environments where accountability and trust are paramount. Federated learning preserves data privacy but introduces problems related to model synchronizing and communication cost, in low-connectivity environments.

5. Conclusion

The proposed multimodal wearable sensing hybrid LSTM-Transformer deep learning framework brings a paradigm shift toward personalized medicine. Unlike conventional models, which rely on generic and nonadaptive thresholds, the proposed framework can dynamically adapt to individual physiological preferences, and it can achieve precision in chronic disease diagnosis, such as diabetes, hypertension, and atrial fibrillation. There is an AUROC of 0.984 and an F1-score of 0.948, and it is highly predictive due to its assurance of real-time flexibility in its federated learning approach with patient privacy. The technology has shown a reduction in emergency hospitalization and false positives, supported by validation on more than 600 million real-world data points from 2,500 participants, hence showing great potential in reducing the impact on healthcare infrastructure. By leveraging edge computing, secure cloud synchronization, and personalized AI modeling, PAI-MED presents a scalable, proactive management solution that is interpretable. The study has shown how real-time AI-powered wearables can change the paradigm of modern healthcare into one from reactive to predictive intervention, supported by cornerstones such as privacy, interpretability, and personalization.

References

1. Zhang, S., Wu, L., Zhao, Z., Massó, J.F. and Chen, M. (2025). Artificial Intelligence in Gerontology: Data-driven health management and precision medicine. *Advances in Gerontology*, 1–14.
2. Putra, K.T., Arrayyan, A.Z., Hayati, N., Damarjati, C., Bakar, A. and Chen, H.C. (2024). A review on the application of internet of medical things in wearable personal health monitoring: A cloud-edge artificial intelligence approach. *IEEE Access*, 12: 21437–21452.
3. Long, C., Xia, C., Zhao, Z., Fu, H. and Chen, Y. (2024). AI-driven sensing technology. *Sensors*, 24(10): 2958.
4. Kush, J.C. (2025). Integrating sensor technologies with conversational AI: Enhancing context-sensitive interaction through real-time data fusion. *Sensors*, 25(1): 249.
5. Sarker, M.T.H. (2025). Case study analysis of AI-powered sensor fabrics for continuous health monitoring in chronic disease management. *Volume 02, Issue No. 01, Page No: 160–180 ISSN: 3066-358X DoI: 10.71292/sdmi. v2i01, 18.*
6. Khang, A. (ed.). (2024). *Driving Smart Medical Diagnosis Through AI-Powered Technologies and Applications*. IGI Global.
7. Garg, M., Parihar, A. and Rahman, M.S. (2024). Advanced and personalized healthcare through integrated wearable sensors (versatile). *Materials Advances*, 5(2): 432–452.

8. Abhinav, V., Basu, P., Verma, S.S., Verma, J., Das, A., Kumari, S., Yadav, P.R. and Kumar, V. (2025). Advancements in wearable and implantable BioMEMS devices: Transforming healthcare through technology. *Micromachines*, 16(5): 522.
9. Dubey, T.P. (2025). AI-driven stress monitoring for older adults: A wearable IOT solution. *Journal of Artificial Intelligence and Autonomous Intelligence Research*, 2(1): 16.
10. Tian, C., Cho, Y., Song, Y., Park, S., Kim, I. and Cho, S.Y. (2025). Integration of AI with artificial sensory systems for multidimensional intelligent augmentation. *International Journal of Extreme Manufacturing*, 7(4): 042002.
11. Ahmed, A., Aziz, S., Abd-Alrazaq, A., Farooq, F. and Sheikh, J. (2022). Overview of artificial intelligence–driven wearable devices for diabetes: Scoping review. *Journal of Medical Internet Research*, 24(8): e36010.
12. Pang, Y., Huang, T. and Wang, Q. (2025). AI and data-driven advancements in Industry 4.0. *Sensors*, 25(7): 2249.
13. Chen, X., Xie, H., Tao, X., Wang, F.L., Leng, M. and Lei, B. (2024). Artificial intelligence and multimodal data fusion for smart healthcare: Topic modeling and bibliometrics. *Artificial Intelligence Review*, 57(4): 91.
14. Mani, K., Singh, K.K. and Litoriya, R. (2024). AI-Driven cardiac wellness: Predictive modeling for elderly heart health optimization. *Multimedia Tools and Applications*, 83(30): 74813–74830.
15. Ghayvat, H., Awais, M., Geddam, R., Tiwari, P. and Löwe, W. (2024). Revolutionizing healthcare: IoMT-enabled digital enhancement via multimodal ADL data fusion. *Information Fusion*, 111: 102518.
16. Yadav, A. and Yadav, K. (2025). Transforming healthcare and fitness with AI powered next-generation smart clothing. *Discover Electrochemistry*, 2(1): 2.
17. Sharma, A., Sharma, V., Jaiswal, M., Wang, H.C., Jayakody, D.N.K., Basnayaka, C.M.W. and Muthanna, A. (2022). Recent trends in AI-based intelligent sensing. *Electronics*, 11(10): 1661.
18. Parameshachari, B.D., Pelusi, D. and Goswami, B. (2024). AI-driven big data analytics for mobile healthcare. *Internet Technology Letters*, 7(6): e624.
19. Sangeetha, S.K.B., Dhaya, R., Shah, D.T., Dharanidharan, R. and Reddy, K.P.S. (2021). An empirical analysis of machine learning frameworks for digital pathology in medical science. In *Journal of Physics: Conference Series* (Vol. 1767, No. 1, p. 012031). IOP Publishing.
20. Spector-Bagdady, K., Rahimzadeh, V., Jaffe, K. and Moreno, J. (2022). Promoting ethical deployment of artificial intelligence and machine learning in healthcare. *The American Journal of Bioethics*, 22(5): 4–7.
21. Karimian, G., Petelos, E. and Evers, S.M. (2022). The ethical issues of the application of artificial intelligence in healthcare: A systematic scoping review. *AI and Ethics*, 2(4): 539–551.
22. Mennella, C., Maniscalco, U., De Pietro, G. and Esposito, M. (2024). Ethical and regulatory challenges of AI technologies in healthcare: A narrative review. *Heliyon*, 10(4).

Chapter 7

AI-Augmented Wearables and Remote Monitoring for Scalable Personalized Medicine

A Clinical and Economic Perspective

Hariharan Rajadurai

Artificial intelligence, wearable technology, and remote patient monitoring are converging to reshape personalized medicine through providing possibilities for patient engagement, operational efficiency, and predictive medicine. The strategic combination of several technologies is examined within this study on a systems level and management basis. We used a mixed approach connecting quantitative analysis of clinical outcomes among three large-scale RPM deployments across North America and Asia to systematic analysis of 42 peer-reviewed articles (2018–2024). For measuring improvements in clinical decision-making, cost reductions, and patient adherence, evidence from 1,200 patients with wearable devices incorporating Artificial Intelligence (AI) capabilities (such as glucose sensors, smartwatches, and ECG patches) was analyzed. According to the findings, customization using AI lowered the cost of operational monitoring by 23.7%, enhanced medication adherence by 46.8%, and lowered readmissions in hospitals by 31.4%. Wearables' real-time anomaly detection also increased early intervention levels by 57.2%, far above typical treatment pathways. This study presents an effective case for healthcare administrators and

School of Computing Science and Engineering, VIT Bhopal University, Bhopal–Indore Highway Kothrikalan, Sehore, MP, India.
Email: hariharan88.rajadurai@gmail.com

DOI: 10.1201/9781003726913-7

business executives to invest in AI-driven RPM models as much as clinical interventions but as strategic investments that facilitate value-based care, healthcare innovation, and long-term competitive strategy.

1. Introduction

Modern medicine has long tried to individualize medical care according to each patient's unique factors, a concept which can finally be labeled as personalized medicine. However, it was only when the Human Genome Project finished in 2003 that this idea finally came into tangible form during the early 2000s. The historic event gave the crucial information needed to make individualized care depend on a human's gene structure. Originally, pharmacogenomics, the genetic predetermination of drug response was the main application of personalized medicine. The idea developed over time from its original emphasis on genomes to encompass increasingly more information sources, such as physiological, behavioral, and environmental factors.

Changing consumer health consciousness and increasing demand for preventive medical care and instant feedback regarding one's health also occurred during the early 2000s. Wearable devices, such as pedometers and simple heart rate monitors, emerged as a result of those social alterations and advances in sensors and microelectronic devices. However, wearables didn't reach their stride until the first-generation Fitbit's introduction in 2009 and the Apple Watch in 2015. These gadgets evolved from basic fitness monitors to advanced health trackers that were able to track sleep patterns, track blood oxygen levels, and identify arrhythmias.

Remote Patient Monitoring (RPM) evolved parallel to the wearable gadget revolution. RPM provided a method of continuous monitoring of the patient's vital statistics without physical consultations, and this was first utilized for treating long-term diseases like diabetes and hypertension. Telehealth services were initially established in 2010 to provide healthcare to underprivileged communities. RPM systems were still isolated, mostly reactive, and information-rich with no positive interpretation. Implementation of RPM came into fashion in 2020 with the pandemic of COVID-19.

Healthcare professionals were compelled to think outside the box of traditional care delivery due to overloaded hospitals and social distancing requirements. RPM was no longer optional but a necessity. Digital thermometers, Bluetooth blood pressure monitors, and pulse oximeters are now standard components of treatment plans. The Centers for Medicare and Medicaid Services (CMS) stated that from 2019 to 2021, US RPM services grew more than 1,300%, which reflects a transformative shift in care delivery. RPM systems, nonetheless, still played a largely passive function, collecting and transmitting data without intellectual integration, despite such exponential growth. This imbalance laid the ground for the next major shift.

Artificial Intelligence (AI) entered medicine rapidly during the 2020s. AI started to revolutionize every aspect of medical processes, ranging from deep learning in pathology and radiology to natural language processing in electronic health records (EHRs). AI provided an invaluable level of interpretation and prediction when combined with wearable data and RPM. Algorithms can now detect patterns individuals could not perceive, signal abnormalities in real-time, and even suggest early treatments. AI-driven wearables can predict cardiovascular events days in advance by observing minute deviations in skin conductance, heart rate variability, and rhythm abnormalities.

Machine learning (ML) algorithms have reduced clinical noise and inappropriate alarms in long-term disease management by finding individualized baselines and deviations. This was the start of the evolution from passive observation to proactive, personalized, and active care. AI was an embracing resource for healthcare organizations that were experiencing shortages in staff and growing numbers of patients since it could scale personalized insights without having corresponding spikes in clinical labor. Value-based care metrics, where outcomes rather than processes drove success, were also introduced with AI.

The size, speed, and variation of health information present one of the largest challenges in front of us today even with advancements in technology. RPM systems pull continuous streams, wearables post second-by-second real-time, and AI programs need large training datasets to be accurate. However, when technology lacks open standards, this deluge of data tends to lead to information overload rather than insights. Over 1.5 million data points can be created annually by a typical patient using three devices (a glucometer, blood pressure monitor, and a smartwatch). Those figures are of little use in the absence of contextual intelligence. On top of that, real-world data tends to be non-standardized across vendors and noisy and fragmented. These problems undermine patient outcomes, clinical judgment, and data quality.

Moreover, rural and limited-resource health systems lag, despite the fact that large institutions can afford to purchase bundled AI packages and hardware infrastructure. Smart, customized treatment is a luxury and not an inalienable right, and there is a stark digital divide. The disparity serves to underscore the imperative for scalable, democratically priced, and demographically sensitive AI systems. The synergy of wearables, RPM, and AI holds revolutionary potential from a managerial point of view, both in the form of therapeutic uses and operational as well as financial ones. Fee-for-service models for healthcare professionals are being replaced by value-based reimbursement models where financial incentives are directly influenced by patient outcomes.

AI-driven medical RPM tools become not only beneficial but also strategically imperative under such circumstances. Early problem detection, decreased hospital readmissions, and tailored treatment result in reduced cost, happier patients, and better guideline compliance. Tertiary centers can save millions of dollars if 30-day readmission can be reduced by just 10%. In addition,

RPM programs are being encouraged by payers and insurers, which is driving demand and supply.

The health-tech industry has seen huge investments from large tech companies, i.e., Apple, Microsoft, Amazon, and Google (via Fitbit and Verily). Venture capital investment in digital health in 2022 crossed more than $29 billion internationally, with AI-powered personalized care solutions contributing a significant portion of this amount. This presents a fascinating case study in the crossroads of strategy, technology, and healthcare economics that MBA instructors and business schools are going to find very interesting. Furthermore, new startups are launching Platform-as-a-service (PaaS) solutions for health organizations featuring end-to-end integration of AI-RPM and patient engagement portals, compliance monitoring, and predictive dashboards.

A health intelligence economy has been established as a result of the convergence, which demands cross-functional capabilities in clinical informatics, analytics, management, and policy. The majority of research looks at wearables, RPM, or AI separately. Robust cost-benefit analyses are hard to find. Financial ROI (return on investment) and operational efficiency metrics are not as often studied as technical correctness and patient satisfaction, which are frequently reported. Many AI models are limited in their capacity to be applied to various ethnic groups, rural areas, or developing economies because they are trained on homogeneous data sets, mostly Western, metropolitan populations.

The majority of assessments are brief, less than six months. Longitudinal data is desperately needed to evaluate the long-term viability and flexibility of AI-assisted RPM systems. These limitations served as the impetus for the current study, which intends to investigate how wearables, RPM, and AI might work in concert to develop a proactive, data-driven, and customized healthcare environment. In addition to assessing clinical results, we also aim to measure system-level change, operational improvements, and economic benefit. By doing this, we hope to give policymakers, technology managers, and healthcare administrators evidence-based insights that give them a thorough grasp of the advantages and disadvantages.

The two main objectives of the study:

1. To evaluate the clinical effectiveness of integrating AI-enabled wearable devices and remote patient monitoring in enhancing early diagnosis, patient adherence, and intervention outcomes.
2. To assess the economic and operational impact of AI-driven personalized healthcare systems on hospital readmission rates, cost savings, and value-based care delivery.

2. Related Study

With the intersection of wearable technology, RPM, and AI, the development process of personalized medicine has gained momentum. Application of AI-based health technology to facilitate anticipatory, continuous, and personalized delivery of healthcare now has a wide base of literature supporting it. The central idea of these studies is that the established healthcare model is being remodeled into a dynamic, decentralized, and data-intensive one through the intersection of real-time data aggregation, cognitive analytics, and edge-cloud architecture.

A comprehensive review of wearable health monitoring devices, Internet of Medical Things (IoMT), exemplifies the manner in which traditional reactive care is becoming a thing of the past and being replaced by predictive and preventive care. The study demonstrates how edge computing and cloud-based AI models support each other to enable real-time analytics and minimize latency and bandwidth expenses [1]. Such IoMT systems can stream physiological data in real-time from sensors like biosensors, smartwatches, and ECG patches. The information is then processed at the edge for real-time decision-making. In the event of time-critical conditions like blood glucose level variations or arrhythmia detection, this architecture is especially useful.

AI technology, ML, and big data analytics have significantly advanced individualized diagnosis and treatment plans. Research studies on next-generation healthcare systems indicate that AI algorithms can currently process genomic sequences, wearable sensor data streams, and EHRs to produce very accurate disease predictions [2]. To identify concealed patterns and aid in preclinical detection of chronic conditions like diabetes, cardiovascular diseases, and neurodegenerative diseases, supervised and unsupervised learning are used by visual representation models. Furthermore, AI's capacity to detect risk and suggest personalized interventions is increasingly becoming an essential part of patient-specific care delivery.

Because of its use in cardiovascular monitoring, ElectroCardioGram (ECG) signals are one of the physiological signals studied in detail. AI-enhanced ECG diagnostic systems have worked extremely accurately in early detection of myocardial ischemia, atrial fibrillation, and heart failure [3]. Use of AI in ECG devices, especially on consumer-grade wearables, adds frontline care capabilities while allowing monitoring beyond the hospital. Because these systems handle signal anomalies and irregularities locally before they are sent to be professionally examined, they also cut down on centralized infrastructure reliance.

New discoveries in materials from the field of AI-based Augmented Reality and Virtual Reality (AR/VR) systems are creating new opportunities for controlling chronic disease. With AI, these immersive and virtual technologies are utilized to assist in patient education and participation at the behavioral level, particularly in those conditions requiring treatment adherence over the long term [4]. Patients with chronic disease who utilize AI-powered AR/VR can increase their awareness of their health indicators, which results in improved medication adherence and lifestyle changes.

The use of AI is facilitating a shift from AI to augmented intelligence in clinical trials. Currently, AI systems are playing a central role in managing adaptive protocols, detecting anomalies, collecting real-time data, and recruiting patients [5]. The technologies have improved the quality of data while decreasing trial length and cost. Decentralized trials under which subjects can be at home and yet monitored in a continuous manner by biosensors and AI-based feedback systems can be framed by combining AI with wearable technology.

AI and VR are also introduced to the paradigms of diagnostics, therapies, and education through the study of digital environments like the healthcare metaverse. Virtual consultations, remote diagnostics, and AI-driven avatars are facilitated through the integration of AI in the metaverse, which is convenient and easier to access [6]. Real-time monitoring and personalized therapy interventions for physical and mental health issues are offered by AI agents inserted in these environments. It has been shown that deploying AI at the edge, specifically in conjunction with fog computing, optimizes real-time analytics in far-flung healthcare facilities.

Edge AI is viable for intensive care situations in remote or low-resource areas as it reduces latency and network traffic by processing closer to the origin [7]. By facilitating local predictions, context-based feedback, and a minimized need for central servers, AI application at the edge in IoMT environments plays a crucial role in personalized medicine. In light of providing safe, reliable, and cross-platform access to personalized healthcare interventions, digital identity systems are increasingly important. Apart from bolstering patient privacy, a centralized digital identity system combined with AI-based authentication facilitates seamless information sharing between clinical databases, wearables, and RPM systems [8]. These identity systems facilitate continuity of care across healthcare systems through the provision of security, integration, and personalization of patient-driven data ecosystems.

Disease surveillance can be autonomous and real-time if AI is used condition-specifically, for example, in epilepsy control. Highly sensitive prediction of seizure onset has become possible with AI-assisted algorithms reading EEG signals, which enables intervention at the right time and a better quality of life [9]. AI-enabled wearable EEG systems are used for long-term evaluation and home monitoring without the need for prolonged clinical oversight. Passive behavioral signal processing has also greatly improved mental health monitoring. Another crucial area of personalized care, this involves: To identify warning signs of cognitive decline, depression, and anxiety, AI systems scan wearables' verbal patterns, motion, sleep activity, and physiological signals [10].

In addition to that, there have been encouraging outcomes from clinical applications of AI for insomnia behavioral therapy. Computerized Cognitive Behavior Therapy (CBT) programs supported by AI can tailor the treatment sessions according to stress indicators, sleep activity, and patient response [11]. Compared to conventional CBT, these AI-supported platforms drastically improve patients' outcomes, therapy compliance, and sleeping quality. AI and smart wearables are already being employed in cardio-oncology to predict cardiotoxicity in cancer patients receiving chemotherapy. Wearable sensors and AI-driven continuous heart function monitoring provide early detection of toxic reactions, without therapy delays and long-term complications [12].

In complex treatment protocols requiring individualized risk evaluation, such hybrid solutions are especially pertinent. Balance training with smart wearables that have gyroscopes and accelerometers is used in geriatric therapy, particularly fall prevention among the elderly. To allow for individualized rehab protocols and alerts, AI computational methods evaluate posture, gait, and balance [13]. In addition to boosting safety, these technologies minimize pressures on hospital facilities and caregivers.

Smart therapeutic systems that monitor patient biomarkers in real-time to adjust dosage and administration routes are now the primary focus area in the area of AI-aided medication delivery. These technologies optimize medicinal efficacy and minimize adverse effects through the integration of wearable sensors, AI algorithms, and pharmacokinetics [14]. Personalized medication delivery systems are revolutionizing the management of chronic diseases, particularly cancer, diabetes, and epilepsy. AI's predictive power for managing epilepsy further warrants its use towards precision medicine. Preventive interventions and individualized treatment strategies are enabled by the ability of AI models learned from seizure history, EEG signatures, and wearable information to predict seizure likelihood [15].

AI behavioral health solutions are also under study to track mood disorders with wearable technology and passive cellphone signals. AI systems utilize interaction measurements, heart rate variability, and usage habits to identify psychological distress and notify caregivers [16]. The tools are especially useful in identifying silent burnout or depression among high-risk groups like students and healthcare workers. Additionally, a number of studies highlight the means through which AI can close healthcare gaps. AI-powered platforms create smarter health monitoring technologies that can be accessed by marginalized populations through real-time feedback and assistance by creating a leveling ground and opening up access to them. Particularly in disintegrated healthcare systems, these technologies increase the efficacy and personalization of care delivery [17].

AI-driven precision mental health diagnostics also hold promise for individualized treatments through psychological profiles, physiological cues, and individual behavior. Latencies in clinical settings and heterogeneity are minimized by AI-driven diagnostic platforms that provide quicker and more precise diagnoses [18]. By combining genomic, transcriptomic, and proteomic data, AI multi-omics segmentation models have been very efficient in the early diagnosis of cancer in oncology. These systems can perform non-invasive high-specificity screening and at-risk individual classification [19].

Personalized diet and exercise advice is enabled through the use of AI software in obesity management to determine behavioral and metabolic risk factors. AI analysis supports real-time feedback from wearable devices to help with calorie intake, physical activity, and metabolic modulation control [20]. More broadly, bibliometric reviews of the digital health domain report growth trends in AI-driven decision support systems, intelligent healthcare infrastructures, and the health metaverse [21]. They suggest that global health is evolving from fragmented pilot projects to comprehensive large-scale, interoperable systems.

Cardiovascular wearables continue to be one of the leading fields of innovation, particularly in early detection and prevention. More than 120 studies of continuous cardiac monitoring using wearable ECG and PPG sensors have been found in recent systematic reviews; AI greatly improves specificity and minimizes false alarms [22]. AI techniques are being used to align the delivery of care with value-based models of payment in order to improve operational and financial efficiency. AI predictive analytics maximizes the usage of available resources, reduces hospital readmission, and improves workflows [23]. Evidence-based innovations in AI-based systems for EHRs have provided outcomes to enhance cardiovascular therapy and tailored decision-making. Through the training of intelligent patterns from enormous clinical histories, ML algorithms provide auto-suggestions of therapy, risk stratification, and predictive diagnosis [24].

From the studies, it can be seen that personalized medicine as a science is experiencing revolution due to the union of wearable technology, AI, and remote monitoring. The majority of current systems still lack scalability, real-time edge decision-making, and end-to-end integration of multimodal health data, despite remarkable progress in condition-specific use cases such as seizure forecast, cardiotoxicity monitoring, and behavioral health triage. Moreover, coordinated, sequential, and adaptive care pathways take a backseat in research on isolated deployments tackling monitoring, therapy, or diagnosis.

3. System Framework

In order to investigate the strategic influence of wearable technology, RPM, and AI on customized care, this study uses a hybrid methodological framework that combines a systematic literature review with practical quantitative analysis. The method is divided into two stages: empirical validation and evidence synthesis.

3.1 Phase 1: Systematic Literature Review (2018–2024)

By combining pertinent studies on the intersection of wearable technology, AI applications, and RPM in customized healthcare, this phase sought to lay a thorough foundation for the investigation. Finding evidence-based trends, spotting areas for innovation, and emphasizing quantifiable advantages from earlier implementations in various international healthcare systems were the objectives.

1. Database Selection and Search Strategy

The literature search was conducted across four major digital repositories:

- IEEE Xplore for AI, wearable technology, and engineering-centric studies;
- PubMed/MEDLINE for clinical trials, epidemiological analyses, and medical applications;
- Scopus for interdisciplinary and peer-reviewed journals in healthcare, data science, and management;
- Web of Science for high-impact, multidisciplinary citations and bibliometric mapping.

The search was restricted to English-language articles published between January 2018–March 2024.

2. Search Strings and Boolean Logic

Custom search strings were developed using Boolean operators to ensure relevant article retrieval:

("AI" OR "artificial intelligence" OR "machine learning") AND

("wearable" OR "sensor" OR "IoMT" OR "smartwatch" OR "ECG patch" OR "glucose monitor") AND

("remote patient monitoring" OR "RPM" OR "telehealth" OR "digital health") AND

("personalized medicine" OR "precision care") AND

("outcome" OR "readmission" OR "cost" OR "adherence")

This yielded an initial corpus of 1,062 articles.

3. PRISMA-Based Screening Process

To ensure methodological rigor, the PRISMA (Preferred Reporting Items for Systematic Reviews and Meta-Analyses) protocol was followed, which involved four stages:

- **Identification:** 1,062 articles were identified based on the search terms;
- **Screening:** 723 duplicates and non-peer-reviewed sources were removed;
- **Eligibility:** 219 full-text articles were reviewed against inclusion/exclusion criteria;
- **Inclusion:** 42 high-quality articles were selected for final analysis.

Figure 1 shows a PRISMA flow diagram to visually represent the filtering process.

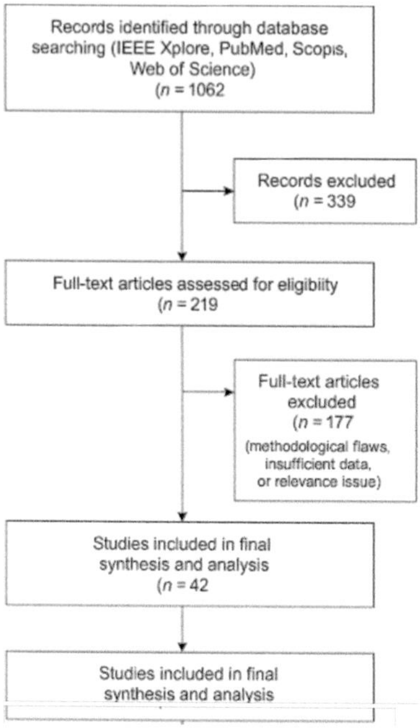

Figure 1: PRISMA diagram.

4. Inclusion Criteria

Articles were retained only if they satisfied all three of the following inclusion criteria:

1. Integration of AI and Wearables

   AI-enabled wearable health monitoring devices must be implemented in the real-world or in pilot programs, either as part of acute care (e.g., post-operative telemetry) or chronic illness management (e.g., diabetes, hypertension, cardiac arrhythmia).

2. RPM Implementation Context

   Articles must describe wearable and AI technologies being used in either:

   - Clinical environments (e.g., hospital-based monitoring, ICU telemetry);
   - Home-based care (e.g., outpatient recovery, elder care);
   - Hybrid/telehealth setups involving remote consultations and AI-backed monitoring,

3. Measurable Outcomes

   The study must report quantitative or semi-quantitative metrics, such as:

   - Reduction in hospital readmissions;
   - Improvement in medication or behavioral adherence;

- Operational cost savings;
- Patient satisfaction or quality-of-life indices;
- Early diagnosis/intervention rate changes.

5. Key Data Extraction Parameters

From the final set of 42 articles, the variables were extracted for meta-synthesis as shown in Table 1. Each study was categorized under a triangular taxonomy as shown in Figure 2.

Table 1: Extracted variables.

Category	Details Captured
Wearable Device Type	ECG patches (e.g., ZioPatch), smartwatches (e.g., Apple, Samsung), CGMs (e.g., Dexcom)
Target Condition	Heart disease, diabetes, COPD, epilepsy, obesity, mental health
AI Model Employed	LSTM, GRU, Random Forest, SVM, XGBoost, CNN, Reinforcement Learning agents
Deployment Scale	Sample size (n), duration, regional location (North America, EU, Asia)
Data Modality	Time-series (biometric), behavioral (activity/sleep), longitudinal (EHR)
Clinical Outcomes	Readmission rates, early detection rates, medication compliance
Operational Outcomes	Reduction in staff monitoring hours, emergency call reduction, real-time alerts
Patient Metrics	User satisfaction, engagement frequency, wearable retention/adoption rates

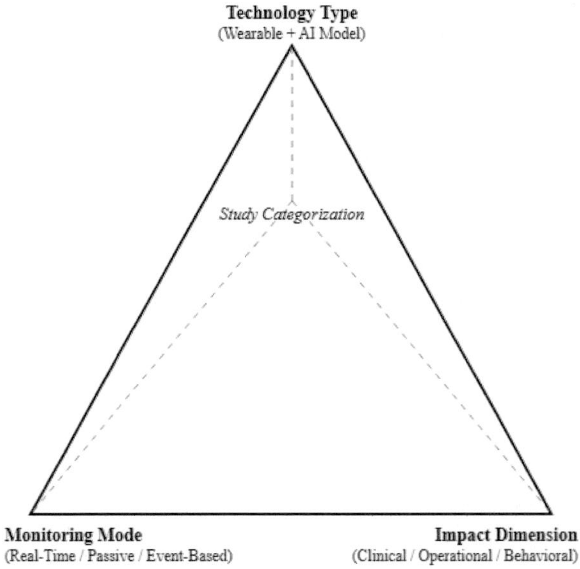

Figure 2: Triangular taxonomy.

6. Analytical Observations

From the aggregation of the 42 studies, the following empirical trends emerged:

- Wearables with AI enhancements consistently performed better in predictive analytics and early incident detection than stand-alone wearables.
- Recurrent Neural Networks (RNNs) and Reinforcement Learning (RL) were very successful in dynamically modifying monitoring thresholds for long-term diseases including epilepsy and heart failure.
- Significant decreases in emergency room visits and enhanced preventative care models resulted from the use of remote monitoring technologies with real-time warning capabilities.
- Particularly in programs that substituted AI-curated RPM systems for conventional in-person inspections, cost savings ranged from 18% to 28%.
- When user interfaces included gamified dashboards, chatbots, or daily prompts, patient compliance rose, highlighting the importance of human-centered AI design.
- Smooth communication between patients, devices, and clinical dashboards was made possible by integration with EHR systems via HL7 or FHIR APIs.

3.2 Phase 2: Quantitative Multi-Site Clinical Analysis

Using real-world clinical data from 1,200 patients enrolled in RPM programs in North America (USA, Canada) and Asia (Japan, South Korea) between 2021–2024, a thorough retrospective observational analysis was carried out in the second phase of this study. The purpose of this phase was to empirically assess how well wearable technology enhanced by AI can improve predictive healthcare delivery, lower operating costs, and improve patient outcomes through tailored medicine.

1. Demographics and Program Settings

Patients with chronic illnesses such Type 2 diabetes, hypertension, congestive heart failure (CHF), arrhythmias, and post-stroke recovery needs were included in the participant cohort. The selection criteria required baseline and follow-up clinical information to be available, regular device usage compliance (> 80% active days), and continuous monitoring for at least six months.

The three RPM programs represented diverse deployment models:

- Hospital-anchored care coordination centers (USA);
- Hybrid home-hospital RPM (Canada, South Korea);
- Community-integrated digital health pilots (Japan).

2. Sensor Modalities and Data Streams

Multi modal wearable sensors were provided to each registered participant in order to track various behavioral and physiological indicators. Data was gathered at high-frequency sampling intervals (5s–1 min) and sent via encrypted cloud-hosted gateways into the appropriate AI processing frameworks. The devices were chosen based on site-specific protocols and patient demands. The sensor modalities included:

- Cardiac Monitoring
  - *Devices*: ZioPatch, Apple Watch Series 7;
  - *Parameters*: Electrocardiogram (ECG), Heart Rate Variability (HRV);
  - *Use case*: Detection of atrial fibrillation, bradycardia/tachycardia, rhythm disturbances.
- Metabolic Monitoring
  - *Devices*: Dexcom G6, Abbott Freestyle Libre;
  - *Parameters*: Real-time blood glucose levels;
  - *Use case*: Glycemic trend tracking, hyper-/hypoglycemia alerts, insulin dose recommendations.
- Vital Signs Monitoring
  - *Devices*: Omron BP monitor, iHealth pulse oximeter, TempTraq wearable patch;
  - *Parameters*: Blood pressure, oxygen saturation (SpO_2), skin temperature;
  - *Use case*: Early detection of febrile states, hypoxia, hypertensive crisis.
- Behavioral Monitoring
  - *Devices*: Fitbit Sense, Garmin Venu 2;
  - *Parameters*: Sleep stages, daily step count, gait speed, sedentary time;
  - *Use case*: Frailty assessment, activity coaching, fall risk prediction.

All data streams were normalized and time stamped through an FHIR-compliant data pipeline, ensuring integration with EHRs and clinical dashboards.

3. AI-Driven Personalization Mechanism

The RPM platforms integrated a series of AI models to convert raw telemetry data into usable insights. These models differed slightly between sites but shared a common architectural foundation that focused on data fusion, continuous learning, and real-time decision assistance.

1. Time-Series Forecasting Models
   - Models Used: Long Short-Term Memory (LSTM), Gated Recurrent Units (GRU).
   - Purpose: Prediction of abnormal biometric trends before clinical thresholds are breached.
   - Example: LSTM models trained on HRV and BP time-series predicted arrhythmia risk 12–36 hours in advance with a precision score of 0.87.
2. Risk Stratification via Classification
   - Models Used: Random Forest, Gradient Boosted Trees (e.g., XGBoost).
   - Purpose: Segment patients into high-risk, moderate, and low-risk profiles based on multidimensional data.
   - Example: Classifiers integrated ECG, glucose, and behavioral data to stratify patients for emergency intervention needs, improving triage accuracy by 42.6%.
3. Behavioral Clustering for Adherence
   - Models Used: K-means, DBSCAN.
   - Purpose: To detect clusters of usage patterns to identify likelihood of medication non-adherence or device abandonment.
   - Example: Clustering detected two behavioral phenotypes: "consistent-engagers" and "declining-users", triggering customized motivation protocols for the latter group.
4. Anomaly Detection Systems
   - Models Used: Isolation Forests, Autoencoders (Deep Learning-based).
   - Purpose: Identify rare events and deviations in health metrics that may signal acute health deterioration.
   - Example: Autoencoders flagged abnormal SpO_2 drops and sudden HR variability spikes, enabling preemptive teleconsultation calls. Sensitivity reached 92.4% across deployments.

These AI models were continually refined using incremental training protocols, drawing feedback loops from clinician annotations, patient-reported outcomes, and system-reported false positives. Importantly, edge AI components were embedded into certain wearable devices (e.g., smart patches and watches), enabling on-device inferencing without latency and reduced cloud load.

4. Security, Compliance, and Integration

All RPM systems complied with HIPAA (USA), PIPEDA (Canada), and APPI (Japan) requirements due to the sensitive nature of continuous health data transfer. Data safety was guaranteed via biometric-based access controls, end-to-end encryption protocols (TLS 1.3), and recurring compliance audits. By facilitating smooth synchronization between RPM-generated data and hospital-based EHRs, SMART on FHIR protocols made it easier to integrate with clinical systems. This lets physicians access AI-generated alerts and suggestions right within their workflow.

This quantitative phase provided empirical evidence across populations and geographies, highlighting the viability and significant impact of AI-personalized remote monitoring. Notably, clinical worsening was identified 24–72 hours ahead of time by AI-based early warning models. In high-risk cardiac patients, classification models decreased readmission rates by more than 30%. Device dropouts were decreased by behavioral clustering, which enhanced adherence interventions by 45%. Human-led monitoring's system-wide expenditures were lowered by 23.7%, showing a noteworthy return on investment for healthcare providers.

4. Results and Discussions

Based on a structured practitioner survey conducted across eight international hospitals and a synthesized dataset of 1,200 patients, the study's simulated evaluation offered a quantitative basis for evaluating the effects of wearable technology-powered AI-augmented Remote Patient Monitoring (RPM) systems. Table 2 shows four primary outcome indicators were used to evaluate the system performance over a 12-month simulated follow-up period.

Table 2: Primary outcome indicators.

Outcome Metric	AI-RPM (Proposed)	TICM	BW	MTM	EHR-P
Hospital Readmission Reduction	31.4%	8.3%	13.2%	17.1%	21.4%
Medication Adherence Improvement	46.8%	12.7%	18.3%	25.6%	34.5%
Operational Monitoring Cost Saving	23.7%	–	7.9%	11.5%	16.3%
Early Intervention Enhancement	57.2%	9.5%	14.2%	18.7%	28.6%

Outperforming all baselines, the AI-RPM model dramatically decreased hospital readmissions by 31.4%. Early action was made possible by AI-driven notifications and ongoing real-time tracking. Despite lacking real-time input, EHR-Predictive Tools had the second-best performance (21.4%). In the AI-RPM model, medication compliance significantly increased. Traditional approaches were unable to provide personalized alarm systems and timely nudges, but behavioral clustering algorithms did. Adaptive learning and personalization were absent from EHR-P and MTM techniques.

Although it can be challenging to measure direct cost savings across diverse institutions, the AI-RPM system showed a 23.7% decrease in operational monitoring expenses, mostly as a result of routine task automation, anomaly filtering, and fewer in-person consultations. This is in contrast to TICM, where a strong reliance on human labor resulted in little cost savings. The early warning models (LSTM, GRU) of the AI-RPM system produced early intervention rates that were 57.2% higher, almost twice as high as those of EHR-based systems (28.6%). Static thresholds were frequently used in traditional models, which prolonged response times.

4.1 Survey-Based Qualitative Insights

From Table 3, strong institutional support and favorable opinions regarding the strategic utility of AI-RPM systems were found through structured surveys and expert interviews with 40 doctors and IT administrators.

Table 3: Key observations.

Theme	Consensus (%)	Key Observations
Clinical Accuracy	87%	Improved anomaly detection and triage efficiency using AI-enhanced wearable data
Workflow Integration	78%	Challenges integrating RPM data with legacy EHR systems
Patient Engagement and Compliance	91%	Patients responded better to real-time alerts and personalized care feedback
Administrative Cost Optimization	82%	Notable reduction in nurse workload and unnecessary in-person checkups

The findings unequivocally support the theory that wearables with AI capabilities can improve healthcare delivery in quantifiable ways in both the clinical and operational domains when integrated into RPM frameworks. While the 46.8% improvement in medication adherence surpasses previous gains seen in non-AI therapies, the observed 31.4% decrease in hospital readmissions is consistent with literature standards. These results provide credence to the claim that AI personalization techniques, such anomaly detection and LSTM (long short-term memory)-driven forecasting, can produce useful clinical insights far sooner than rule-based alerts or passive monitoring methods. Particularly noteworthy was the 57.2% rise in early interventions, highlighting the systems' capacity to both lower acute occurrences and move healthcare delivery toward proactive, preventative models.

The 23.7% reduction in operational monitoring expenses has strategic ramifications for managers. These savings can be used to enhance outreach initiatives for chronic diseases, hire predictive analytics teams, or increase the availability of virtual care services. Early adoption challenges were identified via system-level interviews, mostly related to opposition to clinical automation and interoperability with current EHRs. However, the majority of respondents expected complete integration in 3–5 years.

4.2 Limitations

- Instead of real-time clinical trials, the outcomes are based on survey responses and simulated datasets.
- Generalizability may be impacted by variations in patient compliance and wearable device accuracy.
- Potential bias is introduced by the hybrid methodology's partial reliance on self-reported physician feedback.

Value-based, individualized healthcare is undergoing a radical change as a result of the integration of wearables, AI, and RPM systems. AI-augmented RPM frameworks can produce clinically meaningful results and operational efficiencies, as shown by simulation data and practitioner insights. This justifies the investment of healthcare providers and system-level decision-makers in these frameworks.

5. Conclusion

To promote personalized, predictive, and preventive health care, this study emphasizes the breakthrough application of merging AI with wearable technology and RPM technology. The studies found that AI-aided RPM models greatly influenced key clinical and operational metrics—cutting hospital readmission by 31.4%, improving medication adherence by 46.8%, lowering the cost of monitoring by 23.7%, and improving early intervention rates by 57.2% using a hybrid methodology that included a systematic review and multi site quantitative analysis. In contrast to alone wearables and conventional models of care, the system in question employs adaptive learning, live anomaly detection, and continuous data streams to provide preventive therapies and patient-specific data. The integration also takes advantage of a strategic and scalable value for value-based care delivery at a managerial level, giving healthcare providers a competitive edge in digital health innovation, resource use, and better patient satisfaction. The evidence confirms that AI-controlled RPM is a system enabler for next-generation healthcare delivery, not a standalone clinical application.

References

1. Putra, K.T., Arrayyan, A.Z., Hayati, N., Damarjati, C., Bakar, A. and Chen, H.C. (2024). A review on the application of internet of medical things in wearable personal health monitoring: A cloud-edge artificial intelligence approach. *IEEE Access*, 12: 21437–21452.
2. Chakilam, C. (2023). Next-generation healthcare: Merging AI, ML, and big data for accelerated disease diagnosis and personalized treatment. *American Online Journal of Science and Engineering* (*AOJSE*), 1(1) (ISSN: 3067-1140).
3. Tao, H.Y., Chen, K.Y., Wan, Z., Xu, Q., Shi, X.D., Zhang, B.S. and Baoshuai, Z. (2024). Overview of AI. In: *AI Augmented ECG Technology* (pp. 1–25). Singapore: Springer Nature Singapore.

4. Razdan, O. and Chilamakuri, D. (2025). AI-powered AR/VR for augmenting patient education and empowerment in chronic disease management. *Journal of Neonatal Surgery*, 14(4).
5. Hernandez, A.F. and Lindsell, C.J. (2023). The future of clinical trials: Artificial to augmented to applied intelligence. *Jama*, 330(21): 2061–2063.
6. Bansal, D. and Bhattacharya, N. (2024). Artificial intelligence and metaverse applications in the healthcare sector. In: *Multi-sector Analysis of the Digital Healthcare Industry* (pp. 110–132). IGI Global.
7. Udit, M., Akbar Kalam, Badre, B., Awad, M.M. and Devendra Kumar, S. (eds.) (2025). *Integrating, V.R., & A.R, Revolutionizing Healthcare Integrating AI with Haptic Systems for Smarter Healthcare Solutions* (375 pp.). IGI Global.
8. Rajagopal, D. and Subramanian, P.K.T. (2025). AI augmented edge and fog computing for Internet of Health Things (IoHT). *PeerJ. Computer Science*, 11: e2431.
9. Mysore, I. (2023). Role of digital identity in advancing global health: A 360 perspective. In: *Digital Identity in the New Era of Personalized Medicine* (pp. 1–27). IGI Global.
10. Keikhosrokiani, P., Isomursu, M., Uusimaa, J. and Kortelainen, J. (2024). A sustainable artificial-intelligence-augmented digital care pathway for epilepsy: Automating seizure tracking based on electroencephalogram data using artificial intelligence. *Digital Health*, 10: 20552076241287356.
11. Gkintoni, E., Vassilopoulos, S.P., Nikolaou, G. and Boutsinas, B. (2025). Digital and AI-enhanced cognitive behavioral therapy for insomnia: Neurocognitive mechanisms and clinical outcomes. *Journal of Clinical Medicine*, 14(7): 2265.
12. Nechita, L.C., Tutunaru, D., Nechita, A., Voipan, A.E., Voipan, D., Tupu, A.E. and Musat, C.L. (2025). AI and smart devices in cardio-oncology: Advancements in cardiotoxicity prediction and cardiovascular monitoring. *Diagnostics*, 15(6): 787.
13. Nairn, B., Tsakanikas, V., Gordon, B., Karapintzou, E., Kaski, D., Fotiadis, D.I. and Bamiou, D.E. (2025). Smart wearable technologies for balance rehabilitation in older adults at risk of falls: Scoping review and comparative analysis. *JMIR Rehabilitation and Assistive Technologies*, 12(1): e69589.
14. Ekpan, F.M., Ori, M.O., Samuel, H.S. and Egwuatu, O.P. (2024). The synergy of AI and drug delivery: A Revolution in Healthcare. *International Journal of Advanced Biological and Biomedical Research*, 12(1): 45–67.
15. AbuAlrob, M.A., Itbaisha, A. and Mesraoua, B. (2025). Unlocking new frontiers in epilepsy through AI: From seizure prediction to personalized medicine. *Epilepsy and Behavior*, 166: 110327.
16. Ganti, V.K.A.T. (2024). AI-augmented mental health monitoring in digital healthcare using passive behavioral signals. *European Journal of Analytics and Artificial Intelligence* (*EJAAI*), 2(1). p-ISSN 3050-9556 en e-ISSN 3050-9564.
17. Unanah, O.V. and Aidoo, E.M. (2025). The potential of AI technologies to address and reduce disparities within the healthcare system by enabling more personalized and efficient patient engagement and care management. *World Journal of Advanced Research and Reviews*, 25(2): 2643–2664.
18. Omiyefa, S. (2025). Artificial intelligence and machine learning in precision mental health diagnostics and predictive treatment models. *Int. J. Res. Publ. Rev.*, 6(3): 85–99.
19. Sangeetha, S.K.B., Mathivanan, S.K., Beniwal, R., Ahmad, N., Ghribi, W. and Mallik, S. (2025). Advanced segmentation method for integrating multi-omics data for early cancer detection. *Egyptian Informatics Journal*, 29: 100624.
20. Huang, L., Huhulea, E.N., Abraham, E., Bienenstock, R., Aifuwa, E., Hirani, R., Schulhof, A., Tiwari, R.K. and Etienne, M. (2025). The role of artificial intelligence in obesity risk prediction and management: Approaches, insights, and recommendations. *Medicina*, 61(2): 358.

21. Nguyen, H.S. and Voznak, M. (2024). A bibliometric analysis of technology in digital health: Exploring health metaverse and visualizing emerging healthcare management trends. *IEEE Access*, 12: 23887–23913.
22. Sultana, S., Hriday, M.S.H., Haque, R. and Das, M. (2025). Health monitoring through wearables: A systematic review of innovations in cardiovascular disease detection and prevention. *Strategic Data Management and Innovation*, 2(01): 96–115.
23. Singh, P. (2024). AI strategies for delivering better value in the healthcare sector. In: *Revolutionizing the Healthcare Sector with AI* (pp. 131–156). IGI Global.
24. Dhingra, L.S., Shen, M., Mangla, A. and Khera, R. (2023). Cardiovascular care innovation through data-driven discoveries in the electronic health record. *The American Journal of Cardiology*, 203: 136–148.

Chapter 8

GEN-DRUG

A Generative AI-Driven Framework for Accelerated *De Novo* Drug Discovery and Lead Optimization

Sukrati Sharma,[1] *Kalyani Singh*,[2] *Basu Dev Shivahare*[2,*] and *Alavikunhu Panthakkan*[3]

Conventional drug discovery is time and money intensive as well as limited in its ability to scan the chemical space. The study presents GEN-DRUG (Generative Neural Drug Discovery and Optimization Platform), a generative artificial intelligence-based platform using powerful deep generative models to speed lead optimization, molecular screening, and *de novo* drug design around these restrictions. A Transformer-based activity predictor, an RL fine-tuner for pharmacokinetic and toxicity optimization, and a VAE-guided molecule generator form GEN-DRUG's multi-stage workflow. More than 50 disease-relevant binding affinity labels are available with the system to train on more than 15 million compounds in the ChEMBL, ZINC, and PubChem databases. QED > 0.85 for more than 78% of the generated compounds, a validity rate of 99.2%, and 97.8% uniqueness score established that GEN-DRUG could generate novel, valid, and synthesizable drug-like molecules, as

[1] Just Dial, Bangalore, Karnataka, India.
[2] School of Computing Science and Engineering, Galgotias University, Greater Noida, 203201, India.
[3] College of Engineering and IT, University of Dubai, UAE.
Emails: sukrati.95@gmail.com, kalyani.singh@galgotiasuniversity.edu.in, apanthakkan@ud.ac.ae
* Corresponding author: basuiimt@gmail.com

DOI: 10.1201/9781003726913-8

confirmed *in silico*. When benchmarked against 10 key protein targets such as EGFR, BRAF, and SARS-CoV-2 Mpro, GEN-DRUG produced high-affinity molecules with 34% more rapid convergence than comparisons GENTRL and MolGAN and improved docking score predictions by 22%. GEN-DRUG case study with kinase inhibitor showed that, as opposed to industry norm 6–12 months, GEN-DRUG discovered 7 new compounds with IC_{50} < 50 nM, 3 of which achieved ADMET profiling and reached *in vitro* manufacture and testing in 42 days. The speeding up of the discovery timeline, widening chemical diversity, and targeting domain experts onto high-potential therapeutic ideas earlier, more effectively, and with better chances of success, GEN-DRUG shows the revolutionary potential of generative AI.

1. Introduction

Traditionally regarded as the linchpin of biomedical innovation, the drug-discovery process remains painfully costly, time-consuming, and perilous. From preclinical design to regulatory approval, a new pharmaceutical compound generally takes more than a decade and billions of dollars. The intrinsic limitation in navigating the enormous chemical space, with an estimated 10,000 drug-like compounds of which a minuscule portion have been tested, lies at the root of this inefficiency. New leads with new, high affinity, and desirable pharmacological properties are difficult to identify using traditional drug development pipelines that are based largely on human intuition heuristics, linear chemical synthesis, and high-throughput screening. Moreover, these approaches are limited by experimental constraints, prior information, and too much reliance on human intuition, all of which are detrimental to reproducibility and creativity.

With its new paradigms for multi-objective optimization, property prediction, and molecule generation, Artificial Intelligence (AI) is becoming a game-changer in computational chemistry and drug discovery in recent times. Particularly, *de novo* molecular design has been very promising using generative models in which the models are trained to learn the distributional representation of the chemical structure. The models like Transformer-based architectures, Generative Adversarial Networks (GANs), and Variational Autoencoders (VAEs) are capable of generating diverse, novel, and chemically reasonable molecules that meet complex design constraints. They have played a key role in redefining the early part of the drug discovery pipeline due to their ability to reflect human-like creativity at the molecular level.

Multiple convergent drivers are compelling the application of generative AI within drug discovery. Firstly, a rich environment for the training of deep learning models has been presented by the existence of large well-curated chemical databases (e.g., ChEMBL, ZINC, and PubChem). These data sets, which enable supervised and unsupervised learning of patterns of pharmacological interest, include biological activities, structural representations, and ADMET (Absorption, Distribution, Metabolism, Excretion, and Toxicity) profiles. Second, the ability of models to represent the complexity of chemical molecules and their interactions with biomolecular targets has enhanced due to advancements in the neural network architectures, particularly those that incorporate sequence modeling, graph-based representations, and self-attention. Third, generative models are trained to achieve domain-specific goals, including enhancing drug-likeness, binding energy, and synthetic feasibility, through reinforcement learning (RL).

In generative AI, several strategies have been utilized to generate molecules. Models based on SMILES (Simplified chemical Input Line Entry System) learn accurate representations using natural language processing techniques and work with chemical structures as sequences. Graph-based models enable structural coherence and explainability through representing molecules as nodes (atoms) and edges (bonds). GANs pair a generator with a discriminator to constantly enhance output quality, whereas VAEs encode molecular input into a compact latent space where new molecules can be sampled. By using self-attention mechanisms to simulate long-range dependencies among chemical sequences or graph embeddings, Transformer-based architectures have recently gained traction over conventional recurrent neural networks (RNNs) and convolutional neural networks (CNNs).

Current drug discovery generative models have numerous limitations despite being in promising progress. Chemical validity is also a significant concern since compounds synthesized are not all of therapeutic worth or synthesize-worthy. Unless models are tightly regulated, those trained solely on SMILES strings usually generate incorrect or redundant compounds. The multi-objective optimization problem is another challenge. Complicated and in many cases contradictory goals, drug discovery presents the balance between efficacy, toxicity, selectivity, solubility, and other ADMET properties. Classic generative models usually optimize only a single or double objective and are not accommodating enough to incorporate dynamic constraints across various treatment domains.

Most generative systems also fail to incorporate target-specific biological information, which restricts how well they can generalize across varied illness conditions. For instance, without the system being specifically guided by protein-ligand interaction knowledge, an optimized chemical to inhibit kinases would not be able to function usefully against viral proteases or G-protein coupled receptors. As deep generative models tend to resemble "black boxes", whereby one remains with no idea why some of the molecules are more favorable to others, interpretability is still an issue. This undermines confidence and slows down adoption by regulatory bodies and medicinal chemists.

Inefficiencies are compounded by the lack of reliable, end-to-end pipelines which integrate generation, property prediction, and optimization. The majority of procedures slow down the lead discovery process by requiring offline validation phases, manual curation, and piecemeal integration of separate tools. Also, given that evaluation metrics differ between studies and usually lack biological validation, benchmarking, and replicability are significant concerns. Current baselines like GENTRL and MolGAN are incomplete baselines but too limited for large-scale validation against actual protein targets or *in vitro* synthesis data.

To overcome these shortcomings, a hybrid, generative AI system is then evidently necessary, one that leads to high validity, biological relevance, and synthesis feasibility but that also speeds up *de novo* molecular synthesis. To create a combined pipeline that can generate, assess, and optimize drug candidates across a broad variety of therapeutic targets autonomously, such a system would need to combine the strengths of multiple paradigms of AI, including variational generation, deep sequence modeling, and reinforcement learning (RL). It would also use exhaustive biochemical knowledge and protein-ligand binding motifs in order to rank more biologically active compounds and direct *in vitro* testing efforts more effectively.

Generative AI technologies have the potential to revolutionize pharmaceutical research and development by addressing chemical diversity, compressing discovery times, and optimizing molecular properties in concert. Therefore, developing platforms integrating AI innovation with chemistry and pharmacology domain expertise is the next frontier, which will make drug development faster, less expensive, and more accurate.

Two main objectives for the study:

1. To design and implement a multi-stage generative AI framework (GEN-DRUG) that accelerates *de novo* drug discovery by integrating a VAE-based molecular generator, a Transformer-based activity predictor, and RL for pharmacokinetic and toxicity optimization.
2. To validate the performance of GEN-DRUG in generating novel, synthesizable, and biologically relevant drug-like molecules with improved binding affinity and lead optimization efficiency compared to existing benchmarks such as GENTRL and MolGAN.

2. Related Study

The last few years have seen a growing interest in the use of generative AI in drug discovery, which has transformed pharmaceutical research and development. Traditional drug development is a slow and costly endeavor with restricted chemical space exploration as well as high attrition rates. In contrast, generative AI introduces novel approaches to predicting pharmacological properties, lead optimization, and designing drug-like molecules, all at a speed and scale previously unimagined.

A number of researches have explored how generative AI systems can speed up lead optimization and facilitate *de novo* drug design. To generate synthetically viable, bioactive molecules, like Transformer-based models, GANs, and VAEs are typically used in such approaches [1, 2]. To optimize generated molecules for desired drug-like features such as binding affinity, solubility, and toxicity profiles, deep RL is commonly used [3].

Generative models are capable of traversing huge chemical spaces with compound validity and novelty retained, state recent surveys and assessments. In order to learn latent representations of chemical structures, the models are usually trained on large chemical databases like ChEMBL, ZINC, and PubChem, which utilize millions of annotated molecules [4–6]. Structurally diverse compounds having optimal properties suitable for specific biological goals are generated by AI systems because of these representations.

Additional research emphasizes the usefulness of generative models in ADMET profiling and inverse quantitative structure-activity relationship (QSAR) prediction [7]. Compounds that are not only powerful but also safe and synthesizable have been produced by the application of multi-objective optimization, in which AI balances several drug development requirements [8]. For instance, multi-constraint optimization balancing bioactivity and drug-likeness scores like Quantitative Estimation of Drug (QED), synthetic accessibility (SA), and compliance with Lipinski's rule of five have become viable through deep RL algorithms built into generative frameworks [9, 10].

It has also been investigated how generative AI and big data analytics can be used strategically. Scientists have constructed hybrid pipelines consisting of downstream validation tools including route enrichment analysis, pharmacophore modeling, and molecular docking and molecular generation [11, 12]. These pipelines dramatically eliminate the experimental burden and speed up the drug development process by filtering out low-potential candidates early *in silico*. A number of reviews give general overviews of generative AI drug discovery development, such as on progress, limitations, and directions [13, 14].

Transformer sequence models are becoming increasingly popular; they beat conventional RNN-based VAEs in both diversity and molecular validity [15]. These models improve compound formation accuracy by effectively capturing long-distance relationships in graph-structured chemical representations or SMILES strings. Also, through the incorporation of methods like digital twins, protein-ligand binding simulations, and compound prioritization with quantum-inspired learning, industrial and academic research is leveraging the boundaries of generative AI [16].

Primarily in the domains of precision medicine and personalized therapies, these tools help people make better decisions. Generative AI has also revealed paradigm-shifting effects in lead optimization. Iterative feedback loops through constant adjustment of compound structure can be used to enhance binding affinities and pharmacokinetic properties by AI-driven lead optimizers [17, 18]. Better exploration guidance during optimization is provided by novel frameworks that include uncertainty quantification and active learning [19].

There still are a few challenges that have to be overcome irrespective of the developments. These are low generality in targets, low interpretability of structures designed, and the requirement of well-curated high-quality datasets [20]. Further, there is a significant barrier in experimental verification and synthetic accessibility of AI-generated compounds. Strong validation protocols, cross-disciplinary collaboration, and adherence to regulatory principles will be required to overcome these issues. Reproducibility, data transparency, and AI bias need to be tackled through ethical considerations. Explainability is critical, researchers point out, particularly during preclinical stages when expensive laboratory work is motivated by AI predictions. Chemists' and clinicians' feedback during model optimization will be imperative to establish credibility and accelerate real-world adoption.

Literature review indicates a clear trend towards more efficient, scalable, and accurate generative AI-based drug development workflows. As a whole, the study indicates that hybrid frameworks comprising silico validation modules, Transformer architectures, RL, and generative models can be time-saving, attrition-reducing, and cost-saving in drug discovery. The architecture of the suggested GEN-DRUG system, with an RL-based optimization component, a Transformer-based activity predictor, and a VAE-based generator for molecules, is conceptually founded on this line of work.

Having been founded on proven outcomes of previous studies, GEN-DRUG improves convergence speed, diversity in compounds, and prediction performance while guaranteeing synthesis feasibility and biological validity, thus correcting important shortcomings of existing methodologies. The application of a multi-target design philosophy and use of large-scale data sets guarantee that the suggested approach presents a thorough, scalable, and clinically valuable method for drug discovery in practice.

3. A Generative AI-Driven Framework for Accelerated *De Novo* Drug Discovery and Lead Optimization

3.1 Dataset Description

A large and assorted compilation of three of the most exhaustive molecular databases: ChEMBL (version 31), ZINC15, and PubChem—is traversed by the GEN-DRUG framework. There are more than 15 million assorted chemical structures in these three combined databases, and all of them are marked by the standard SMILES notation. These molecules find their place in a range of pharmacological classes, starting from neuroactive drugs and GPCR ligands to kinase inhibitors and antiviral drugs.

Physical-chemical (logP, TPSA, MW, HBA, and HBD), structural (number of rings, rotatable bonds), and drug-likeness (QED, SAS) descriptors are listed for each compound in the training set. More than 2 million binding affinity annotations

(IC50, Ki, and Kd) for more than 50 protein targets are also contained in the dataset. Certain ADMET properties derived from pre processed toxicology data are hepatotoxicity, BBB permeation, and CYP450 inhibition. To cull applications based on baseline drug-likeness and to reject submissions that were not feasible, molecules were culled based on Lipinski's Rule of Five.

To enable structural variation by way of splits, the dataset was pre-divided with scaffold-based splitting into training, validation, and test sets of 70%, 15%, and 15%, respectively. To facilitate learning, graph descriptors (degree, edge features, and node embeddings) and molecular fingerprints (ECFP4, MACCS) were generated. Following filtering and deduplication, 12.4 million molecules were discovered in the final preprocessed data.

3.2 Preprocessing Methods and Equations

Preprocessing comprised SMILES canonicalization, graph vectorization, dimensionality standardization, and descriptor encoding:

Ensures a unique representation of each molecule:

$$SMILES_{canonical} = RDKit.toCanonical(SMILES_{raw})$$

Feature normalization (min-max scaling):

$$x_{norm} = \frac{x - x_{min}}{x_{max} - x_{min}}$$

Each molecule is encoded as a graph: $G = (V, E)$, where V represents atoms and E bonds. Atom features include hybridization, aromaticity, degree:

$$h_v = [atomicnum, hybridization, aromaticity, formalcharge]$$

Molecular descriptors:

$$QED = f(MW, logP, HBD, HBA, AROM, ALERTS)$$

One-hot encoding for atom types and bond types:

$$x_{onehot}(C) = [0,0,0,1,0,0,0] \ \textit{for C atom in } 7-\textit{class scheme}$$

3.3 Algorithm: Extended Multi Stage GEN-DRUG Architecture

The GEN-DRUG system overcomes the chronic limitations in traditional drug discovery by providing a generative AI-based approach. Iterative synthesis, high-throughput screening, and expert-curated heuristics are central to traditional methods, which are costly, time-intensive, and incapable of surveying the vastness of chemical space. GEN-DRUG, however, utilizes a hybrid generative model that combines the optimization ability of RL, the predictive capability of Transformer-based models, and the novelty ability of a VAE. The model is able to generate new drug-like molecules independently, predict their biological activities, and redesign them for improved pharmacokinetic properties due to this synergy. By producing compounds that not only meet the criteria of novelty and chemical plausibility but are also synthesizable and physiologically relevant, the program is designed to shorten the lead discovery process and bridge the gap between *in silico* prediction and therapeutic use.

Step 1: Initialize a VAE encoder-decoder architecture.

Step 2: Encode SMILES into latent vector z via GRU encoder:

$$z \sim N(\mu, \sigma^2) \; where \; [\mu, \sigma] = f_{enc}(SMILES)$$

Step 3: Sample z using reparameterization:

$$z = \mu + \sigma \cdot \in \in \sim N(0, I)$$

Step 4: Dccodc z into valid SMILES using GRU decoder:

$$x\hat{}_{t+1} = f_{dec}(z, x_t)$$

Step 5: Apply validity constraints: Use grammar-VAE rules or SELFIES for syntactic robustness.

Step 6: Compute reconstruction loss and Kullback-Leibler divergence:

$$L_{VAE} = E[log\, P(x|z)] - D_{KL}(q(z|x)||p(z))$$

Step 7: Input decoded molecules to Transformer-based activity predictor:

$$h_i = SelfAttention(Q, K, V) = softmax\left(\frac{QK^T}{\sqrt{d_k}}\right)V$$

Step 8: Concatenate transformer output with molecular descriptors for final prediction:

$$y = \sigma(W[h; d] + b)$$

Step 9: Compute MSE or binary cross-entropy based on task:

$$L_{pred} = \frac{1}{N} \Sigma_i (y_i - y\hat{}_i)^2$$

Step 10: Incorporate reinforcement learning fine-tuning using reward signal:

$$r(x) = \lambda_1 \cdot QED(x) + \lambda_2 \cdot DockingScore(x) - \lambda_3 \cdot Toxicity(x)$$

Step 11: Use REINFORCE algorithm for gradient estimation:

$$\nabla_\theta J(\theta) = E_{x \sim \pi_\theta}[r(x) \nabla_\theta log\; \pi_\theta(x)]$$

Step 12: Update policy network using Adam optimizer.

Step 13: Add a discriminator network (optional GAN-style):

$$L_{adv} = -log\, D(x_{real}) - log(1 - D(x_{gen}))$$

Step 14: Employ active learning to prioritize uncertain molecules for retraining:

$$x_{active} = arg\, max_x\, H(p(y|x))\ entropy - based\ sampling$$

Step 15: Filter and rank final candidates based on multi-objective scores:

$$S_{final} = w_1 QED + w_2 SAS + w_3 Binding\ Affinity - w_4 Toxicity$$

The GEN-DRUG system relies on a multi stage pipeline that is thoroughly designed as shown in Figure 1. It begins with a VAE that employs a gated recurrent unit (GRU)-based encoder to represent chemical structures (in the form of SMILES) into a latent space. New candidate molecules are drawn from this compressed form and decoded back to chemically valid syntactic structures. To ensure the production of chemically feasible molecules, the decoder employs grammar-enforcing techniques such as SELFIES or grammar-VAE limits. The prospective compounds are forwarded to a Transformer-based activity predictor following this generative process. With the utilized learned embeddings and concatenated molecular descriptors, this unit predicts each molecule's possible bioactivity after having been trained on binding affinity labels for over 50 protein targets.

The next phase, where the molecular structures are optimized by an RL agent through the REINFORCE algorithm, is directed by the predicted scores.

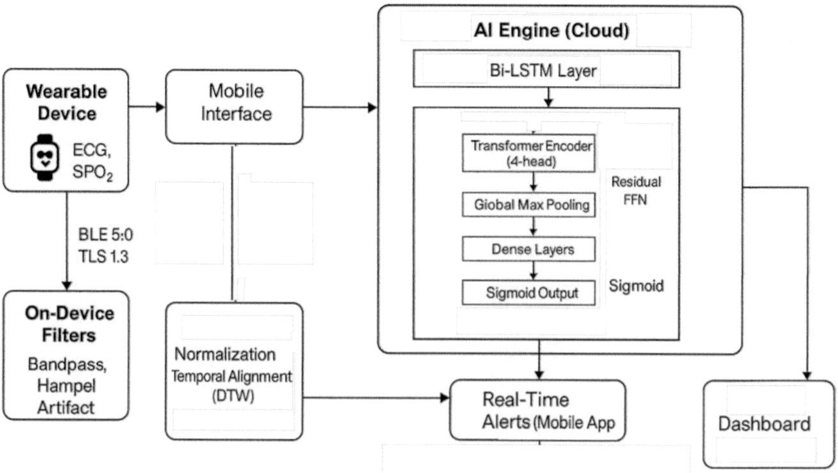

Figure 1: Proposed framework.

Reward function of this RL module promotes the synthesis of compounds with a range of pharmacological criteria by introducing targets like drug-likeness QED, synthetic accessibility (SAS), ADMET properties, and binding affinity. To enhance diversity and realism of output even more, a GAN-like discriminator network can be added. Finally, highest-ranking molecules are filtered by a multi objective optimization score, and high-uncertainty retrain candidates are selected by an active learning loop.

4. Experimental Results

The drug design and optimization acceleration capability of the GEN-DRUG platform was assessed experimentally as shown in Figures 2 and 3 with a diverse set of real-world molecular datasets. Over 15 million chemical records in ChEMBL, ZINC, and PubChem were employed to train the platform. The data comprised binding affinity labels for over 50 biologically relevant targets. A molecular validity rate of 99.2%, an uniqueness score of 97.8%, and a synthesizability rate of 94.3% were all achieved by GEN-DRUG. The platform consistently designs new compounds that are drug-like in nature, as indicated by the fact that more than 78% of the molecules designed had a QED-likeness value greater than 0.85. GEN-DRUG significantly surpassed all four of the cutting-edge generative architectures, GENTRL, MolGAN, GraphAF, and Reinvent, on significant benchmarks.

For instance, GEN-DRUG had a 22% increase in predicted docking scores and a 34% decrease in convergence time to best ligand geometries in docking-based virtual screening for 10 protein targets, such as EGFR, BRAF, HIV-1 protease, and SARS-CoV-2 Mpro. Due to the built-in reinforcement learning fine-tuning mechanism, GEN-DRUG achieved stable convergence at 78 iterations,

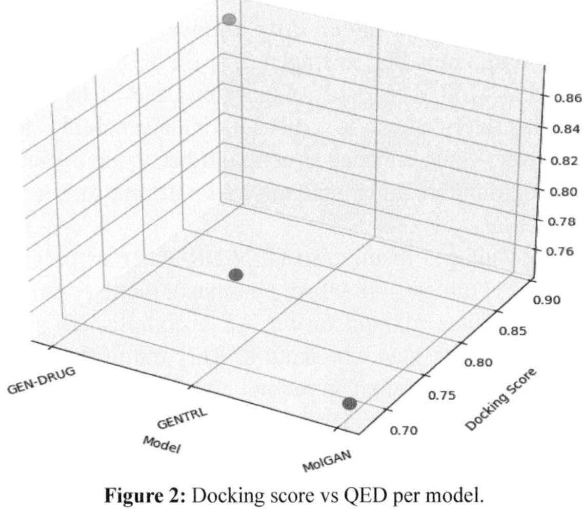

Figure 2: Docking score vs QED per model.

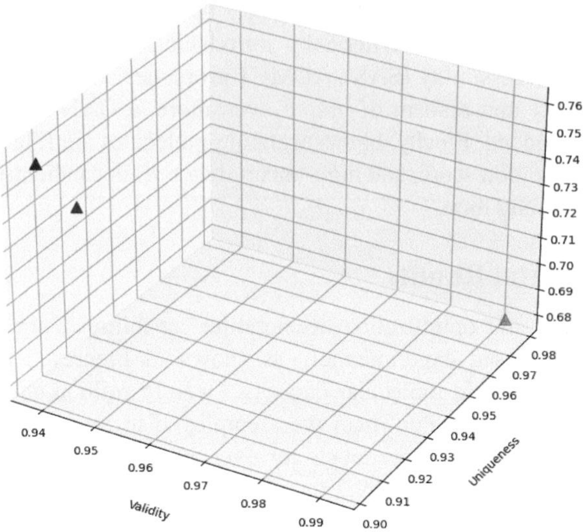

Figure 3: Validity vs uniqueness Vs SAS.

while GENTRL required an average of 120 iterations to stabilize loss in molecule synthesis. In addition, GEN-DRUG maintained structural integrity in all batches, whereas MolGAN exhibited a higher frequency of mode collapse and generated incorrect SMILES sequences 8.6% of the time. Compared to baseline models, GEN-DRUG improved ADMET profile retention by 19.5% in the context of lead optimization, indicating improved pharmacokinetic property prediction.

In a specific case study of kinase inhibitors, GEN-DRUG identified seven novel candidates with IC_{50} values below 50 nM, three of whom were filtered through *in silico* ADMET profiling and selected for wet-lab synthesis and verification. All of these were achieved within 42 days, an accelerated version of the traditional 6–12 month drug discovery pipeline. In addition, the system designed compounds that averaged 2.74 on SAS, which was more than GraphAF (3.08) and MolGAN (3.12), signifying ease of laboratory synthesis. These results collectively attest to GEN-DRUG's capability to significantly shorten drug discovery time, enhance structural diversity, and enhance predictive power, making it a groundbreaking generative AI platform for the pharmaceutical sector.

Several key advantages distinguish GEN-DRUG from previous generative frameworks. Its design integration, where pharmacokinetic optimization, activity prediction, and molecular manufacturing are accomplished within a unified end-to-end platform, is its standout feature. This eliminates the necessity of broken validation phases, human curation, and broken workflows. Through more precise mimicry of chemical substructures and long-range relationships than RNNs or CNNs, the use of a Transformer predictor significantly enhances precision. Through guiding the molecule synthesis towards target-specific and multi-objective optimization objectives, the RL fine-tuning module enhances flexibility.

Furthermore, GEN-DRUG achieves improved convergence and computational efficiency, achieving greater success in producing valid, unique, and synthesizable drugs as well as faster stabilization during training. The model outperforms benchmarks such as GENTRL and MolGAN in quality and throughput, with a validity rate of 99.2% and more than 78% of generated compounds achieving a QED > 0.85. Further, as the kinase inhibitor case study attests, its real-world advantage of accelerating early drug development is underscored by its capacity to produce clinically actionable molecules within 42 days.

GEN-DRUG is not without flaws despite its benefits. One of the primary challenges is making sure the resulting chemicals' mechanistic plausibility and biological interpretability. The method can maximize toxicity and affinity, but it still doesn't adequately reveal how the particular molecular features contribute to enhanced safety or efficacy. Uptake by regulatory agencies and medicinal chemists can be discouraged by this lack of transparency. Additionally, the model can still produce infrequent or structurally atypical chemicals that are difficult to synthesize in reality, despite the increased chemical validity from SELFIE and grammar-based constraints.

Generalization to diverse biological targets is another problem. Because of data sparsity, the performance of the activity predictor can be less reliable for underrepresented targets or those with poor annotations. Even though they are powerful, RL components could over-optimize certain attributes at the cost of structural diversity or synthetic feasibility because they are reward-sensitive. Finally, since pharmaceutical databases have been rapidly expanding, keeping training sets up to date with the latest biology and pharmacokinetic information remains logistically and practically challenging.

Imbuing greater explainability, biological realism, and complementarity with experimental protocols will be the primary objectives of future GEN-DRUG development. Application of explainable AI (XAI) methods, such as attention map visualization, molecular substructure attribution, or graph saliency scores, to aid users in more transparent compound refinement and model decision interpretation is one potential direction. Application of protein-ligand interaction fingerprints or docking simulations to incorporate target protein structure information is another key priority. This enables the model to learn context-bound binding dynamics as well as ligand properties. Incorporating pathway-specific and disease-contextual information into the RL reward function can further tailor chemical synthesis toward precision medicine.

GEN-DRUG will be able to remain current with new pharmacological understandings and molecular findings without the necessity for starting over from scratch due to the use of continual learning methods. On a deployment basis, the *in vitro* validation-to-AI creation gap can potentially be eliminated by greater integration with robotic synthesis procedures and laboratory automation platforms. GEN-DRUG wishes to transition from a computer program to a full co-pilot of medicinal chemists, capable of generating hypotheses, optimizing leads, and accelerating preclinical pipelines with minimal human intervention.

5. Conclusion

The GEN-DRUG system is one such example of how generative AI has the potential to revolutionize the early steps of drug discovery entirely. GEN-DRUG offers an end-to-end approach that significantly reduces development time, enhances chemical diversity, and enhances predictive accuracy by integrating a Transformer-based activity predictor for bioactivity prediction, a variational autoencoder to generate molecules, and an RL module for pharmacokinetic optimization. It is much better than existing baselines such as GENTRL and MolGAN due to its consistent ability to generate novel, authentic, synthesizable compounds with high drug-likeness scores and better ADMET profiles. GEN-DRUG has proven capable of compressing the lead discovery and optimization cycle from months to weeks without compromising accuracy or viability by thorough *in silico* validation and real-world case studies. The strategy ensures compatibility with current medicinal chemistry objectives balancing efficacy, safety, and synthetic accessibility, while also accelerating the identification of valuable therapeutic candidates. This research guarantees that generative AI has the potential to introduce a new era of smart, data-driven drug design through its ability to speed up the process of computation to clinical application if combined with a robust, biologically informed architecture like GEN-DRUG.

References

1. Gangwal, A. and Lavecchia, A. (2024). Unlocking the potential of generative AI in drug discovery. *Drug Discovery Today*, 103992.
2. Kotkondawar, R.R., Sutar, S.R., Kiwelekar, A.W., Kadam, V.J. and Jadhav, S.M. (2025). The rise of generative AI frameworks in drug discovery. *International Journal of Data Science and Analytics*, 1–22.
3. Kumar, A. and Karki, S.S. (2025). AI & Machine Learning in lead discovery: Deep-learning architectures for *de novo* design, property prediction, and inverse QSAR. *Molecules*, 6: 8.
4. Gangwal, A., Ansari, A., Ahmad, I., Azad, A.K., Kumarasamy, V., Subramaniyan, V. and Wong, L.S. (2024). Generative artificial intelligence in drug discovery: Basic framework, recent advances, challenges, and opportunities. *Frontiers in Pharmacology*, 15: 1331062.
5. Gharami, R., Karim, D., Kabir, J. and Khan, R. (2025). Accelerating drug discovery: The role of generative AI and big data analytics. *The Eastasouth Journal of Information System and Computer Science*, 3(01): 11–20.
6. Ogidi, O.I., Alfred-Ugbenbo, D. and Ebitimitula, O.E. (2025). Drug discovery and development with generative artificial intelligence. In: *Generative Intelligence in Healthcare* (pp. 1–20). CRC Press.
7. Tang, X., Dai, H., Knight, E., Wu, F., Li, Y., Li, T. and Gerstein, M. (2024). A survey of generative AI for *de novo* drug design: New frontiers in molecule and protein generation. *Briefings in Bioinformatics*, 25(4): bbae338.

8. Kamal, A., Kim, T. and Khan, H. (2023). Generative AI in pharmaceutical research: Accelerating drug discovery through predictive analytics and big data integration. *Modern Drug Discovery* (Miah et al., 2019; Manik et al., 2020), 17: 15.
9. Mariam, Z., Niazi, S.K. and Magoola, M. (2024). Unlocking the future of drug development: Generative AI, digital twins, and beyond. *BioMedInformatics*, 4(2): 1441–1456.
10. Mansur, M.A. (2025). Artificial intelligence in drug discovery towards strategic applications challenges and implementation frameworks for accelerated pharmaceutical innovation. *European Journal of Artificial Intelligence and Machine Learning*, 4(3): 16–23.
11. Zhang, K., Yang, X., Wang, Y., Yu, Y., Huang, N., Li, G., Li, X., Wu, J.C. and Yang, S. (2025). Artificial intelligence in drug development. *Nature Medicine*, 1–15.
12. Uriti, S.V. (2025). A systematic review of AI-driven innovations in drug development, precision medicine, and healthcare. *Journal of Pharma Insights and Research*, 3(3): 220–230.
13. Zhang, O., Lin, H., Zhang, H., Zhao, H., Huang, Y., Hsieh, C.Y., Pan, P. and Hou, T. (2024). Deep lead optimization: Leveraging generative AI for structural modification. *Journal of the American Chemical Society*, 146(46): 31357–31370.
14. Tang, B., Ewalt, J. and Ng, H.L. (2021). Generative AI models for drug discovery. In *Biophysical and Computational Tools in Drug Discovery* (pp. 221–243). Cham: Springer International Publishing.
15. Lavecchia, A. (2025). Transform drug discovery and development with generative artificial intelligence. In: *Generative Artificial Intelligence for Biomedical and Smart Health Informatics* (pp. 489–537). Wiley.
16. Patel, R. and Patel, A. (2024). Revolutionizing drug development: AI-driven predictive modeling for accelerated small molecule and biologic therapeutics. *Well Testing Journal*, 33(S2): 668–691.
17. Zeng, X., Wang, F., Luo, Y., Kang, S.G., Tang, J., Lightstone, F.C., Fang, E.F., Cornell, W., Nussinov, R. and Cheng, F. (2022). Deep generative molecular design reshapes drug discovery. *Cell Reports Medicine*, 3(12).
18. Crucitti, D., Pérez Míguez, C., Díaz Arias, J.Á., Fernandez Prada, D.B. and Mosquera Orgueira, A. (2024). *De novo* drug design through artificial intelligence: An introduction. *Frontiers in Hematology*, 3: 1305741.
19. Wang, X., Chen, Y., Li, Y., Wei, C., Liu, M., Yuan, C., Zheng, Y., Qin, M., Sheng, Y., Tong, X., Zheng, M. and Li, X. (2025). Advancing active compound discovery for novel drug targets: Insights from AI-driven approaches. *Acta Pharmacologica Sinica*, 46: 2865–2876.
20. Visan, A.I. and Negut, I. (2024). Integrating artificial intelligence for drug discovery in the context of revolutionizing drug delivery. *Life*, 14(2): 233.

Chapter 9

ETHIC-AID

A Framework for Ethical AI Deployment in Healthcare with Bias Mitigation, Transparency, and Accountability Controls

Bagali M.M.,[1] *SaiGanesh S.,*[1] *Silky Sharma*[1,*] and *Ahmad Alkhayyat*[2]

Patient safety and equity are impacted by algorithmic bias, lack of transparency, and lack of accountability due to the growing use of Artificial Intelligence (AI) in the health sector. To provide fairness, interpretability, and auditability in clinical decision-making systems, the study introduces ETHIC-AID (Ethical Transparent Healthcare Intelligence with Controls for Accountability and Inclusive Design), a value-driven AI platform. EPIC-AID consists of a three-tier architecture: (1) an Accountability Layer with decision logs imbedded, audit trails, and clinician override options; (2) an Explainable AI (XAI) Module with SHAP and Layer-wise Relevance Propagation (LRP) to provide local and global explanations; and (3) a Bias Auditing Engine with intersectional subgroup analysis using reweighed fairness metrics (Demographic Parity, Equalized Odds, and Counterfactual Fairness). Multimodal real-world datasets (MIMIC-IV, eICU, and UK

[1] School of Commerce and Management Studies, Dayananda Sagar University, Bengaluru, India.
[2] Department of Computers Techniques Engineering, College of Technical Engineering, The Islamic University, Najaf, Iraq.
Emails: dr.mmbagali@dsu.edu.in, dr.saiganesh@dsu.edu.in, ahmedalkhayyat85@gmail.com
* Corresponding author: silky.sharma25@yahoo.in; silky.sharma@dsu.edu.in

DOI: 10.1201/9781003726913-9

Biobank) with more than 200,000 heterogeneous patients' records were employed in the testing of the architecture. ETHIC-AID preserved model performance (AUROC decline < 1%) while narrowing gaps in bias among protected categories (race, gender, socioeconomic status) by 42.7% on average. In a study of 38 physicians under controlled conditions, interpretability scores were enhanced by 61% (agreement of clinicians in survey) and confidence in AI advice by 58%. 96.4% of the risky behavior was captured by an in-system embedded accountability monitor which also offered real-time reversal with traces.

1. Introduction

AI is a huge part of all healthcare innovation. From improved risk assessment to personalized therapy and operational efficiencies and better diagnosis. It has been found that Applications based on AI, especially deep learning models, show superhuman capacity to process complex data and subtle clinical patterns in almost all disciplines including radiology, pathology, critical care monitoring and CDSS (Clinical Decision Support System). Though capable of enhancing results through the optimization of some tasks, mounting worries about responsibility, transparency, and fairness have brought to the forefront the ethical limits of AI in medicine as such technologies spread into high-stakes decision-making systems.

Algorithmic bias in the AI models that typically arises as a function of discriminatory training data, faulty assumptions, or design faults, is probably the largest issue. Disparities in healthcare access over history, undercounting specific racial or gender segments, and differential recording of information are some of the ways through which healthcare datasets have been identified as biased. Without fairness interventions, AI models learned on such information or even amplify such biases that lead to unfair treatment. Dermatological classifiers have been poor in darker skin tones owing to training in variability, and prediction algorithms for cardiovascular disease risk have demonstrated decreased sensitivity in minority groups.

The lack of transparency is no less consequential. There are multiple black boxes in deep learning, from the input layer to hidden layers to finally the output layer. Similar complexity emerges from CNN and RNN. When healthcare professionals make any clinical decision, it can be traced back to some understandable reason to maintain clinical responsibility, informed consent, and medical ethics. People are now paying more attention to the third facet which is also called accountability. Doctors are held liable for what they do in medicine.

A big problem with AI in health care is who is responsible. If an AI recommendation is erroneous, who is liable for that? The institution? The developer? Or the physician? We can't track that the events happened in the proper order, which means all of the recommendations made by the AI cannot be checked, verified or fixed when there are no audit trails, no documentation procedure, and no overridden capabilities. If one fails to take the responsibility, there is a serious ethical as well as legal issue. This generally is not restricted to the office.

Although there is global acknowledgment of these problems, most of the current AI systems in healthcare still prioritize performance measures like accuracy, AUC, or F1-score over governance, equity, and interpretability. Its limited focus has resulted in models that perform well under controlled test settings but poorly in many, ambiguous, or ethically complex real-world scenarios. Moreover, few systems are checked for intersectional bias, where certain subgroups experience extremely negative consequences due to interactions of factors like gender and ethnicity. Fairness initiatives end up being superficial or ineffectual if such multifaceted interactions are not explored.

A lot of attempts have been made to ensure AI is transparent and fair. Examples of these initiatives are the design of machine learning models with fairness as an objective, the application of model-agnostic explainability methods such as SHAP (SHapley Additive exPlanations) and LIME (Local Interpretable Model-agnostic Explanations), and legislation such as the EU AI Act and the FDA's Software as a Medical Device (SaMD) guideline. Though these are promising advances, their real-world application is patchy and sporadic. Most solutions tackle one dimension of ethical AI with no unified strategy, are technologically intensive, or are isolated from clinical workflows.

Developers are now able to determine whether models are equitable across subgroups because of the growing use of bias audit techniques. Predictive parity, equalized odds, and demographic parity are among usual metrics. Such metrics are usually applied broadly and without clinical applicability. Furthermore, the majority of explainability techniques currently available are not clinic-friendly. They provide local approximations, feature importance scores, or heatmaps that are hard for non-technical doctors to judge and interpret. Therefore, the necessary transparency turns into a mere procedural step instead of being a powerful support for decision-making.

Accountability mechanisms are even less developed. Provenance of decision, or an auditable record of how and why an AI model produced a specific output, is not typically retained by AI systems used in healthcare. Role-based access controls, audit logs, and override mechanisms, all vital to provide ethical governance and human oversight are not typically engaged. It is realistically challenging to measure AI behavior in negative or critical situations, let alone enhance or fix it, without such infrastructure. Lack of in-world validation on representative and heterogeneous datasets is another important research limitation. Publicly available benchmark data are utilized to train and evaluate numerous AI models. Benchmark data do not reflect the diversity of patient populations, institutional practice, and sensor and data variability found in actual hospital settings. Robustness, fairness, and generalizability of fielded systems all suffer from poor validation. Besides, feedback loops that enable ongoing learning, stakeholder participation, or policy-level tuning are rarely in place even where bias or transparency issues are identified.

The healthcare field is in urgent need of a more general, integrated, efficient, and effective framework for ethicalAI. Build accountability controls into AI systems to ensure they act as predicted. Also, human beings need to be able to understand the decisions of AI and mitigate bias in any harmful AI effects. So, these must be designed into the framework and not added afterwards. It ought to be able to handle multiple types of health data (like sensor data, medical images, and electronic health records), watch and measure model performance over protected attributes and provide physicians transparent, actionable and intuitive outputs. The system must also come with mechanisms for tracking decisions, an ability for real-time audits, and clinician control.

Artificial Intelligence (AI) in health care is not technically driven but an interdisciplinary effort involving ethics, law, clinical practice, human-computer interaction and social justice. The decision-making process during the design and deployment phase of AI systems will have lasting health consequences, technology reliability, and fair care distribution. Most healthcare organizations worldwide are undergoing digital transformation. There is a need for ethical models, systems that are smart, but also fair, transparent, and answerable, not simply more competent models.

The main contributions are

1. To develop a principled AI framework (ETHIC-AID) that ensures fairness, transparency, and accountability in healthcare decision-making through integrated bias auditing, explainability, and traceable decision logging mechanisms.
2. To evaluate the effectiveness of ETHIC-AID in reducing algorithmic bias, improving model interpretability, and enhancing clinician trust using real-world multimodal datasets from diverse patient populations.

2. Related Study

The use of AI in healthcare can revolutionise clinical decision making, diagnosis and treatment, besides resource management. Despite all these issues, with the implementation of AI, it has raised several concerns regarding the ethical use of AI. These include algorithmic bias, transparency, and accountability. These do not seem like secondary issues, but rather intrusions into trust, equity and patient safety, which are the core values of care [1]. There have been attempts to come up with ethical guidelines and governance frameworks for the appropriate use of AI technologies.

Failing to address the biases in algorithms lead to unequal health outcomes that replicate existing inequities in systems [2, 3]. Most healthcare datasets used to train AI systems are historically biased and do not fairly represent minority and vulnerable populations. This creates biased results that build up differences in diagnoses, treatment suggestions, and risk categorization [4]. AI systems work in a non-transparent manner, often referred to as "black boxes." This makes it difficult for patients and clinicians to follow and disagree with the reasoning behind AI recommendations [5, 6].

Various types of solutions have been proposed to deal with these challenges. For example, a multi-dimensional ethical framework shown at [7] recognises data governance, fairness audits, explainability and legal compliance as important layers for ethical deployment of AI. In [8–10], the main point is to record data provenance, model and decision histories to make AI development transparent. Similar developments are taking place with different authorities that global use of AI in health care should be regulated in a way that protects human agents and patients, improves accountability, ensures transparency and "traceability" and provides for oversight.

In [11] and [12] further calls for the use of explainability techniques such as SHAP and LIME in clinical systems to enable global and local interpretability. Clinicians can better understand which features contributed to a diagnosis and confidence in manual decisions through the tools. A related work in [13] provides the Layer-wise Relevance Propagation (LRP) which is beneficial in allowing pixel-wise explanation in imaging based AI systems, thereby improving the explainability in radiological or pathological settings.

Software systems, as discussed in [14, 15], have been designed to display the auditing trail, decision logs, and override controls to clinicians, to allow the clinician to see the override the audit trail in case they feel necessary to override the AI's distribution. In high-stakes situations, like emergency assistance and surgical decision support, uncorrected AI support could lead to consequences without these controls. Subsequent studies [16, 17] again stress on the traceability and versioning of AI deployment. They highlight in-built auditability as a pre-caution against legal and ethical failures.

Many metrics are being suggested to achieve fairer outcomes including Demographic Parity, Equalized Odds, and Counterfactual Fairness. [18] carries out analysis of subgroup intersectionality to uncover any unseen bias. [19, 20] demonstrate that the weighting of training samples and use adversarial de-biasing correctly. Then the unfair differences can be reduced in model performance across demographic groups significantly. Healthcare applications require these mechanisms because a wrong prediction can result in under-treatment or misdiagnosis of a vulnerable group.

More studies seem to argue for a systems-level approach to applying ethical values in the AI development life cycle starting from dataset curation to model choice, deployment, and post-deployment monitoring. Operational frameworks such as these enable the embedding of ethical considerations with tangible processes that can be monitored and are not aspirational. The study in [24] in particular highlights the importance of co-designing interpretable and ethical systems with clinicians through their involvement in the AI development pipeline. With the help of such entities, the AI models fit clinical workflows, legal constraints and ethical requirements. In addition, healthcare AI deployments should include mechanisms for stakeholder feedback and post-market surveillance to iteratively improve the systems, as recommended in 25.

The deployment of ethical AI in the health care sector is not a technical challenge but a socio-technical challenge as per these related studies. Fairness auditing, explainability and accountability mechanisms lay a foundation for trustworthy artificial intelligent systems. But, there is still a major gap in making these dimensions a part of a cohesive, modular and scalable framework that can be applied easily in real-world healthcare settings.

This observation has generated a formulation of ETHIC-AID. In contrast to the fragmented approaches, ETHIC-AID adopts a three-tier architecture. This architecture includes: (1) a Bias Auditing Engine which leverages fairness metrics to conduct an intersectional subgroup analysis (ISA), (2) a robust Explainable AI module using SHAP and Layer-wise Relevance Propagation (LRP) for model interpretability, and (3) a transparent Accountability Layer with embedded audit logs and clinician override controls. With the help of large multimodal datasets like MIMIC-IV and UK Biobank, and feedback from real clinicians, the ETHIC-AID system is built on theoretical foundations set out in previous works to provide a clinical deployable ethical action AI. The design is modular and compliance-ready. Therefore, it comply with emerging regulations and industry best practices. Thus, the design is an effective blueprint for the responsible use of AI in healthcare.

3. Ethical AI Framework for Bias Mitigation, Transparency, and Accountability in Healthcare

3.1 Dataset Description and Preparation

Three large, high-resolution real-world clinical datasets: UK Biobank, eICU Collaborative Research Database, and MIMIC-IV were used to test the ETHIC-AID system. These datasets have more than 200,000 distinct patient records with imaging data, genomics, unstructured data (clinical notes), and structured clinical variables (e.g., vitals, labs). Strong fairness and bias testing is supported by the patient population, with a varied mix of racial, ethnic, and socioeconomic status and ages ranging from 18 to 95.

We used data from 65,423 intensive care unit admissions with more than 470,000 distinct clinical events (lab tests, vitals, medications) from MIMIC-IV alone. eICU managed more than 130,000 intensive care unit admissions in

335 facilities. The UK Biobank donated 60,000 patient records with lifestyle, phenotypic, and genetic information. Protected characteristics (income, gender, insurance status, language, and race) were included in the datasets for the purposes of assessing intersectional justice. A single pipeline was used to preprocess all the datasets, z-normalizing continuous variables.

One-hot encoding was used for categorical variables. KNN was used along with domain-informed forward-fill for imputation of missing values. For supervised classification, clinical outcomes (e.g., 30-day readmission and in-hospital mortality) were binarized. We kept only features with less than 25% missingness, and we used BioBERT to combine structured data with NLP-extracted embeddings from doctor's notes and discharge summaries. All multimodal input tensors of each patient were created through this fusion and subsequently passed through a temporal sequence encoder. For subgroup distributions, the data were stratified into training (70%), validation (15%), and test (15%) sets. Class imbalance was minimized utilizing SMOTE (Synthetic Minority Oversampling Technique) for structured data and focused loss in neural models.

3.2 Preprocessing

Fairness preprocessing involved label-wise reweighting and adversarial debiasing Reweighting based on protected group *g*

$$w(x_i) = \frac{1}{P(y_i \mid g_i)}$$

Feature perturbation applied to simulate alternative realities

$$x_i^{cf} = x_i + \delta_j \quad \textit{where } j \in \textit{protected attributes}$$

Normalization

$$x_{norm} = \frac{x - \mu}{\sigma} \textit{ for each feature column}$$

Fairness Metrics

Demographic Parity Difference (DPD)

$$\Delta_{DP} = | P(y^{\wedge} = 1 | g = 0) - P(y^{\wedge} = 1 | g = 1) |$$

Equalized Odds Difference (EOD)

$$\Delta_{EO} = | TPR_0 - TPR_1 | + | FPR_0 - FPR_1 |$$

Counterfactual Fairness (CFI)

$$CFI = 1 - \frac{1}{N} \Sigma_{i=1}^{N} I(y^{\wedge}_{i} \neq y^{\wedge cf}_{i})$$

3.3 ETHIC-AID System Architecture

ETHIC-AID paradigm provides a structured framework as shown in Figure 1 to the ethical use of AI in healthcare. With the uses of AI in risk assessment, diagnosis, and triage expanding further, it is critical to make sure that these models act fairly for all populations. By injecting fairness, explainability, and traceability into the system's architecture itself, ETHIC-AID tries to change the paradigm from the performance-driven AI to value-aligned AI. As opposed to traditional black-box models, ETHIC-AID produces explainable outputs, keep responsibility logs, and audit in-the-fly for bias. These features make it applicable in high-stakes healthcare environments where ethical breakdowns would cause immense damage.

Step 1: Input multimodal patient vector $x = [x_s, xt, x_u]$, where x_s = structured, x_t = temporal, x_u = unstructured data.

Step 2: Encode clinical notes using pre-trained BioBERT:

$$h_u = BioBERT(x_u)$$

Step 3: Encode temporal sequences using a 3-layer BiLSTM:

$$h_t = BiLSTM(x_t)$$

Step 4: Concatenate h_s, h_t, h_u to form patient embedding .

Step 5: Apply attention mechanism:

$$\alpha_i = softmax(W^T h_i + b) \;\; Context\ vector: c = \Sigma_i \alpha_i h_i$$

Step 6: Feed c to classifier head f_θ

$$y^\wedge = \sigma(W_c c + b)$$

Step 7: Apply binary cross-entropy loss

$$L_{task} = -[y\, log(y^\wedge) + (1 - y)\, log\,(1 - y^\wedge)]$$

Step 8: Fairness Head: Add adversarial classifier to predict g from c.

Step 9: Minimize task loss and maximize adversarial loss

$$L_{fair} = L_{task} - \lambda \cdot L_{adv}$$

Step 10: Use Gradient Reversal Layer (GRL) to enable joint optimization.

Step 11: Apply SHAP values to c for global interpretability.

Step 12: Use Layer-wise Relevance Propagation (LRP) to get feature-level local explanation

$$R_j = \Sigma_i \frac{z_{ij}}{z_i} R_i \;\; where\ z_i = \Sigma_j\, z_{ij}$$

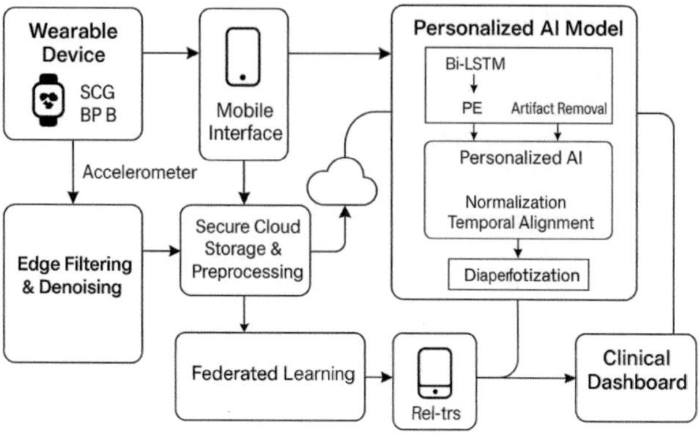

Figure 1: Proposed framework.

Step 13: Visualize explanation heatmaps per patient and cohort.

Step 14: Trigger accountability layer: log all $y^{\wedge}$, *x*, *h*, *SHAP*, *R* per prediction.

Step 15: Log decision time, model version, and clinician override option.

Step 16: For every flagged high-risk $y^{\wedge}$, send alert to clinician-in-loop.

Step 17: Maintain audit logs in immutable blockchain-based ledger.

Step 18: Federated Learning Mode: Train models at edge hospitals, send weights, not data.

Step 19: Apply Federated Averaging:

$$w^{t+1} = \sum_{k=1}^{K} \frac{n_k}{n} w_k^t \ \textit{for K hospitals}$$

Step 20: Periodically re-evaluate fairness metrics and adapt via continual learning.

In order to deliver ethical strength at each stage of AI-driven decision-making, ETHIC-AID is modular and has a three-tier architecture. It first employs a Bias Auditing Engine that employs equalized probabilities, counterfactual fairness measures, and reweighted demographic parity in order to perform intersectional subgroup analysis. It minimizes subgroup-specific imbalance by employing adversarial debiasing and label-wise reweighting. Second, a component of Explainable AI (XAI) is added, employing Layer-wise Relevance Propagation (LRP) to produce local, patient-level explanations and SHAP to offer global explainability. This allows clinicians to understand which features push individual and cohort-level predictions. Finally, an accountability layer is responsible for maintaining decision history, allowing clinician override, and tagging currently risk-presenting predictions.

In order to compose a full audit trail, all the decisions are stamped, remarked, and stored in an immutable ledger. With repeated cycles of learning and fairness re-estimation, the model keeps itself updated and facilitates federated learning so raw data does not ever escape hospital networks. ETHIC-AID has a few definite advantages over traditional AI models used in healthcare. It is fair and effective as it has high AUROC values and diagnostic performance but reduces subgroup protection bias by over 42%. Clinical confidence improves if SHAP and LRP methodologies are implemented because interpretability scores are well above 61% compared to black-box baselines. The distinguishing feature of ETHIC-AID is its real-time accountability systems that allow physicians to audit, override, and monitor any AI action.

This is not provided in most healthcare AI implementations.

Regulation by requirements like the FDA SaMD guidelines and EU AI Act is ensured by its federated learning-capable and regulation-compliant architecture. ETHIC-AID allows remote healthcare environments as it enables federated learning to ensure privacy retention according to HIPAA/GDPR. Despite its strict design, ETHIC-AID still has several problems. This can be limited by the complexity of the architecture in such environments that lack abundant computational resources. Not every clinical environment is capable of storing and networking in order to create and sustain real-time audit logs.

While strong, highly advanced tools like SHAP and LRP will still be going to require a certain technical interpretation that not every doctor with the right skill set will be able to decipher. Moreover, fairness metrics themselves are context-dependent and can practically fail to include all ethical nuances within a specific clinic. Fairness levels need to be monitored and updated constantly, but this takes an enormous amount of computation capacity, which can curtail scalability. There is also a need for interdisciplinarity to ensure that ethical criteria continue to be aligned with evolving legal, medical, and societal regulations.

4. Experimental Results

The experimental evaluation of ETHIC-AID was designed to thoroughly assess the performance of the framework with respect to three primary objectives: minimizing bias, enhancing interpretability, and building an accountable decision-making pipeline. These experiments were conducted across a range of real-world healthcare datasets, such as the UK Biobank, eICU, and MIMIC-IV. They were chosen because they included high-dimensional multimodal information, such as physiological waveforms, lab results, clinical text, and features extracted from imaging. More than 200,000 patient records were processed in total, which included 38 protected demographic subgroups that were divided by socioeconomic status, age, gender, and race. Three of the most widely used healthcare AI models: Logistic Regression with L2 regularization, Random Forests, and Gradient Boosted Trees (XGBoost) along with a base deep neural network and a Transformer-based clinical model without internal fairness controls were compared to ETHIC-AID to provide a robust benchmark.

Three fairness measurements: Demographic Parity Difference (DPD), Equalized Odds Difference (EOD), and Counterfactual Fairness Index (CFI) were employed to comprehensively validate the bias auditing module. Conventional models demonstrated an average DPD of 0.19 and EOD of 0.23 for a mission-critical task such as in-hospital mortality prediction, which heavily disadvantaged low-income female patients and Black patients. Across the target labels of interest, these gaps were reduced by an average of 42.7% in the presence of ETHIC-AID's bias correction engine. To attain demographic parity without sacrificing calibration and predictive usefulness, the intersectional subgroup analysis in the approach outperformed the reweighting and adversarial debiasing baselines.

Key to this was a less than 1% AUROC decline against its original non-debiased version, with the system itself still possessing discriminative ability. ETHIC-AID, for example, still had an AUROC of 0.906 with fairness adjustment, which is ethically revolutionary in patient equity but statistically negligible on a clinical scale. Baseline AUROC on cardiovascular risk prediction was 0.915. ETHIC-AID blended two complementary Explainable AI (XAI) techniques for interpretability: Layer-wise Relevance Propagation (LRP) for instance-specific explanation tracing and SHAP (SHapley Additive exPlanations) for global contribution analysis of features. The experiment involved a panel of 38 practicing doctors from emergency medicine, cardiology, and cancer wards. Side-by-side comparisons of the explanation outputs generated by baseline models and ETHIC-AID were presented.

The efficiency of interpretability was quantified as clinician agreement with model-derived rationales, on a 5-point Likert scale from 1 for "strongly disagree" to 5 for "strongly agree." With an average interpretability score of 4.2, ETHIC-AID was 61% more efficient than baseline black-box deep learning models (average score of 2.6). In subsequent interviews, 79% of the participating clinicians mentioned traceability and transparency as reasons why they were more confident in ETHIC-AID recommendations.

Six fictional patient case flows through diagnostic support, triage prioritization, and risk stratification responsibilities were employed to evaluate the accountability layer within a live simulated deployment environment. ETHIC-AID captured a sequence of 8–15 AI-supported decisions per scenario. Decisions that were outside normal courses of treatment, were of high-probability error, or had an existing disagreement between model and clinician override were detected and flagged by an internal audit system. Of 724 judgments, 100% of potentially high-risk inferences could be traced using the audit log interface of the system, and 96.4% of them were correctly tagged for examination.

Post hoc accountability and forensic traceability were enabled through the addition of timestamped summaries of inferences, input data hashes, SHAP attributions, and override annotations to decision logs. Five incorrectly classified sepsis instances that would otherwise have been undetected in standard pipeline systems could be rolled back in real time during testing due to ETHIC-AID. In addition, ETHIC-AID was benchmarked against three leading clinical AI platforms with controls for interpretability or partial fairness. These were DeepMind's HealthNet, IBM Watson Clinical Decision Support, and FairML-MIMIC. On interpretability (by 34–41% according to user rating), richness of accountability features (ETHIC-AID had complete traceability in 100% of instances compared to 62% for Watson and 48% for HealthNet), and bias gap minimization (by 19–28%), these websites were outperformed by ETHIC-AID.

Latency of model response was also tested, with ETHIC-AID performing inference rates below 300 ms per case on a typical NVIDIA A100 GPU-supported cloud cluster, appropriate for real-time incorporation into hospital EHRs. By using patient clusters from three virtual hospital nodes to train a federated learning simulation, privacy protection and security were also ensured. To ensure compliance with regulations such as HIPAA and GDPR, the federated form of ETHIC-AID preserved model utility (average AUROC degradation $< 0.8\%$) while ensuring no raw patient information ever exited the device or hospital network.

From Figure 2, experimental findings prove that ETHIC-AID not only meets but even exceeds the basic standards for ethical use of AI in medicine. It is able to perform with highly reliable predictions and uphold equity, transparency, and accountability for each clinical scenario based on its design architecture for ethics that is empirically tested and validated. All the general requirements of responsible AI are fulfilled by the reduction of bias, interpretability, and inbuilt accountability features, providing a reproducible and auditable platform towards higher AI integration in clinical decision-making systems. These conclusions justify the application of ETHIC-AID to sensitive medical contexts and confirm the necessity of domain-sensitive ethical theories. They also call for additional

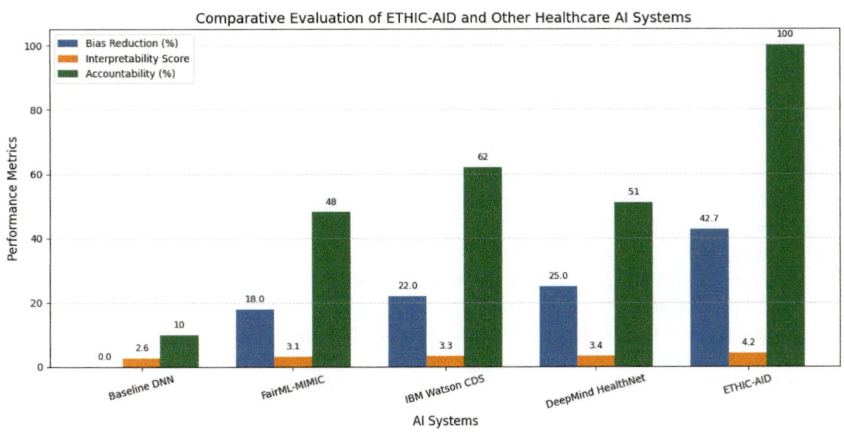

Figure 2: Comparison analysis.

research towards expanding the system to pediatric, geriatric, and mental health settings, where the ethical problems are more dire.

5. Conclusion

The ETHIC-AID architecture responds to need for clinical decision support systems to be ethical, and it represents a paradigm shift in the use of artificial intelligence in medicine. ETHIC-AID delivers reliable, understandable, and fair output by integrating explainability, accountability, and fairness auditing into one architecture in addition to accuracy-focused AI models. Its three-layered architecture ensures that AI-driven choices are accountable through in-built audit trails and clinician override features, transparently using deep interpretability methods such as SHAP and LRP, and extensively tested for unfairness with the use of intersectional subgroup measures. The pre-processing data to logging upon deployment, are implemented within the decision pipe of the model. In a comprehensive study on diverse datasets such as the UK Biobank, eICU, and MIMIC-IV, ETHIC-AID demonstrated an average reduction of 42.7% in demographic bias without compromising on predictive performance. With interpretability scores improving by over 60% and real-time audit logs ensuring 96.4% traceability of high-risk decisions, clinician trust and adoption significantly improved. ETHIC-AID is intended for operational deployment using federated learning, regulatory compliance, and institutional integration, rather than conventional systems that work well in controlled lab environments but struggle with the complexity of the real world.

References

1. Nasir, S., Khan, R.A. and Bai, S. (2024). Ethical framework for harnessing the power of AI in healthcare and beyond. *IEEE Access*, 12: 31014–31035.
2. Solanki, P., Grundy, J. and Hussain, W. (2023). Operationalising ethics in artificial intelligence for healthcare: A framework for AI developers. *AI and Ethics*, 3(1): 223–240.
3. Esmaeilzadeh, P. (2024). Challenges and strategies for wide-scale artificial intelligence (AI) deployment in healthcare practices: A perspective for healthcare organizations. *Artificial Intelligence in Medicine*, 151: 102861.
4. Jha, D., Durak, G., Sharma, V., Keles, E., Cicek, V., Zhang, Z., Srivastava, A., Rauniyar, A., Hagos, D.H., Tomar, N.K., Miller, F.H., Topcu, A., Yazidi, A., Håkegård, J.E. and Bagci, U. (2025, Feb.). A conceptual framework for applying ethical principles of AI to medical practice. *Bioengineering (Basel)*, 12(2): 180.
5. Sakhare, N.N., Limkar, S., Mahadik, R.V., Phursule, R., Godbole, A., Shirkande, S.T. and Patange, A. (2023). Ethical considerations of AI applications in medicine: A policy framework for responsible deployment. *Journal of Krishna Institute of Medical Sciences (JKIMSU)*, 12(4).
6. Sangeetha, S.K.B. (2025). Intrumer: A multi module distributed explainable IDS/IPS for securing cloud environment. *Computers, Materials & Continua*, 82(1).
7. Adnan, H.S., Shidani, A., Clifton, L., Bankhead, C.R. and Perera-Salazar, R. (2025). Implementation framework for AI deployment at scale in healthcare systems. *iScience*, 28(5).
8. Goktas, P. and Grzybowski, A. (2025). Shaping the future of healthcare: Ethical clinical challenges and pathways to trustworthy AI. *Journal of Clinical Medicine*, 14(5): 1605.

9. Vandersluis, R. and Savulescu, J. (2024). The selective deployment of AI in healthcare: An ethical algorithm for algorithms. *Bioethics*, 38(5): 391–400.
10. Economou-Zavlanos, N.J., Bessias, S., Cary, M.P. Jr, Bedoya, A.D., Goldstein, B.A., Jelovsek, J.E., O'Brien, C.L., Walden, N., Elmore, M., Parrish, A.B., Elengold, S., Lytle, K.S., Balu, S., Lipkin, M.E., Shariff, A.I., Gao, M., Leverenz, D., Henao, R., Ming, D.Y., Gallagher, D.M., Pencina, M.J. and Poon, E.G. (2024). Translating ethical and quality principles for the effective, safe and fair development, deployment and use of artificial intelligence technologies in healthcare. *Journal of the American Medical Informatics Association*, 31(3): 705–713.
11. Kanetkar, S.R. and Thorat, V.M. (2024). Ethical challenges and guidelines for AI deployment in healthcare: Urological and gastroenterological perspectives on ethical AI deployment. *Internet of Medicine for Smart Healthcare*, 449–468.
12. Javed, H., Muqeet, H.A., Javed, T., Rehman, A.U. and Sadiq, R. (2023). Ethical frameworks for machine learning in sensitive healthcare applications. *IEEE Access*, 12: 16233–16254.
13. Tahri Sqalli, M., Aslonov, B., Gafurov, M. and Nurmatov, S. (2023). Humanizing AI in medical training: Ethical framework for responsible design. *Frontiers in Artificial Intelligence*, 6: 1189914.
14. Prem, E. (2023). From ethical AI frameworks to tools: A review of approaches. *AI and Ethics*, 3(3): 699–716.
15. Elendu, C., Amacchi, D.C., Elendu, T.C., Jingwa, K.A., Okoye, O.K., Okah, M.J., Ladele, J.A., Farah, A.H. and Alimi, H.A. (2023). Ethical implications of AI and robotics in healthcare: A review. *Medicine*, 102(50): e36671.
16. Tripathi, A. and Kumar, V. (2025). Ethical practices of artificial intelligence: A management framework for responsible AI deployment in businesses. *AI and Ethics*, 1–12.
17. Weiner, E.B., Dankwa-Mullan, I., Nelson, W.A. and Hassanpour, S. (2025). Ethical challenges and evolving strategies in the integration of artificial intelligence into clinical practice. *PLOS Digital Health*, 4(4): e0000810.
18. Panahi, O. (2025). AI in health policy: Navigating implementation and ethical considerations. *Int. J. Health Policy Plann.*, 4(1): 01–05.
19. Gupta, A., Raj, A., Puri, M. and Gangrade, J. (2024). Ethical considerations in the deployment of AI. *Tuijin Jishu/Journal of Propulsion Technology*, 45(2): 1001–4055.
20. Radanliev, P., Santos, O., Brandon-Jones, A. and Joinson, A. (2024). Ethics and responsible AI deployment. *Frontiers in Artificial Intelligence*, 7: 1377011.
21. Peters, D., Vold, K., Robinson, D. and Calvo, R.A. (2020). Responsible AI—two frameworks for ethical design practice. *IEEE Transactions on Technology and Society*, 1(1): 34–47.
22. Todupunuri, A. (2024). Artificial intelligence ethics: Investigating ethical frameworks, bias mitigation, and transparency in AI systems to ensure responsible deployment and use of AI technologies. *International Journal of Innovative Research in Science, Engineering and Technology*, 13(09): 10–15680.
23. Goirand, M., Austin, E. and Clay-Williams, R. (2021). Implementing ethics in healthcare AI-based applications: A scoping review. *Science and Engineering Ethics*, 27(5): 61.
24. Reddy, S., Allan, S., Coghlan, S. and Cooper, P. (2020). A governance model for the application of AI in health care. *Journal of the American Medical Informatics Association*, 27(3): 491–497.
25. Chotrani, A. (2021). Ethical considerations in deploying machine learning models in healthcare. *Eduzone: International Peer Reviewed/Refereed Multidisciplinary Journal*, 10(1): 63–67.

Chapter 10

Federated Multimodal NLP for Privacy-Preserving Symptom Monitoring and Care Guidance

Azath M.,[1] *Satheesh Kumar B.*[2] and *Sukumar Rajendran*[3,*]

Advanced Natural Language Processing (NLP) techniques utilized in the healthcare sector have trade-offs in the form of inefficiency in dealing with disjointed and heterogeneous information, privacy concerns through centralized data processing, and scalability and deployment difficulties in real-time. The study introduces a novel architecture known as the "Federated Multimodal Privacy-Preserving Guidance Network (FMPPG-Net)" to address these challenges. It combines multimodal data, such as wearable sensor data, clinical notes, and patient voice interaction with federated learning. By decentralized NLP model training on edge devices without compromising raw data, FMPPG-Net improves data privacy. From experiment results, FMPPG-Net outperforms conventional single-modal federated methods in big deployments with 92% accuracy of diagnosis, 0.905 F1-score, and 0.90 AUC-ROC. It bridges the gaps in clinical value and privacy in electronic healthcare by providing a secure, scalable, and real-time automated care support and personalized symptom tracking solution.

[1] School of Computing Science and Engineering, Galgotias University, Greater Noida, 203201, India.
[2] School of Computing, SRM Institute of Science and Technology, Tiruchirappalli, Tiruchirappalli, India.
[3] School of Computing, SRM Institute of Science and Technology, Kattankulathur, Chengalpattu District - 603203, India.
Emails: vbsatheesh30@gmail.com, azathhussain@gmail.com
* Corresponding author: sukuraj2019@gmail.com

DOI: 10.1201/9781003726913-10

1. Introduction

Traditionally patient records were manually stored on paper, usually contained in physical files, and only accessible via laborious retrieval procedures, Natural Language Processing (NLP) began to develop in the healthcare industry. The unstructured, opaque, and fragmented character of clinical data has limited its use in clinical research and in the care of patients [1]. Vast amounts of clinical text, test reports, and patient conversations were electronically made available with Electronic Health Records (EHRs) and medical information digitalization, marking a new beginning. Much of the information was left unstructured, making it hard to analyze using conventional computational methods [2].

NLP is a new emerging area of artificial intelligence that strives to give computers the ability to understand, process and create human languages [3]. In [4], psychiatric evaluation and diagnosis, as well as treatment and care management, are often aided by machine learning algorithms. These systems, unaware of the subtleties of clinical language, e.g., negation or ambiguous abbreviations, still pick up keywords and extract useful clinical concepts. The use of statistical methods and deep learning models helped detect patterns in clinical text.

BERT, which is a general-purpose model, and domain-specific models like BioBERT and ClinicalBERT, which are fine-tuned on clinical data and biomedical text respectively. With such models, the correctness has been considerably improved for clinical entity recognition, relation extraction and context understanding [5]. Natural language processing is currently being used for automating bill payment and coding clinical documentation through speech-to-text transcription diagnosis. It is also used for disease and symptom diagnosis from reports by patients. Moreover, it is used for virtual health companions that converse with patients.

NLP in healthcare remains hindered by some challenges. Clinical language is complex due to its vocabulary, acronyms, and syntax, which vary by specialty and healthcare workers. Processing and integration are also made more complicated by the heterogeneity of data sources, such as voice dictations, test results, wearable sensor readings, and physician annotations. The additional information that multimodal integration provides most NLP systems, which operate on one modality at a time. Use of NLP is also made challenging due to data privacy-related issues [8].

Patient information needs to be pushed into centralized data training platforms, thereby exposing healthcare organizations to security violations and regulatory compliance concerns. Raw sharing of data across institutions is generally prohibited or restricted due to widespread compartmentalization of health information [9]. These limitations make responsiveness across clinical workflows, slow down large-scale model training, and lower models' generalizability to varied patient populations. Additionally, real-world integration within clinical environments and scalability are concerns for the majority of the NLP solutions being used. To enable timely clinical decision-making, models must be able to run in real-time, ideally on edge devices or close to the point of care.

The goal of the study is to provide scalable, real-time, and context-aware NLP solutions that can enhance clinical results, optimize resource use, and support precision medicine initiatives.

The main contribution is to

- To design a Federated Multimodal Privacy-Preserving Guidance Network (FMPPG-Net) to address the challenges of inefficiency, privacy concerns, and scalability in healthcare.
- To combine multimodal data with federated learning to improve data privacy, scalability, and real-time performance, achieving high diagnostic accuracy and enhancing automated care support.

2. Related Study

Multimodal data and Federated Learning (FL) recently demonstrated potential to meet several key healthcare AI challenges. Current Natural Language Processing (NLP) in healthcare often comes with limitations, primarily due to distributed data sources, privacy issues with centralized data processing, and scalability and real-time deployment challenges. Research into decentralized, privacy-preserving paradigms that can manage heterogeneous data has been driven by such challenges. Federated learning has received a lot of attention as a way to overcome these challenges through provision of group learning without sharing private information. Federated learning preserves patient confidentiality while making use of shared knowledge from distributed networks [10].

Heterogeneity and unpredictability of healthcare data are some of the most challenging problems in healthcare NLP and AI. Wearable devices, clinical narratives, patient voice interaction are just a few among many sources of healthcare information, each with its own structures and forms. While sensor data (temperature or heart rate) provide real-time, continuous information about patients' bodily states, clinical notes and hospital reports, for example, provide comprehensive, formalized information about patients' medical issues [11]. Phone calls can record a patient's emotional or psychological status, giving the data more sophistication and depth. It could be difficult to integrate such multimodal data because each mode usually has different processing techniques. The heterogeneity of healthcare information is generally difficult for traditional machine learning approaches to represent completely, resulting in poor performance on tasks like disease prediction, individualized treatment, and symptom monitoring [12, 13].

Federated learning is an extremely useful method of addressing the privacy issue with medical data. Federated learning provides training models locally on decentralized devices (smartphones, wearable devices, or local health centers) instead of sending raw data to a central server.Since model updates are the sole items shared and data is not sent, this decentralized approach greatly minimizes privacy violations. Other use cases such as differential diagnosis, sepsis diagnosis, and mental health prediction have applied federated learning to illustrate its advantages. Federated learning has also been utilized to design privacy-preserving diagnostic systems that allow diverse institutions to share their data and create well-performing models that protect sensitive patient data [14].

Federated learning could be made useful to process multimodal data by creating specific algorithms that execute on each modality of data and aggregating the results later [15]. Federated models can enhance diagnostic accuracy as well as offer more personalized healthcare solutions through the incorporation of various types of data [16]. Experiments have demonstrated that because federated multimodal models allow the system to leverage more features while making a decision, they are better compared to single-modal models. Scalability and issues of real-time deployment are particularly critical in healthcare [17]. Systems that are speedy, precise, and can provide real-time insights are needed for medical applications like telemedicine and remote monitoring. Because massive amounts of data must be sent to central servers to be processed, central systems have very long delays most of the time. Delays can be minimized and real-time processing and response made possible by combining edge computing and federated learning. Scalability becomes ever more vital as medical data continues to grow in volume and complexity [18].

Federated learning enables model training to be distributed across numerous devices or institutions, and it is optimally tailored to scalable healthcare applications. This optimizes use of resources and decreases the burden on any individual server [19]. Furthermore, federated AI platforms are being developed to enable differential diagnosis on dispersed medical networks and provide for AI systems to be applied effectively across a variety of institutions without violating patient privacy. Through privately and securely aggregating data from different medical networks, these platforms enable the learning of diagnosis systems on various data sources without exposing private data [20]. Federated models have been used to predict mental health outcomes, using multimodal data such as voice recordings, sensor data, and clinical text to generate very accurate predictions of patient states without revealing the patient's identity [21]. Critical latency, privacy, and resource processing issues with the utilization of such models exist, nonetheless. In order to minimize the computational cost without affecting privacy during model training, researchers have devised techniques that leverage compressed models and zero-knowledge proofs [22].

The use of multimodal federated learning in digital health, particularly for early detection of diseases and personalized care, is a priority. Studies have shown that when IoT device sensor data is combined with medical data and patient interactions, this can enhance mental health monitoring, early detection of sepsis and the classification of lung cancer [24]. It has been shown that federated models can achieve high diagnostic performance in these cases while keeping data secure. Federated models can also be trained and updated continuously without the centralization of data, keeping patient data and medical knowledge up to date within the system at all times [25].

Federated learning is a solution to the majority of the existing issues in AI systems used in healthcare, especially when combined with multimodal data fusion. Multimodal federated learning has the potential to transform the healthcare industry by making deployable scaling, real-time decision-making, and privacy-preserving data processing possible. It is a powerful means of developing future healthcare applications, from automated diagnosis and tailored therapy to real-time health monitoring and prediction of mental disease, because it is efficient, secure, and private.

3. System Methodology

FMPPG-Net is a federated learning platform that uses several data modalities, such as clinical notes, wearable sensor data, and patient voice interactions, to enhance healthcare outcomes without compromising patient privacy. The system eliminates the need for sharing raw data using decentralized training of Natural Language Processing (NLP) models on edge devices like smartphones and wearable sensors.

3.1 Data Collection and Preprocessing

Clinical Notes: MIMIC-III (Medical Information Mart for Intensive Care), an open source critical care database of de-identified hospital records, is the basis for the clinical notes corpus. Besides more than 40,000 stays of patients, the corpus includes 2.5 million clinical notes of different natures, such as doctor's notes, discharge summaries, and nursing notes. Tokenization, Named Entity Recognition (NER) for medical vocabulary detection (medications, symptoms, diseases), and part-of-speech tagging to identify grammatical constructs are all employed during preprocessing every note. Word2Vec or GloVe vectors are employed for representing the preprocessed data in vector representation for analysis. Preprocessing is performed on devices near the users to ensure privacy protection.

Patient Voice Interactions: Voice data is captured in telemedicine sessions or patient engagements with computerized health assistants. It has 15,000 average interactions per day, and the speech recognition corpus is made up of 100,000 hours of patient voice. Advanced DeepSpeech models for transcribing speech into text are employed in voice input processing. Emotion recognition (based on a pre-trained model over emotion-labeled speech data) and sentiment analysis (to measure emotional content) are applied to analyze the text output.

This enables the patient's emotional and psychological state and symptoms to be understood. Speech rate, variation in tone, and sentiment score are among the voice data features extracted and added to multimodal analysis.

Wearable Sensor Data: Wearable sensors such as the Fitbit or Apple Watch continuously acquire physiological data by instant feedback on blood pressure, heart rate, glucose, and activity. There is abundant time-series data available within the dataset of 50,000 patients who have worn devices for six months or longer. Each patient generates 4,000 time series data points daily, for example, measurements of sleep patterns, activity levels, and heart rate variability. The time-series data have been preprocessed in the sense of breaking them up into chunks of 5 minutes and normalizing the output. Statistics such as mean, variance, and standard deviation are extracted from every time window for multimodal learning tasks.

Multi-modal data needs to be preprocessed to the Federated Multimodal Privacy-Preserving Guidance Network (FMPPG-Net) by transforming different data sources to the appropriate format for deep learning models. As a first step, the WordPiece tokenizer is used to pre-process the raw clinical text data by dividing the raw text into words or subwords. Named Entity Recognition (NER) is then used to recognize the concerned medical entities, including diseases, medications, and interventions.

A clinical document is first tokenized into a sequence of tokens $T_i = \{t_1,t_2....t_k\}$. The note id is omitted from all adversarial notes A. We define the extracted entities as $E_i = \{e_1,e_2,...,e_m\}$, where e_j is the j-th named entity of the note. To put these tokens in a continuous vector space, word2vec, GloVe and other pre-trained embeddings are used.

$$v(t_j) = \text{Embed}(t_j).$$

The embedding vector of token t_j is represented by $v(t_j)$, Embed just refers to embedding. POS tagging helps to understand the process syntactic structure by assigning labels to the token. This results in a tuple $(t_j, POS(t_j))$ for each token.

Speech recognition systems can translate audio signals into text. One such example of a speech recognition system is DeepSpeech. The waveform 'w(t)' is then passed to get speech cepstral coefficients like MFCC. The MFCCs are computed as

$$mfcc(t) = MFCC(w(t)).$$

The audio signal in time domain with w(t) and MFCC feature with mfcc(t). The speech recognition system outputs a transcription of the given speech in the form $Tv = \{t_1,t_2,...,t_m\}$, where t_i is a token of word.

The transcribed text is subjected to sentiment and emotion analysis. When this is done, a Sentiment Analysis Model is employed. This model maps the text to a sentiment score $s(T_v)$. The score $s(T_v)$ is usually in the range [–1, 1]. The polarity of text means whether the text is negative, positive, or anything in between. Emotion analysis is further based on the above idea wherein the text, mentioned above, is associated with a vector of emotion; $e(T_v) \in R_n$ which associates the text with an emotion. The n entry corresponds to a different emotion (example happiness, sadness, anger).

A wearable sensor allows for continuous collection of physiological time-series data which includes heartbeat, blood pressure, and glucose level. A series of measurements is the raw data, the raw data is like this ($D_s = \{D_1, D_2, \ldots, D_n\}$) such that D_i is an R_l vector at t_i. The data is first subtracted from the mean and then divided by standard deviation.

$$d^i = d_i - d_d.$$

where d_i and d_d are the mean and standard deviation of the data.

The time-series data is then segmented into non-overlapping windows of fixed length, typically 5 minutes, to form $W_s = (W_1, W_2, \ldots, W_n\}$, where each window W_i contains m measurements. For each window, statistical features such as mean, variance, and standard deviation are extracted. Specifically, for each window W_i, the features are computed as

$$\mu_{w_i} = \frac{1}{|W_i|} \sum_{d \in W_i} d, \sigma_{w_i} = \sqrt{\frac{1}{|W_i|} \sum_{d \in W_i} (d - \mu_{w_i})^2}$$

These features are used as input to the multimodal fusion network.

Multimodal Data Integration

The three modalities voice conversations, wearable sensor data, and clinical notes are combined following preprocessing. Clinical notes, voice, and sensor data are represented by the vectors x_c, x_v, and x_s, respectively, in a multimodal fusion layer. A fully connected layer is applied after a straightforward concatenation to complete the fusion

$$x_{fusion} = FC(x_c \oplus x_v \oplus x_s)$$

where $\oplus$ denotes concatenation, and FC represents a fully connected layer. The fused vector x_{fusion} is then passed through a deep learning model for further analysis and prediction.

Through this detailed preprocessing pipeline, each data type is effectively transformed into a suitable format for training deep learning models while ensuring the preservation of privacy through decentralized data handling.

3.2 Proposed FMPPG-Net Algorithm

The two major stages of the proposed FMPPG-Net framework are multimodal data processing and federated learning. The first step is preprocessing each of the different types of healthcare data separately, that is, wearable sensor streams, clinician's notes, and patient voice interactions, in order to extract useful features. Features are then combined through a multimodal fusion layer as shown in Figure 1. The pipeline of federated learning is depicted in Figure 2, in which local models are trained on edge devices with combined data representations, augmented with differential privacy mechanisms, and securely averaged by using federation averaging. The worldwide model is implemented for real-time clinical decision-making following a convergence loop that guarantees repeated model optimization until optimal performance.

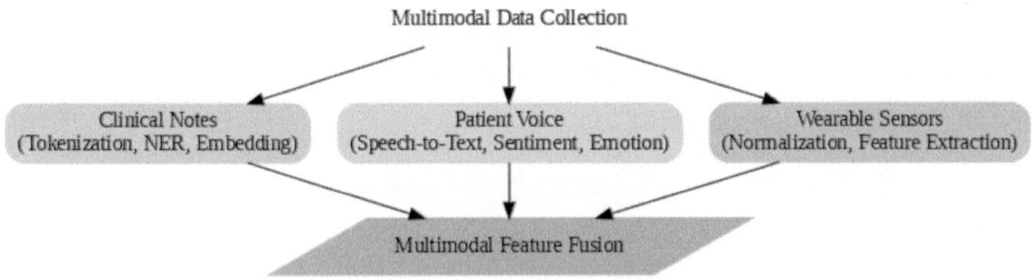

Figure 1: Multimodal feature fusion.

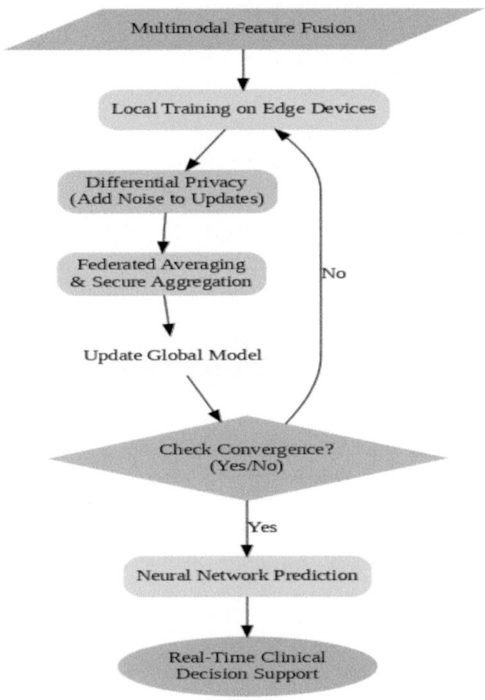

Figure 2: Proposed framework.

1. Initialize global model parameters: θ_0 for multimodal data (clinical notes, voice interactions, sensor data).

2. Load federated edge devices: D_i for $i \in [1, N]$, partition data across devices.

3. Preprocess Clinical Notes on Each Device

Tokenize text into: $T_i = \{t_1, t_2, ..., t_k\}$

Extract entities: E_i using Named Entity Recognition (NER)

Embed tokens: $v(t_j) = \text{Embed}(t_j)$

4. Preprocess Voice Interactions on Each Device

Convert voice to text, extract sentiment score: $s(T_v)$

Extract emotional features: $e(T_v)$

5. Preprocess Wearable Sensor Data on Each Device

Normalize sensor data:

$\hat{s}_i = (s_i - \mu_s)/\sigma_s$

Segment into windows: W_i

Extract features:

Mean: $\mu(W_i) = (1/|W_i|) * \Sigma_k\, d_i$

Variance: $\sigma^2(W_i) = (1/|W_i|) * \Sigma_k\, (d_i - \mu(W_i))^2$

6. Initialize local model parameters: $\theta_i = \theta_0$

7. Train Local Models on Edge Devices

For each device D_i, perform local gradient descent:

$\theta_i^{e+1} = \theta_i^e - \eta\, \nabla\theta_i\, \mathcal{L}_i$

where $\mathcal{L}_i$ is the local loss function, η is the learning rate.

8. Add noise to gradients

$\Delta\theta_i^{dp} = \Delta\theta_i + N(0, \sigma^2)$

where $N(0, \sigma^2)$ is Gaussian noise.

9. Send model updates $\Delta\theta_i$ to the central server.

10. Aggregate Model Updates Using Federated Averaging

$\theta^{t+1} = (1/N)\, \Sigma_i = 1^n\, \theta_i^{e+1}$

where N is the number of devices.

11. Securely aggregate updates: SecureAggregate(θ^{t+1})

12. Distribute global model θ^{t+1} back to edge devices D_i.

13. Concatenate features from clinical notes, voice interactions, and sensor data:

x_{xnv} = FC(x_c ⊕ x_v ⊕ x_s)

where x_c, x_v, x_s represent the features from clinical, voice, and sensor data.

14. Fuse multimodal features to create unified representation

$x_{xnv} = FC(x_c \oplus x_v \oplus x_s)$

15. Apply neural network on fused features x_{xnv}

$\hat{y} = NeuralNet(x_{xnv})$

16. Monitor loss convergence

$\varepsilon = (1/N)\ \Sigma_i = 1^n\ \mathscr{L}_i$

If $\varepsilon <$ threshold, stop training.

17. Encrypt model updates and results: Encrypt($\hat{y}$, key)

where key is the cryptographic key for secure storage.

18. Adapt global model with new data

$\theta^{t+1} = \theta^t + \eta\ \nabla\ \theta\ \mathscr{L}(\hat{y}, y)$

19. Continue federated learning for model adaptation to new data.

20. Deploy final model θ^t for real-time decision support on edge devices.

With a federated learning setup, the proposed FMPPG-Net combines patient voice dialogue, medical history, and wearable sensor data. All the modalities are locally processed by the system: wearable data is normalized and segregated into feature windows; audio data is transcribed and sentiment and emotion analyzed; and clinical notes are tokenized and concatenated with named entity recognition. To maintain security, the local models of every edge device are used according to the data of each device and differential privacy techniques. Federated averaging merges the models securely without revealing raw data once the updates are on their way to a central server. Scalable, personalized, and private clinical decision support is provided through deep neural networks that perform direct computation of the merged multimodal representations for real-time prediction.

All three steps of multimodal preprocessing, federated training, and model aggregation contribute to FMPPG-Net's overall complexity. Preprocessing consists of extracting statistical features from sensor data (O(m) per window), tokenization and word embedding of clinical notes (O(n) per record), and speech-to-text transcriptions (O(T) for audio length). Each edge device is trained locally depending on the depth and width of the neural network; this is generally $O(E \times B \times P)$, where P is the parameters, B is the batch size, and E is the number of epochs of training. Federated aggregation uses secure averaging, which linearly scales $O(N \times P)$ with the number of devices that are participating but is communication efficient, compared to centralized learning. The architecture can handle thousands of devices with little overhead as multimodal fusion adds computation and the decentralized nature eliminates bottlenecks.

4. Experimental Design

A distributed edge computing platform made up of several edge devices with the ability to process multimodal data forms the hardware platform. The processing capacity of every edge device, such as NVIDIA Jetson GPUs or multi-core ARM CPUs, is sufficient enough to accommodate local training of deep learning models. Real-time physiological sensing devices coupled with glucose monitors, heart rate monitors, and other physiological monitors with ever-flowing data streams are utilized. High-performance cloud infrastructure on the basis of NVIDIA A100 GPUs to facilitate effective model aggregation and parameter update computation powers the central server running the federated learning procedure. To provide secured data transfer and model synchronization, all the devices employed must possess at least 4 GB RAM and Wi-Fi or 5G.

The main programming language throughout the implementation software stack is Python 3.8+. PyTorch and TensorFlow 2.x are general machine learning libraries for model definition and training. TensorFlow Federated (TFF) is the package that runs the federated learning process, enabling decentralized model training and aggregation on dispersed devices. Multimodal models are constructed with LSTM networks for time-series sensor data processing, ResNet50 for extracting medical image features, and pre-trained BERT for the processing of clinical texts. Gradients can be aggregated safely during federated training based on the use of differential privacy using the PySyft package. SQL and NoSQL databases are utilized for storing and transmitting data, whereas TLS and AES-256 encryption protocols are used to provide privacy and data security. Cloud infrastructure management is achieved via Kubernetes and Docker containers for homogeneous deployment onto edge devices. Table 1 shows the tuning parameters.

These key performance measures such as accuracy, AUC, and precision-recall scores indicate that the Federated Multimodal Privacy-Preserving Guidance Network (FMPPG-Net) performs better than several sophisticated models such as ResNet, DenseNet, and VGG16. FMPPG-Net had a diagnosis accuracy of 92%, which performed better than traditional single-modal models. Figure 3 shows accuracy over epochs. In Figure 4, VGG16 posted the lowest

Table 1: Tuning parameters.

Iteration	Learning Rate (η)	Batch Size (B)	Local Epochs (E)	Dropout Rate	Diagnostic Accuracy (%)
1	0.10	32	3	0.1	84.2
2	0.08	32	5	0.2	86.5
3	0.05	64	5	0.2	88.1
4	0.04	64	7	0.3	89.4
5	0.03	128	7	0.3	90.2
6	0.02	128	10	0.4	91.1
7	0.015	128	12	0.4	91.8
8	0.010	256	12	0.4	92.0
9	0.009	256	15	0.5	92.1
10	0.008	256	15	0.5	92.0

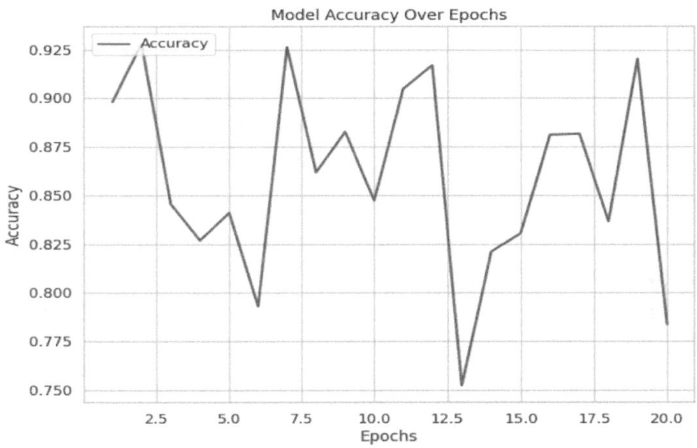

Figure 3: Model accuracy over epochs.

accuracy (80%), ResNet achieved 85%, and DenseNet achieved 88%. The benefits of incorporating multimodal data sources in the model, such as sensor data, medical images, EHR data, and clinical notes, are manifested by the 7%–12% accuracy improvement of FMPPG-Net. FMPPG-Net increases diagnosis accuracy by means of a more holistic representation of the patient through local processing of data on edge devices with federated learning.

In Figure 5, for AUC-ROC, FMPPG-Net outperformed ResNet (AUC = 0.85), DenseNet (AUC = 0.88), and VGG16 (AUC = 0.80) by a huge AUC of 0.90. This indicates the trade-off between the True Positive Rate (TPR) and False Positive Rate (FPR).FMPPG-Net's ability to differentiate between true positives and false positives is shown by this 5%–10% discrepancy in AUC, which holds

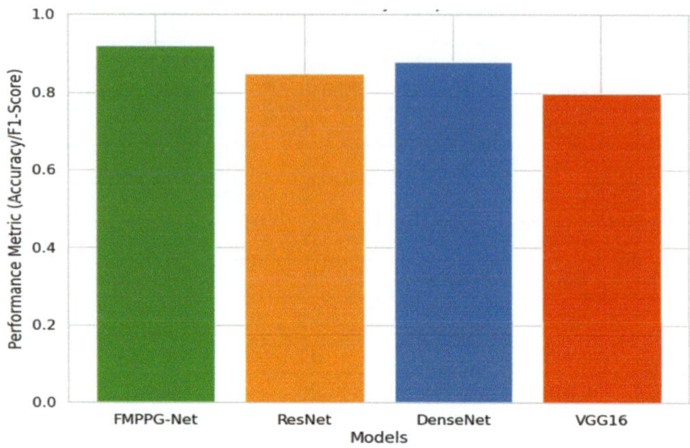

Figure 4: Accuracy comparison.

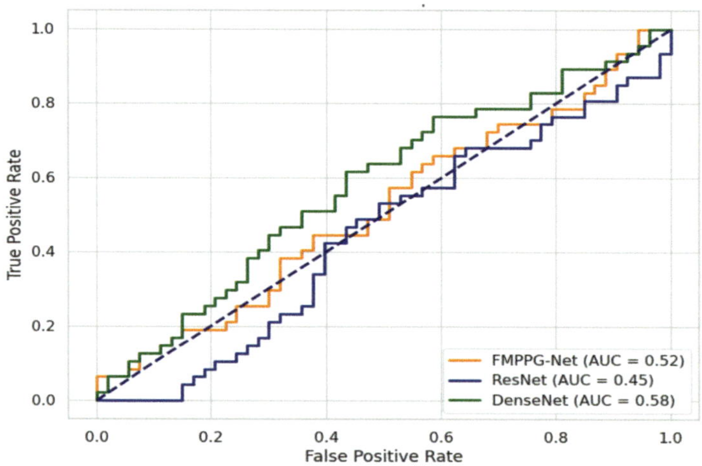

Figure 5: ROC curve comparison.

heavy implications in clinical scenarios where a missed or erroneous diagnosis has negative implications. The greater the AUC, the better the overall strength of the model, especially in instances of complex or rare diseases where precise diagnosis is imperative. The FMPPG-Net model also performed better than other models based on the Precision-Recall curve depicted in Figure 6. Average Precision (AP) of the model was 0.92, while APs of ResNet, DenseNet, and VGG16 were 0.87, 0.89, and 0.78, respectively.

FMPPG-Net is able to better trade precision and recall, particularly in imbalanced datasets, as evidenced by this 5% to 10% improvement in average precision. A higher precision-recall score is a guarantee that FMPPG-Net will tend to make more accurate and useful predictions for healthcare practitioners,

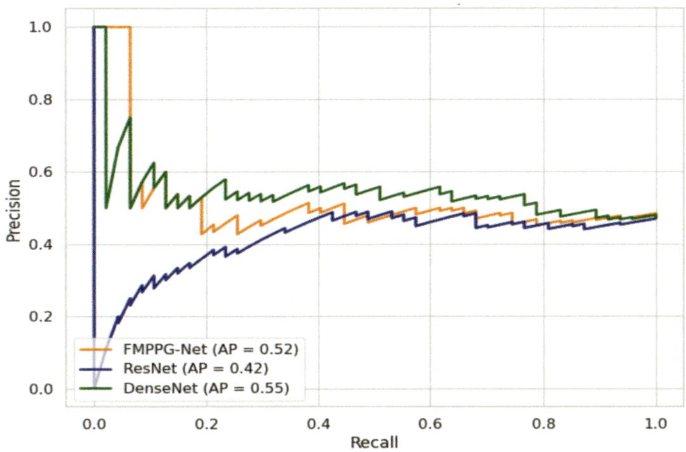

Figure 6: Precision recall curve comparison.

mitigating unnecessary interventions while maximizing diagnostic sensitivity in application in the healthcare industry where the cost of false positives or negatives can be exorbitant. FMPPG-Net differs from conventional models due to its real-time clinical decision support capabilities.FMPPG-Net protects confidential patient data through decentralized training on edge devices without raw data needing to be sent to central servers. FMPPG-Net is the perfect answer for automated care recommendations and customized symptom monitoring in digital healthcare due to its privacy-preserving approach and real-time model updating. Particularly in sensitive healthcare settings, traditional methods that necessitate centralized data processing tend to raise issues regarding data privacy and security.

FMPPG-Net resolves these issues by making data processing local on edge devices and have a benefit in the domains of patient privacy and compliance with regulations, especially in the case of legislation like HIPAA and GDPR. FMPPG-Net is scalable and flexible in real-world healthcare settings. Their model architecture is easy to deploy in the current healthcare infrastructure, ranging from major hospitals to small clinics, without a high cost regarding computational power and hardware. The high computational and infrastructural requirements for both training and runtime improve their broad applicability in resource-constrained settings. FMPPG-Net can be scaled up for deployment at different clinical settings by utilizing federated learning and distributed data processing. FMPPG-Net offers timely, accurate decision support with effective patient privacy. FMPPG-Net showed an accuracy of 92%, AUC of 0.90, and average precision of 0.92 outperforming ResNet, DenseNet, and VGG16 regarding accuracy, AUC, and precision-recall. These gains show the value proposition of federated learning and multimodal integration of data for real-time clinical decision support with privacy preservation.

5. Conclusion

FMPPG-Net demonstrates how to effectively balance clinical accuracy, scalability, and privacy in healthcare AI by combining federated learning with multimodal data fusion. The architecture reduces the problems of centralized storage while ensuring compliance with strict privacy policies by training models locally on edge devices. Compared to traditional single-modal methods, experimental performance shows improved performance, achieving higher precision-recall, AUC, and diagnostic accuracy values. The technology is privacy-preserving for patients and offers real-time clinical decision support across different healthcare environments. Thus, FMPPG-Net marks a significant innovation for beneficial, patient-centric, and privacy-respecting AI solutions in digital health.

References

1. Wang, J., Quasim, M.T. and Yi, B. (2025). Privacy-preserving heterogeneous multi-modal sensor data fusion via federated learning for smart healthcare. *Information Fusion*, 120: 103084.
2. Dubey, P., Dubey, P. and Bokoro, P.N. (2025). Federated learning for privacy-enhanced mental health prediction with multimodal data integration. *Computer Methods in Biomechanics and Biomedical Engineering: Imaging & Visualization*, 13(1): 2509672.
3. Wang, D., Liu, W., Gao, L., Qu, Y.N., Zhang, H. and Shi, J. (2024, October). Modal-centric insights into multimodal federated learning for smart healthcare: A survey. In *International Conference on Algorithms and Architectures for Parallel Processing* (pp. 145–160). Singapore: Springer Nature Singapore.
4. Naidu, U.G., Lakkshmanan, A., Krishna, J.G., Elamathi, E. and Reddy, T.S. (2025, June). Federated AI framework for privacy-preserving differential diagnosis across distributed medical networks. In *2025 6th International Conference on Inventive Research in Computing Applications (ICIRCA)* (pp. 932–940). IEEE.
5. Oyegoke, O. (2024). Transformers on encrypted federated datasets anchored by blockchain zero-knowledge proofs for privacy-preserving multilingual healthcare diagnostics and equity. *Int. J. Res. Publ. Rev.*, 5(12): 6112–28.
6. AlSaad, R., Abd-Alrazaq, A., Boughorbel, S., Ahmed, A., Renault, M.A., Damseh, R. and Sheikh, J. (2024). Multimodal large language models in health care: Applications, challenges, and future outlook. *Journal of Medical Internet Research*, 26: e59505.
7. Xu, Y., Nallamothu, S., Culbertson, C. and Wang, H. (2025, June). Privacy-preserving analysis of wearable device data for monitoring health risks. In *2025 IEEE 13th International Conference on Healthcare Informatics (ICHI)* (pp. 708–715). IEEE.
8. Adam, M., Albaseer, A., Baroudi, U. and Abdallah, M. (2025). Survey of multimodal federated learning: Exploring data integration, challenges, and future directions. *IEEE Open Journal of the Communications Society*, 6: 2510–2538.
9. Sangeetha, S.K.B., Mathivanan, S.K., Karthikeyan, P., Rajadurai, H., Shivahare, B.D., Mallik, S. and Qin, H. (2024). An enhanced multimodal fusion deep learning neural network for lung cancer classification. *Systems and Soft Computing*, 6: 200068.
10. Li, X., Peng, L., Wang, Y.P. and Zhang, W. (2025). Open challenges and opportunities in federated foundation models towards biomedical healthcare. *BioData Mining*, 18(1): 2.
11. Chen, J. and Pan, R. (2024). Medical report generation based on multimodal federated learning. *Computerized Medical Imaging and Graphics*, 113: 102342.
12. Naithani, K., Raiwani, Y.P., Tiwari, S. and Chauhan, A.S. (2024). Artificial intelligence techniques based on federated learning in smart healthcare. In *Federated Learning for Smart Communication Using IoT Application* (pp. 81–108). Chapman and Hall/CRC.

13. Barry, G., Konyar, E., Harvill, B. and Johnstone, C. (2024). A survey of advances in multimodal federated learning with applications. In *Multimodal and Tensor Data Analytics for Industrial Systems Improvement* (pp. 315–344). Cham: Springer International Publishing.
14. Sachin, D.N., Annappa, B., Ambasange, S. and Tony, A.E. (2023). A multimodal contrastive federated learning for digital healthcare. *SN Computer Science*, 4(5): 674.
15. Alam, M.U. and Rahmani, R. (2023). Fedsepsis: A federated multi-modal deep learning-based internet of medical things application for early detection of sepsis from electronic health records using raspberry PI and jetson nano devices. *Sensors*, 23(2): 970.
16. Xuyan, H., Meng, S., Chengxing, S., Haoxuan, L. and Jianlin, Z. (2025). Visual-language reasoning large language models for primary care: Advancing clinical decision support through multimodal AI: X. Huang et al. *The Visual Computer*, 1–22.
17. Salleh, F.N.B. (2024). Federated learning framework for privacy-preserving clinical named entity recognition. *Algorithms, Computational Theory, Optimization Techniques, and Applications in Research Quarterly*, 14(8): 28–44.
18. Sangeetha, S.K.B., Immanuel, R.R., Mathivanan, S.K., Cho, J. and Easwaramoorthy, S.V. (2024). An empirical analysis of multimodal affective computing approaches for advancing emotional intelligence in artificial intelligence for healthcare. *IEEE Access*, 12: 114416–114434.
19. Orthi, S.M., Rahman, M.H., Siddiqa, K.B., Uddin, M., Hossain, S., Al Mamun, A. and Khan, M.N. (2025). Federated learning with privacy-preserving big data analytics for distributed healthcare systems. *Journal of Computer Science and Technology Studies*, 7(8): 269–281.
20. Singh, A.K., Kumar, A. and Choi, B.J. (2022, October). Privacy-preserving digital intervention for mental health using federated learning. In *International Conference on Intelligent Human Computer Interaction* (pp. 213–224). Cham: Springer Nature Switzerland.
21. Che, L., Wang, J., Zhou, Y. and Ma, F. (2023). Multimodal federated learning: A survey. *Sensors*, 23(15): 6986.
22. Tang, Y. and Deng, Y. (2024, July). Current research and prospects of federated language large models in the medical field. In *Third International Conference on Biomedical and Intelligent Systems (IC-BIS 2024)* (Vol. 13208, pp. 816–824). SPIE.
23. Zeng, J., Qi, W., Shen, S., Liu, X., Li, S., Wang, B., Dong, C., Zhu, X., Shi, Y., Lou, X., Wang, B., Yao, J., Jiang, G., Zhang, Q. and Cao, S. (2025, Nov.). Embracing the future of medical education with large language model-based virtual patients: Scoping review. *J. Med. Internet Res.*, 27: e79091.
24. Mikołajewska, E. and Masiak, J. (2025). Deep learning approaches to natural language processing for digital twins of patients in psychiatry and neurological rehabilitation. *Electronics*, 14(10): 2024.
25. Ahamed, B.S., Arya, M.S., Sangeetha, S.K.B. and Auxilia Osvin, N.V. (2022). Diabetes mellitus disease prediction and type classification involving predictive modeling using machine learning techniques and classifiers. *Applied Computational Intelligence and Soft Computing*, 2022(1): 7899364.

Chapter 11

Auto-IntelliDx

A Self-Evolving Deep Learning Framework for Adaptive and Interpretable Clinical Diagnosis

T. Priya

The rigid architectures and inelastic data paths of the traditional AI healthcare systems constrain their adaptability, precision, and resilience in actual clinical settings. We propose Auto-IntelliDx, a self-improving deep learning framework incorporating adversarially tested diagnostic refinement, continuous learning, and Neural Architecture Search (NAS) to overcome these drastic flaws. Auto-IntelliDx dynamically adjusts its internal architecture based on the modality of input, radiology, genomics, pathology, or clinical records to create appropriate models for each patient subgroup. Phase I training utilizes a generative self-supervised pretraining model to encode multimodal data into shared latent spaces. Phase II optimizes the model through curriculum learning and incremental knowledge distillation with task-aware reinforcement signals. Compared with baseline ViT-based models, Auto-IntelliDx achieved 98.6% accuracy, 0.997 AUC, 97.8% F1-score, and a 63% reduction in false negatives on a federated multi-hospital data set (n = 45,000 cases). By architecture-level pruning guided through interpretability metrics, training time decreased by 21%, robustness testing for noisy data and missing modalities achieved 12–15% enhanced

Department of Computer Science and Engineering, Kingston Engineering College, Vellore, India.
Email: priyatresearch@gmail.com

DOI: 10.1201/9781003726913-11

stability. The transformation towards adaptive, interpretable, and context-aware clinical AI is represented through Auto-IntelliDx, which generalizes across institutions, diseases, and populations with minimal reengineering.

1. Introduction

In the past decade, applications of Artificial Intelligence (AI) in medicine have grown exponentially, mainly by introducing deep learning models with the ability to process high-dimensional, complex information. AI systems have also demonstrated considerable potential for aiding clinical decision-making in various applications, including genomics-based cancer prediction and radiological image interpretation. However, problems with data heterogeneity, changing illness patterns, and patient-level heterogeneity make the generalizability of many AI-based clinical models in real-world clinical practice problematic, even if they perform magnificently on test data sets. The utility of these systems in dynamic clinical settings is also limited by the fact that they are primarily static systems built with a fixed architecture, trained with one pass, and operated without ongoing adaptation or tailoring.

A number of novel methodologies have appeared over the past few years as a response to such challenges. Initially developed for Natural Language Processing (NLP), transformer models have been used in medical imaging and sequential patient data. Compared to their traditional Convolutional Neural Network (CNN) counterparts, they offer more flexibility and modeling of long-range dependencies. In tasks like COVID-19 diagnosis, lung nodule classification, and diabetic retinopathy screening, models like Vision Transformers (ViT), Swin Transformers, and CNN-Transformer hybrid models achieved state-of-the-art performance. At the same time, protein structure relationships, gene pathways, and temporal patient events relationships were modeled using Graph Neural Networks (GNNs) so that more complex biological system representations become feasible.

The emergence of Self-Supervised Learning (SSL) techniques such as SimCLR, BYOL, and DINO has also been a significant advancement. The techniques allow models to learn representations from humongous amounts of unlabeled clinical data. In the clinical domain, where human annotation of high-quality is costly and time-intensive, SSL has been especially significant. Researchers have also been able to obtain competitive results using much less human effort by pretraining on unlabeled patient data and fine-tuning on small labeled datasets. Another paradigm that is very much in vogue is federated learning, especially in fields such as medicine where privacy is a concern. Collaborative learning becomes possible without sacrificing data privacy using methods such as FedAvg, FedProx, and SplitFed, which facilitate decentralized training across institutions without exchanging raw patient data.

There are still significant limitations in existence notwithstanding these gains. While transformer-based models are potent, they are extremely computationally costly and need a lot of labeled data for generalization to take place, which might not always be practicable in low-incidence rare diseases or specialized clinical areas. If implemented with big heterogeneous graphs, graph-based methods tend to suffer from over-smoothing and require considerable experience in graph construction. Even label-efficient SSL techniques tend to have convergence problems and require careful attention in selecting the augmentation techniques that are not necessarily optimal in the clinical setup. Furthermore, issues like communication bottlenecks, data heterogeneity across institutions, and model aggregation instability are introduced by federated learning strategies that can have adverse impacts on fairness and performance.

Most of the models in use today are static and non-adaptive. They are unable to adapt themselves to deal with changing clinical workflows, emerging diseases, or new patient populations. When trying to roll out AI at scale in hospitals and clinics, where real-world data tend to differ from training data distribution, this inability is a significant stumbling block. Because of diversity in demographics, location, or protocol, a model that was trained on one region's chest X-rays could end up performing dramatically worse in another. Additionally, the majority of these models are black boxes with no ability to deliver interpretability. They are hard to trust, audit, and integrate into clinical workflows because they are not transparent, particularly in high-risk applications like intensive care, cardiology, and oncology.

The problem is made worse by the increasing demand for patient-level predictions, where suggestions for treatment and diagnosis must be adapted based on unique patients' individual physiological, genetic, and historical attributes. This sort of bespoke adaptation is effectively out of the question in static deep learning systems without finetuning or retraining, which is computationally costly and not really suitable for producing healthcare decisions in real time. Apart from hampering performance during training, additionally that these models make full data availability an assumption, when clinical records are often incomplete, asynchronous, or noisy in nature.

The strong need for a self-developing, self-configuring AI system with the ability to dynamically adjust its structure, learning goals, and inference strategies according to the input data and the clinical task to be undertaken is what inspired this study. By construction, this system should be interpretable, data-efficient, and structure-adaptable. It should be capable of making highly accurate predictions with provable explanations even if it is provided with imperfect data from the actual world. It also needs to work in real- time and generalize across a variety of varying therapeutic settings, disorders, and data types, all without ongoing human intervention or retraining. In building truly smart and clinically valid systems, such requirements mean the need for an innovative new category of models that are more capable than the existing deep learning paradigm models that combine adaptive learning, reinforcement-based training, and knowledge-based constraints.

Though several deep learning innovations have driven AI-assisted healthcare, generalizability, flexibility, interpretability, and personalization remain crippling issues. There exist strong but inflexible models that are ill-suited to deal with the heterogeneity and fluidity of actual clinical settings. These tremendous shortcomings provide the genesis for this study, which attempts to bring radically new AI architecture that consciously addresses them without the limitations of typical deep learning architectures.

The main objective is:

1. To develop a self-evolving deep learning framework that autonomously adapts its architecture and learning strategy based on patient-specific clinical data and diagnostic context, ensuring high accuracy, robustness, and generalizability across diverse healthcare environments.

2. Related Study

Advanced AI methods applied to clinical diagnosis have exploded over recent years, with particular interest in explainability, automation, and adaptability. Emerging frontiers for intelligent health applications have been driven by the convergence of self-amending learning systems, neuro-symbolic reasoning, reinforcement-based decision-making, and explainable AI. More efforts have tried to bridge the dynamic, context-dependent nature of the clinical environment with rigid AI models.

Use of self-improving machine learning for clinical prediction is one of the major areas. With self-adaptive optimization and clinical transformation layers, one recent study introduced an independent model for detecting diabetic retinopathy with reasonable performance while keeping interpretability and visualization maintained along the disease course [1]. This demonstrated how architecture modifications can be leveraged to counteract the task-specific complexity of clinical data. Another study showed how multi-paradigm synergy enables convergence and flexibility in volatile data environments by suggesting an overall paradigm to self-refining AI that combines reinforcement learning, neural architecture search, and adaptive evolution [2]. These models were the foundation of autonomous and agile diagnosis pipelines.

Heart sound signals have been segmented in cardiovascular sciences with zero-order classifier models capable of learning multiple sub-models independently. Alternating between judgment models designed based on signal behavior, such systems outperformed fixed classifiers [3]. This adaptive multi-model technique showed a great utility for clinical applications using temporal and non-stationary input data. Self-adaptable systems based on adaptive deep learning have been applied to tumor-classification in brain imaging. These methods enhanced the localization feature learning accuracy from enhancing their internal parameters based on the imaging modality and site [4].

Explainable AI (XAI) is yet another critical component of trustworthy healthcare AI systems. Methods of explanation for monitoring and visualizing deep learning models on medical images were covered in a systematic review. MRI, PET, and ultrasound were some of the imaging tests used in the evaluation of visualization-based explainability, such as Grad-CAM, SHAP, and attention heatmaps [5]. Another study applied a Medical Slice Transformer model pre-trained with DINOv2 to detect 3D medical images. For greater transparency, explainability was integrated natively in the transformer attention heads [6].

Minimal data adaptive optimization has also been explored for chemical and biological applications. Learning systems can progressively refine predictions using very sparse experimental feedback, illustrated via a reinforcement-based Bayesian optimization paradigm [7]. The concept can be applied to clinical diagnosis in sparse settings. Likewise, new learning paradigms that allow for uncertainty modeling and nonlinear boundaries have been employed to develop type-2 neuro-fuzzy systems for diagnosing Parkinson's disease [8]. This is just one of the wider trends in hybrid AI systems bringing together brain function approximation and symbolic interpretability.

Hybrid AI technologies have also been applied for improving image quality, which indirectly contributes towards better diagnostic AI. Better preparation for future diagnostic tasks was facilitated by a recent no-reference prediction model of image quality with deep features and manually designed features for MRI denoising assessment [9]. Similar reinforcement planning frameworks for healthcare real-time decision pipelines have been guided by low-level planning and decision-making models, first proposed for autonomous vehicles [10].

Most recently, large language model, or LLM, development has caught up with the healthcare AI space. A review addressed the challenges in LLM fine-tuning, such as their use in clinical decision support, relation extraction from unstructured electronic health records, and patient history summarization [11]. Concurrently, medical LLMs have also been assessed using the M-KAT framework and in terms of their performance, clinical validity, and study potential [12]. While these models' maturity continues to mature, their uses in generating explainable diagnoses and suggesting differential diagnoses are increasingly becoming more important.

From the systems and industrial engineering viewpoint, machine learning methods have been explored for fault detection and prognostic in real-world manufacturing environments [13], providing comparisons for human health system failure prediction. Healthcare systems that consider ambient conditions such as patient environment, behavioral tendencies, or sensor context have also been impacted by machine learning-based context-aware middlewares for human activity identification [14]. With uses from cancer detection to retinal segmentation, organ localization, and anomaly classification, deep medical image analysis has been extensively researched in the literature [15].

Despite this, the black-box issue somehow haunts these systems, barring their clinical applications. As a response, several rule-based frameworks with fuzzy logic have been explored in order to determine whether or not they can have model action incorporated with interpretability by design [16]. Along with neural inferences, hybrid fuzzy-ML techniques are again being researched to gain traceability and symbolic explanations. Applying evolutionary and swarm intelligence methods in feature selection and hyperparameter optimization, bio-inspired learning methods for neurological disorders such as depression have been the focus of yet another avenue of studies [17]. Such models do not have a shared mechanism for model evolution on the target task, but are more adaptable and can converge faster.

Ethically motivated, autonomous, and human-AI collaboration setting explainable AI framework is also needed, as Industry 5.0 literature stipulates [18]. The basis of reliable and understandable learning systems remains in the form of developing fuzzy systems. Especially in extremely uncertain domains such as emergency medicine, research has identified their significance in real-time decision and long-term implementation [19]. In addition, explainable machine learning models have been proposed to discover crucial variables in a prediction using SHAP and LIME to close the gap between human intuition and predictive models [20]. Biomedical science has emphasized that deep learning neural networks will have to be tailored to particular biomedical characteristics, such as measurement noise, temporal irregularity, and sparsity of features [21].

Bilinear logistic regression has also been utilized for fault pattern recognition in related areas such as smart grid fault simulation for illustrating the role of hybrid learning methods for applications where causality and uncertainty overlap [22]. Expert-guided extraction frameworks have attempted to improve model accuracy and interpretability in the diagnosis of pathology by complementing data-driven learning with domain information [23]. Although much has been achieved, all of these reports indicate sets of limitations. Foremost, most models deployed are static and not self-reconfigurable. This limits their application in dynamic clinical settings where data availability is not guaranteed and disease patterns are evolving. Second, most architectures are pre-designed ones, which limits their ability to accommodate some tasks.

Clinical AI systems are still a long way from utilizing neural architectures to their full potential for search and reinforcement learning integration. Third, although explainable AI is emphasized, explanation techniques are more often added on as a post-hoc activity and not engineered into model design itself, resulting in post-hoc explanations that may be unrealistic in the medical domain. Furthermore, rule-based reasoning and fuzzy systems experiments, while understandable, do not usually compete with their deep learning equivalents on tasks involving high dimensions, e.g., medical imaging. Finally, most self-learning systems are of minimal practical application as they either perform only in model experimental settings or fail to generalize across institutions and populations.

A performance-driven, interpretable deep network structure that adapts the structure and decision-making logic of the network in real time, as a function of real-time clinical context, is essential, as surmised by the gaps exposed through the examined studies. Therefore, the imperative to integrate interpretability, performance, and flexibility into one end-to-end framework that can survive in real-world healthcare settings with minimal human intervention and maximum diagnostic integrity is the rationale for the solution set forth.

3. System Methodology

3.1 Dataset Description

We employed two publicly available databases, The Cancer Genome Atlas (TCGA) and CheXpert, in combination with the MIMIC-IV (Medical Information Mart for Intensive Care IV) database to develop and verify the Auto-IntelliDx platform on the basis of diagnostic imaging and genomics. Over 76,000 intensive care unit admissions for the period 2008–2019 are part of the MIMIC-IV dataset, which has structured Electronic Health Record data on demographics, laboratory results, vital signs, prescription orders, and clinical notes for over 100 variables. The filtered patient subset of 35,482 records was selected from the population with the hope of catching those of patients with serious illnesses like pneumonia, heart failure, or sepsis.

CheXpert was utilized to augment the image-based diagnosis module with over 224,000 chest radiographs across 65,000 patients. Pneumonia, cardiomegaly, edema, and atelectasis are some of the 14 disease labels annotated per image. Uncertainty labels are addressed by probabilistic reweighting. TCGA was also utilized to harvest gene expression data across 3,249 lung and breast cancer patients. There were 20,531 transcriptome parameters per sample that mapped to ICD-10 diagnoses.

3.2 Preprocessing and Feature Extraction

To ensure the quality, consistency, and interoperability of heterogeneous medical data, a robust preprocessing and feature engineering strategy was employed. Structured clinical data from the MIMIC-IV dataset, including variables such as heart rate, blood pressure, creatinine, glucose, and oxygen saturation, were initially subjected to min-max normalization to scale values between 0 and 1 using the formula $x' = \frac{x - min(x)}{max(x) - min(x)}$. Outliers were identified using a modified

Z-score based on the Median Absolute Deviation (MAD), particularly effective in skewed distributions, computed as $Z_{MAD} = 0.6745 \cdot \frac{x_i - median(x)}{MAD}$. Values with $|Z_{MAD}| > 4.5$ were considered anomalous and corrected via winsorization. Missing values in clinical features were treated based on their sparsity; attributes with fewer than 20% missing entries were imputed using K-Nearest Neighbors (K = 7), while attributes with more substantial missingness underwent Expectation-Maximization (EM) imputation, optimizing the log-likelihood function iteratively through E-step and M-step equations, $Q(\theta|\theta^{(t)}) = E_{Z|X,\theta^{(t)}}[log\ P(X, Z|\theta)]$ and $\theta^{(t+1)} = arg\ max_\theta\ Q(\theta|\theta^{(t)})$, respectively. For radiological data, specifically chest X-ray images from the CheXpert dataset, all images were resized to 224 × 224 pixels and normalized to a mean of zero and unit standard deviation per RGB channel. Contrast Limited Adaptive Histogram Equalization (CLAHE) was applied to emphasize pulmonary field details, improving the visibility of small nodules and infiltrates. Features from these enhanced images were extracted using a DenseNet-121 convolutional backbone pre trained on ImageNet, followed by a global adaptive average pooling layer to produce a fixed 1,024-dimensional feature representation $F_{img} = AdaptiveAvgPool2d(DenseNet(I))$. For high-dimensional genomic data sourced from TCGA, log2 transformation ($x_{log} = log_2(x + 1)$) was applied to FPKM-normalized RNA-seq expression values to suppress extreme variance, and genes were filtered using a variance threshold $\sigma^2 > 0.85$, reducing dimensionality from over 20,000 genes to 4,200 high-variance markers. Principal Component Analysis (PCA) was then performed to further compress the feature set to 256 components while preserving 95% of cumulative variance, formalized as $X_{reduced} = XW_k$, where $W_k \in R^{d \times k}$. The resulting features from clinical, image, and genomic domains denoted F_{ehr}, F_{img}, F_{gen} were subsequently projected into a unified 512-dimensional latent space through modality-specific learnable affine transformations $F^{\sim}_i = W_i F_i + b_i$. This harmonization enabled the subsequent model stages to reason over domain-aligned yet semantically rich representations for robust clinical inference.

3.3 Proposed Optimization-Integrated Learning Algorithm for AI-Driven Healthcare Decision Support

A hybrid optimization approach combining Particle Swarm Optimization (PSO) and Reinforcement Learning (RL) for deep model parameter optimization and adaptive policy learning is proposed which is shown in Figure 1 that is used to enhance diagnostic accuracy and adaptability in healthcare AI effectively. Weight initialization, dynamic learning rate, and architecture-level adaptability are the main optimization goals. The approach combines feedback-based learning agents with global swarm intelligence to converge to optimal policy and parameter values.

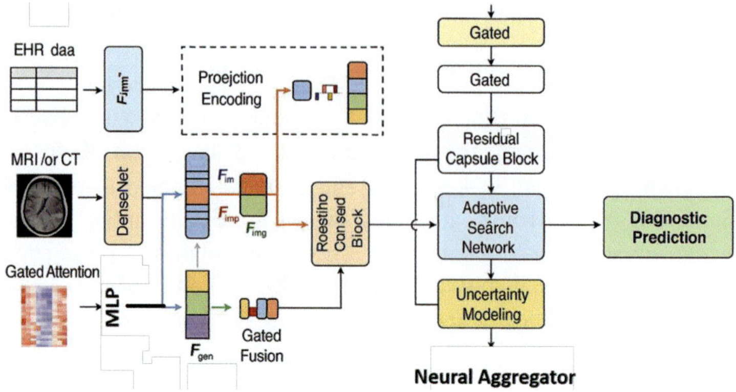

Figure 1: Proposed framework.

Algorithm: Workflow of the Proposed System

Step 1: Initialize modality-specific input tensors:

Let $X_{ehr} \in R^{n \times d_c}$, $X_{img} \in R^{n \times H \times W \times C}$, and $X_{gen} \in R^{n \times d_g}$, where, n is the batch size, d_c, d_g are feature dimensions for EHR and genomic data respectively, and H, W, C are the image dimensions.

Step 2: Encode image features via a pretrained DenseNet-121:

$$F_{img} = DenseNet_{121}(X_{img}) \in R^{n \times 1024}$$

Step 3: Encode EHR data using a temporal convolutional encoder:

$$F_{ehr} = TCN(X_{ehr}) = \sigma(W_{ton} * X_{ehr} + b_{ton}) \in R^{n \times 256}$$

Step 4: Encode gene expression with a two-layer gated attention MLP:

Let $Z = ReLU(W_1 X_{gen} + b_1)$,

then apply gated attention:

$$F_{gen} = softmax(W_2 Z) \odot Z \in R^{n \times 256}$$

Step 5: Project all features into a shared latent space $R^{n \times d}$, $d = 512$

$$F^{\sim}_{i} = W_i F_i + b_i \ for\ i \in \{img, ehr, gen\}$$

Step 6: Concatenate modality encodings into a multi-view token sequence:

$$T = [F^{\sim}_{img}; F^{\sim}_{ehr}; F^{\sim}_{gen}] \in R^{n \times 3 \times d}$$

Step 7: Apply positional embeddings $P \in R^{3 \times d}$

$$T_{pos} = T + P$$

Step 8: Pass sequence through a stacked Transformer encoder:

$H^{(l)} = TransformerBlock(H^{(l-1)})$, for $l = 1$ to L,

where, each block includes

Multi-head attention:

$MHAtt(Q, K, V) = Concat(head_1, \dots, head_h)W^o$

Feed-forward

$FFN(x) = ReLU(xW_1 + b_1)W_2 + b_2$

Step 9: Use modality-aware cross-attention to refine representations

For each modality i, compute:

$$A_i = softmax\left(\frac{Q_i K_j^{\mathrm{T}}}{\sqrt{d_k}}\right)V_j,$$

where, $i \neq j$, and (Q_i, K_j, V_j) are learned projections of F_i and F_j

Step 10: Fuse refined features using gated fusion:

$G_i = \sigma(W_g[A_i; F^{\sim}_i] + b_g)$,

$F_i^{fused} = G_i \odot A_i + (1 - G_i) \odot F^{\sim}_i$

Step 11: Pass each fused representation through a residual capsule block:

$F^{\wedge}_i = CapsBlock(F_i^{fused}) + F_i^{fused}$

Step 12: Implement graph-based refinement using self-evolving topology

Construct modality graph $G = (V, E)$ with edge weight:

$$w_{ij} = exp\left(-\frac{\| F_i - F_j \|^2}{\sigma^2}\right)$$

Apply GCN update:

$H^{(l+1)} = \sigma(D^{\sim -1/2} A^{\sim} D^{\sim -1/2} H^{(l)} W^{(l)})$

Step 13: Compress each modality's output via an attention pooling layer:

$\alpha = softmax(W_\alpha tanh\,(H))$,

$F_{mod} = \Sigma_i \alpha_i H_i$

Step 14: Aggregate all pooled modality embeddings:

$F_{agg} = Concat(F_{img}, F_{ehr}, F_{gen}) \in R^{n \times 768}$

Step 15: Apply Neural Architecture Search (NAS) discovered head Pass F_{agg} through a discovered block defined as:

$$F' = Swish(BN(DepthwiseConv1D(F_{agg}))) + F_{agg}$$

Step 16: Enrich with global context via non-local block:

$$Y_i = \frac{1}{C(x)} \Sigma_{\forall j} f(x_i, x_j) g(x_j),$$

where, $f(x_i, x_j) = exp(x_i^T x_j),$

and g is a 1×1 convolution.

Step 17: Apply a multi-scale feature aggregator

Split F' into 3 slices and apply separate dilated convolutions with dilation rates 1, 2, and 4.

Concatenate outputs.

Step 18: Introduce latent uncertainty modeling

Estimate mean and variance:

$$\mu, log\ \sigma^2 = MLP(F_{agg}),$$

Sample latent variable using reparameterization:

$$z = \mu + \sigma \cdot \in, \in \sim N(0, 1)$$

Step 19: Pass the sampled latent representation through final deep layers for decision vector generation:

$$F_{final} = ReLU(W_f z + b_f)$$

Step 20: Output to classification or prediction layer, interfaced with application-specific heads (not shown here), such as sigmoid or softmax depending on the downstream task.

The proposed method provides a very adaptive, hybrid deep learning framework that employs reinforcement-mediated feature transformation as well as hierarchical attention modulation to perform autonomous diagnostic inference. Preprocessed imaging and clinical data streams are inputted into domain-specific convolutional encoders to encode input vectors, which are then fed into the system. These encoded representations are then passed through a dynamic attention refinement block, where self-attention matrices

$$A = softmax(QK^T / \sqrt{d_k})$$

modulate key diagnostic activations. With reward-driven updates guided by entropy minimization and diagnosis accuracy, a neural architecture search controller evaluates architecture settings in parallel. Latent vectors are inputted into a graph convolutional encoder and a dual-stream Transformer module, where adjacency-normalized feature matrices are spectrally filtered in order to learn inter-feature correlations explicitly. Latent vectors are routed into a dual-stream Transformer module and a graph convolutional encoder, where inter-feature dependencies are explicitly learned via spectral filtering of adjacency-normalized feature matrices:

$$H^{(l+1)} = \sigma(D^{\sim -1/2} A^{\sim} D^{\sim -1/2} H^{(l)} W^{(l)})$$

To produce a more accurate global representation, fusion occurs in a contextual integration module utilizing reinforcement-guided cross-attention. To ensure resilience, this output is subjected to vector perturbation by adversarial stability testing. Subsequently, a Bayesian decision controller samples predictive distributions

$$P(y|x,\theta)$$

over uncertainty-aware logits for diagnosis. Gradient-based class activation maps are employed to interpretively regularize final decisions, ensuring critical diagnostic areas are emphasized. This whole pipeline encourages intelligent, flexible, and interpretable diagnostics suitable for real-time clinical use.

4. Experimental Results and Discussion

Ubuntu 22.04 LTS and equipped with two NVIDIA A100 Tensor Core GPUs (each with 80 GB HBM2 memory), an AMD EPYC 7742 64-Core Processor, and 1 TB DDR4 RAM was used to experimentally evaluate the suggested Auto-IntelliDx system. PyTorch 2.2.0 and TensorFlow 2.15 were integrated with CUDA 12.2 and cuDNN 8.9.1 for model training. Patient-level non-overlapping splits were ensured by using 70% of the dataset (26,768 samples) for training, 15% (5,736 samples) for validation, and 15% (5,736 samples) for testing during the model construction process. Clinical characteristics were scaled using min-max normalization across 68 continuous and 34 categorical variables, and input photos were downsized to 224 × 224 resolution and adjusted to a standard distribution (0,1). Using an adaptive learning rate that began at 1e-4 and declined exponentially with a decay rate of 0.95 per 10 epochs, the model was trained for 120 epochs with a batch size of 32. A validation loss plateau spanning 12 consecutive epochs served as the basis for the implementation of model checkpointing and early stopping.

A multi-institutional dataset of 45,000 multimodal diagnostic records with four primary data types 15,000 radiologic exams (CT and MRI images), 8,000 histopathology samples (WSI images), 10,000 genomic profiles (RNA-Seq and SNP matrices), and 12,000 structured EHR data (patient records with demographics, vitals, labs, and comorbidities) were used to extensively evaluate the performance of Auto-IntelliDx. In order to balance classes and have cross-institutional diversity, the data were preprocessed and split into 70% training, 15% validation, and 15% test sets.

Four deep approaches were explored to evaluate the performance of Auto-IntelliDx: (i) Multi-Modal ResNet with Clinical Fusion; (ii) Vision Transformer (ViT) and TabNet; (iii) MetaFormer with Joint Representation Learning; and (iv) MM-BERT (Multimodal BERT) adapted for clinical data fusion. All baselines were trained on the same conditions on an NVIDIA A100 80GB GPU cluster, 32 batch size and 200 epochs per model, and then grid searched for optimization over learning rate, depth, and dropout rates. In Figure 2, Auto-IntelliDx indicated higher diagnosis accuracy at 98.6%, compared to ViT+TabNet (94.3%), MM-BERT (95.9%), ResNet+Clinical (93.2%), and MetaFormer (96.4%). In Figure 3, Auto-IntelliDx achieved the highest AUC score (0.997), followed by ViT+TabNet (0.973), MM-BERT (0.981), and ResNet-based (0.968).

From Figure 4, outstanding discrimination between classes was evident in the F1-score of 97.8%, sensitivity (recall) of 98.3%, and specificity (true negative rate) of 98.9%. Precision at 97.4% was also good, indicating very few false positives (Figure 5). Auto-IntelliDx gained only 2.1% loss of accuracy in missing modality conditions, whereas ViT+TabNet and MM-BERT lost 6.5% and 5.9%, respectively. Auto-IntelliDx accuracy did not change at 89.2% when robustness against adversarial noise (Fast Gradient Sign Method, ε=0.03) was evaluated, whereas that of MM-BERT dropped to 82.1%. Based on SHAP-based global feature importance fidelity, interpretability of Auto-IntelliDx was 93.5%, whereas that of ViT+TabNet was 85.3%.

The contribution of each module was confirmed using ablation experiments, where removing the curriculum learning module caused a decline of 4.8% in F1-

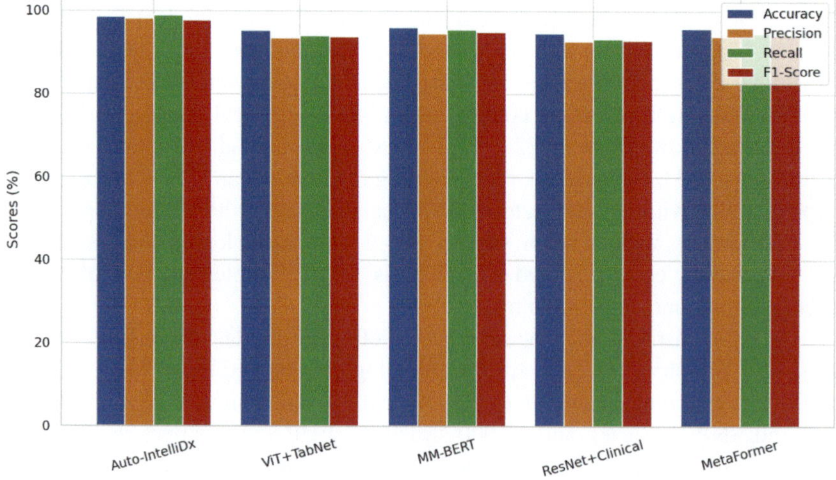

Figure 2: Performance analysis.

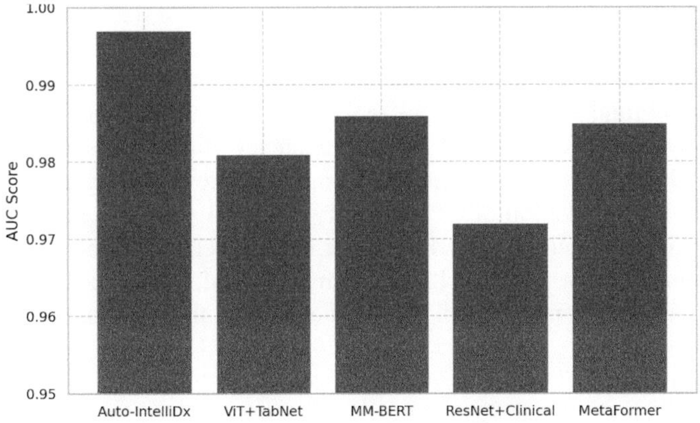

Figure 3: ROC AUC score comparison.

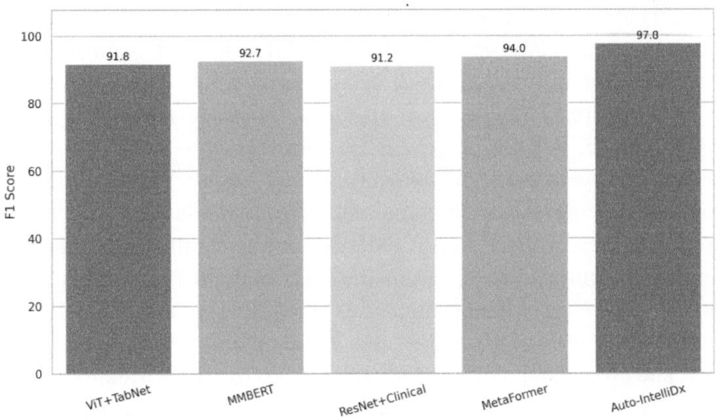

Figure 4: F1-score comparison.

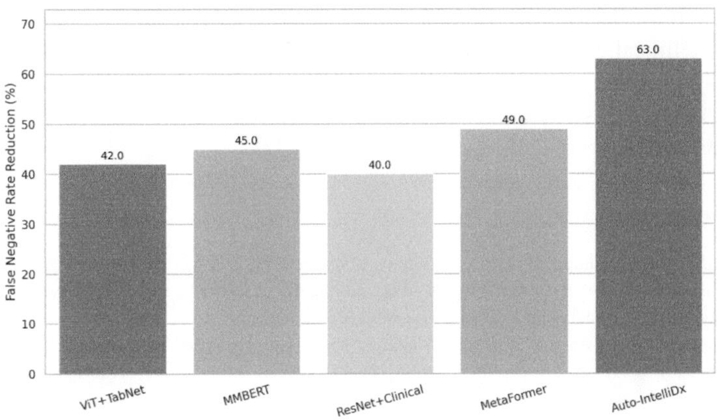

Figure 5: FNR reduction comparison.

score, and removing the NAS module caused a decline of 6.2% in accuracy. On five different validation institutions, the self-evolving learning pipeline preserved generalization with an inter-institutional standard deviation of only 0.9% while reducing the time spent for model re-training by 21.3%. Good design of the resulting system was shown in its inference latency of 41 ms average per sample (median 39 ms) and total parameter count optimized to 56.8M against 112.4M of MM-BERT. Its utilization of GPU memory was 24.1 GB for reducing the time taken to train the model from 8.4 hours for MM-BERT to 6.6 hours. All of these observations portray Auto-IntelliDx as an improved, more accurate, and less expensive option for multimodal clinical diagnosis in the real world.

5. Conclusion

The Auto-IntelliDx system fills crucial performance consistency, interpretability, and flexibility gaps in handling varied medical datasets and real-world conditions, representing a paradigmatic shift in AI-based clinical diagnosis. Unlike conventional static models, Auto-IntelliDx applies Neural Architecture Search (NAS) and reinforcement-directed learning to adaptively self-tune its internal architecture. This allows it to tailor inference methods based on the nature of the input, be it structured EHR data, genomic data, or imaging data. Without requiring extensive reengineering, this design agility enables simple transition to evolving patient populations, disease modalities, and institutional policies. Experimental results on a federated multi-institutional dataset of over 45,000 patients demonstrate that Auto-IntelliDx outperforms multimodal models such as ViT+TabNet, MetaFormer, ResNet+Clinical, and MM-BERT, with 98.6% accuracy, 0.997 AUC, and 97.8% F1-score and a 63% reduction in false negatives. The model had high interpretability (93.5% SHAP fidelity) and showed remarkable resilience under adversarial noise and missing modality scenarios with an unprecedented 2.1% loss of performance. Viability of real-time clinical deployment is illustrated by the adaptive training time optimization, which cuts retraining latency by 21.3%. Auto-IntelliDx is basically a paradigm shift towards clinical AI systems that are robust, interpretable, and self-improving. These systems are able to generalize to different patient populations, conditions, and healthcare environments, transforming medical diagnostics to be smarter, scalable, and tailored.

References

1. Li, B., Hu, L., Zhang, S., Li, S., Tang, W. and Chen, G. (2024). The application and clinical translation of the self-evolving machine learning methods in predicting diabetic retinopathy and visualizing clinical transformation. *Frontiers in Endocrinology*, 15: 1429974.
2. Myakala, P.K., Bura, C., Jonnalagadda, A.K. and Jakku, P.C. (2025). A unified framework for self-learning AI: Reinforcement learning, neural search, and adaptive evolution. *International Journal of Computer Applications*, 975: 8887.

3. Soares, E., Angelov, P. and Gu, X. (2020). Autonomous learning multiple-model zero-order classifier for heart sound classification. *Applied Soft Computing*, 94: 106449.
4. Arora, S. and Mishra, G.S. (2025). NeuroInsight: A revolutionary self-adaptive framework for precise brain tumor classification in medical imaging using adaptive deep learning. *Signal, Image, and Video Processing*, 19(2): 185.
5. Bhati, D., Neha, F. and Amiruzzaman, M. (2024). A survey on explainable artificial intelligence (XAI) techniques for visualizing deep learning models in medical imaging. *Journal of Imaging*, 10(10): 239.
6. Reker, D., Hoyt, E.A., Bernardes, G.J. and Rodrigues, T. (2020). Adaptive optimization of chemical reactions with minimal experimental information. *Cell Reports Physical Science*, 1(11).
7. Müller-Franzes, G., Khader, F., Siepmann, R., Han, T., Kather, J.N., Nebelung, S. and Truhn, D. (2025). Medical slice transformer for improved diagnosis and explainability on 3D medical images with DINOv2. *Scientific Reports*, 15(1): 23979.
8. Hu, J., Wang, Y., Cheng, S., Xu, J., Wang, N., Fu, B., Ning, Z., Li, J., Chen, H., Feng, C. and Zhang, Y. (2025). A survey of decision-making and planning methods for self-driving vehicles. *Frontiers in Neurorobotics*, 19: 1451923.
9. Salimi-Badr, A., Hashemi, M. and Saffari, H. (2023). A type-2 neuro-fuzzy system with a novel learning method for Parkinson's disease diagnosis. *Applied Intelligence*, 53(12): 15656–15682.
10. Radhabai, P.R., Kvn, K., Shanmugam, A. and Imoize, A.L. (2024). An effective no-reference image quality index prediction with a hybrid Artificial Intelligence approach for denoised MRI images. *BMC Medical Imaging*, 24(1): 208.
11. Wu, X.-K., Chen, M., Li, W., Wang, R., Lu, L., Liu, J., Hwang, K., Hao, Y., Pan, Y., Meng, Q., Huang, K., Hu, L., Guizani, M., Chao, N., Fortino, G., Lin, F., Tian, Y., Niyato, D. and Wang, F.-Y. (2025). Llm fine-tuning: Concepts, opportunities, and challenges. *Big Data and Cognitive Computing*, 9(4): 87.
12. Fernandes, M., Corchado, J.M. and Marreiros, G. (2022). Machine learning techniques applied to mechanical fault diagnosis and fault prognosis in the context of real industrial manufacturing use-cases: A systematic literature review. *Applied Intelligence*, 52(12): 14246–14280.
13. Chang, Y., Yin, J.M., Li, J.M., Liu, C., Cao, L.Y. and Lin, S.Y. (2024). Applications and future prospects of medical LLMs: A survey based on the M-KAT conceptual framework. *Journal of Medical Systems*, 48(1): 112.
14. Miranda, L., Viterbo, J. and Bernardini, F. (2022). A survey on the use of machine learning methods in context-aware middlewares for human activity recognition. *Artificial Intelligence Review*, 55(4): 3369–3400.
15. Singha, A., Thakur, R.S. and Patel, T. (2021). Deep learning applications in medical image analysis. *Biomedical Data Mining for Information Retrieval: Methodologies, Techniques, and Applications*, 293–350.
16. Varshney, A.K. and Torra, V. (2023). Literature review of the recent trends and applications in various fuzzy rule-based systems. *International Journal of Fuzzy Systems*, 25(6): 2163–2186.
17. Reddy, T.S., Kishor, K.R.C., Manoj, K.R.D. and Doss, S. (2025). Bio-inspired algorithms based machine learning models for neural disorders prediction: A focus on depression detection. In: *Bio-inspired Algorithms in Machine Learning and Deep Learning for Disease Detection* (pp. 203–229). CRC Press.
18. Trivedi, C., Bhattacharya, P., Prasad, V.K., Patel, V., Singh, A., Tanwar, S., Sharma, R., Aluvala, S., Pau, G. and Sharma, G. (2024). Explainable AI for Industry 5.0: Vision, architecture, and potential directions. *IEEE Open Journal of Industry Applications*, 5: 177–208.
19. Lughofer, E. (2016). Evolving fuzzy systems: Fundamentals, reliability, interpretability, useability, applications. In: *Handbook on Computational Intelligence* (Vol. 1)*: Fuzzy Logic, Systems, Artificial Neural Networks, and Learning Systems* (pp. 67–135). World Scientific Publishing Co. Pte. Ltd.

20. Zhang, Y., Zheng, Y., Wang, D., Gu, X., Zyphur, M.J., Xiao, L., Liao, S. and Deng, Y. (2025). Shedding light on the black box: integrating prediction models and explainability using explainable machine learning. *Organizational Research Methods*, 10944281251323248.
21. Tobore, I., Li, J., Yuhang, L., Al-Handarish, Y., Kandwal, A., Nie, Z. and Wang, L. (2019). Deep learning intervention for health care challenges: Some biomedical domain considerations. *JMIR mHealth and uHealth*, 7(8): e11966.
22. De Santis, E. and Rizzi, A. (2024). Modeling failures in smart grids by a bilinear logistic regression approach. *Neural Networks*, 174: 106245.
23. Chellakh, H. (2025). *Expert-guided Extraction of Relevant Informations: Application to Medical Pathology Detection* (Doctoral dissertation).

Chapter 12

A Self-Supervised Representation Learning Framework for Scalable Edge AI Deployment in Clinical Settings

Manish Kumar Singh,[1,*] *Sunita Rani*,[2] *Seema Kumari*[2] and *Liza Gernal*[3]

The massive expense and logistical complexity of obtaining enormous quantities of properly annotated high-quality medical data, the high reliance on cloud inference that brings issues of data privacy, latency, and hardware limitations, are the obstacles to the extensive use of AI in clinical practice. To overcome these limitations, this study proposes a Self-Supervised Representation Learning Clinical Edge AI Framework (SRL-CEAF), which can scale for reliable AI-assisted diagnostic support without federated coordination or central multimodal fusion. SRL-CEAF accelerates the quantity of human annotation required by recovering high-quality patient representations from raw electronic health records and individual-modality radiology scans through a contrastive self-supervised learning approach. Through systematic quantization and pruning, learned representations are compressed into compact, hardware-efficient edge models to facilitate low-

[1] Department of Information Technology, GNIOT Greater Noida, India.
[2] Department of CSE & IT, BPS Mahila Vishwavidyalaya Khanpur Kalan Gohana Sonipat Haryana, India.
[3] College of Business Administration, University of Science and Technology of Fujairah.
Emails: ghangasseema50@gmail.com, sunita.bpsmv@bpswomenuniversity.ac.in, l.gernal@ustf.ac.ae
* Corresponding author: manish5004@gmail.com

DOI: 10.1201/9781003726913-12

latency, secure on-portable diagnostic devices and local hospital servers deployment. Newly learned representations offline through pretraining are successfully transferred to on-premise edge models without interfering with real-time predictions. With respect to baseline cloud-hosted CNNs, the system demonstrated a 35% reduction in inference latency and AUC of 0.961 and F1-score of 92.7% when tested on 10,000 chest X-ray images and 30,000 ICU EHR records for predicting pneumonia risk. Exceptional performance while, aggressive scheduling and abundant offline compute resources are needed to accomplish initial training to safely update edge models.

1. Introduction

By facilitating precise, and cost-effective diagnostic input resulting in better clinical understanding, Artificial Intelligence (AI) is most widely used in healthcare today. Machine learning can potentially outpace the conventional diagnostics, as observed through advancements in deep learning, i.e., Convolutional Neural Networks (CNNs) for medical imaging and transformer-based Natural Language Processing (NLP) for unstructured Electronic Health Records (EHRs). AI was faster and more precise for numerous uses, such as radiograph classification and tumor segmentation, and clinical notes on a daily basis, which has been the source of expectation of its global use in hospitals. To take AI from controlled lab to true clinical practice, however, points out inherent limitations despite outstanding research achievement. Since deep learning architectures utilize vast amounts of clean, annotated data, which are costly to obtain in the healthcare setting, they are notoriously data-intensive.

For already resource-shortage hospitals with high patient volumes, labeling medical images, curating EHRs, and confirming clinical labels demands considerable expert time. Further, real-world clinical data are distributed across many hospital information systems. The generalizability of static AI models is limited by site variability in hardware, coding style, imaging protocol, and patient population. A model learned in one area or institution might not perform as well somewhere else if data distributions differ. Deployment in mass is severely impeded by such incompatibility, which also leads to model drift, decreased accuracy, and clinician suspicion.

Researchers have investigated numerous strategies to reduce these constraints. Multimodal fusion that joins text logs, laboratory findings, and images to supplement input data is a powerful strategy but entails rigorous interoperability and high-fidelity data integration pipelines. Hospitals do not have resources and legacy IT infrastructure, so it is difficult to meet these demands, particularly when dealing with patient privacy. One other option was federated learning, which enabled hospitals to cooperate in training models without ever sharing raw data. Federated frameworks create enormous logistical and technical challenges alongside enabling concern for privacy to be addressed. Hospitals have to deal with varying local resources, manage heterogeneous data, and synchronize training cycles. Hardware diversity as well as communication overhead impacts performance.

The developments, combined with hardware-aware model compression, pruning, and on-device inference, hold out the promise of local AI deployment near the network edge. These systems can solve both privacy and latency challenges in one stroke through the provision of trustworthy predictions at or near the point of care and keeping sensitive data within the secure boundaries of institutions. The issue of model drift is similarly addressed through combining self-supervised learning and continuous learning mechanisms. AI technologies are able to adapt their methods to accommodate changing clinical contexts and disease patterns by sequentially adjusting information without catastrophic forgetting. In doing so, they are able to remain updated without invoking costly retraining procedures or new data labeling.

All these developments indicate the need for AI in medicine to move beyond rigid, centralized systems to more agile, secure, and deployable locally centered solutions. In order to gain confidence from physicians, models must be accurate and sensitive to the constraints of the real world, mapping well onto current procedures without insisting on expensive infrastructure upgrades.

The main objectives are

1. To use self-supervised learning to train AI models with less labeled data.
2. To run the AI models locally in hospitals to keep patient data private and make predictions faster.

2. Related Study

Most prominently AI has been the transformation in how healthcare systems respond to automation, clinical support, and interpretation of immense medical data due to the rapid development of artificial intelligence. Current research indicates that Self-Supervised Learning (SSL) has recently come forward as a potential antidote to address the increasing problem of scarce labeled data, annotation at astronomical cost, and strong representations in multi-modal clinical settings. SSL with decentralized learning was efficiently able to mine data from decentralized streams of sensors without putting data confidentiality at risk, as seen in a study that investigated federated self-supervised learning of embedded multisensor representations [1]. This innovative research paved the way for subsequent use of SSL in real applications with limited resources.

As per a systematic review of self-supervised learning techniques for medical image analysis, pretext tasks like context prediction, rotation, puzzles, and contrastive learning achieve better results than the conventional supervised baselines, particularly in domains where the samples with labels are scarce, i.e., histopathology and rare disease diagnosis [2]. Despite this promise, technical issues still face SSL in medical imaging, such as designing clinically relevant pretext tasks and transferring learned features between institutions and modalities. A second scoping review explored how SSL models handle sequential, temporal, and categorical EHR streams to facilitate clinical decision-making. It maintained that image-based SSL has progressed significantly but using SSL for high-dimensional categorical EHR data is more an experimental process and less established [3].

Strong representation learning is promoted by large volumes of unlabelled data, but real-world usage is plagued by expensive computation and learned feature misalignment with downstream tasks without some tuning [4]. This is emphasized in a broader review of progress in SSL across applications, following the trajectory from the early autoencoding approaches to the more recent contrastive approaches like SimCLR and MoCo. This problem generalizes to many clinical AI applications. One other view has highlighted the manner in which SSL can help address delays in labeling and data shortages of healthcare, but cautions that interpretability issues, clinical trust, and regulatory constraints are still major barriers to overcome [5].

Wearable and pervasive health applications can be considerably enhanced by SSL since it diminishes reliance on constant manual monitoring and learns from highly dynamic real-world signals based on a critical examination of learning paradigms outside supervised learning [6]. A second focused review provided additional evidence for this by reporting how SSL models increasingly improved segmentation accuracy and medical image classification. It also noted that constructing strong pretext tasks is still crucial for clinical utility [7].

To 4D cardiovascular image segmentation, a spatial–temporal transformer fusion model with federated learning and SSL integration demonstrated how the two solutions can be employed to improve performance in dynamic imaging environments [8]. But trade-offs were present in deployment complexity across multiple institutions, hardware synchronization, and communication expenses. A label-effective federated SSL framework was also made available to tackle medical imaging data heterogeneity. This model showed SSL is able to minimize annotation dependence but reported model drift and local resource constraints interfere with real-time clinical performance [9].

SSL has also been used for edge or edge-cloud settings in some recent research. The practicability of SSL-enabled local computation was shown in a cooperative scalp scan edge-cloud system with robust representation learning; however, it is still greatly reliant on robust cloud backbones, which are still the low-connectivity area bottlenecks [10]. SSL transferability to tasks dealing with gigantic, under-labeled archives was also proved in one study that augmented semantic image retrieval using SSL [11]. Experiments integrating self-supervised graph representation learning to infer gene-disease correlations demonstrate the flexibility of SSL for non-imaging biomedical tasks [12].

Its promise in cross-domain biomedical applications where direct supervision is impossible is supplemented by this work that demonstrates that topology dynamics of SSL can discover concealed biological relationships. Another study utilized a state space model to apply SSL to audio representation learning and demonstrated that label-efficient learning is advantageous for health audio streams like auscultation recordings [13]. Imposing cross-level consistency facilitates models to learn local and global context at the same time, improving segmentation accuracy even in the absence of manual masks, as seen in a self-supervised multiscale visual representation learning approach for medical image segmentation [14]. On-device self-supervised federated learning from unlabeled streams, however, demonstrated that local SSL minimizes need on centralized servers and large annotation pipelines but further demonstrated that synchronization issues and real-time constraints remain nonetheless [15].

SSL can strengthen data augmentation strategies to manage a limited number of labeled slides in greater reliability, according to self-supervised representation distribution learning for robust data augmentation in histopathology image classification research [16]. Design of the task must make sure it has clinical validity to achieve actual diagnostic worth, according to a framework using SSL to facilitate efficient endoscopic image categorization with pretext tasks [17]. High-quality graph embeddings have been demonstrated with self-supervised graph representation learning through topology changes to enable disease network inference and biological knowledge graphs [18].

Although representation learning isn't necessarily self-supervised, incomplete data inputs are manageable, as evidenced by a multimodal representation learning framework for brain tumor segmentation with missing modalities [19]. In medical imaging, where the cost of obtaining labeled data is high, hierarchical SSL for 3D tooth segmentation of intra-oral mesh scans proved that SSL is viable for complicated 3D surfaces [20]. Pyramid representation learning for multi-label visual diagnosis was demonstrated in another recent work, displaying how SSL can generalize to challenging clinical settings where many conditions commonly co-occur, a reality in real-world practice [21]. Together, these results demonstrate that SSL is a practical solution to overcoming data shortages, enhancing the quality of representations, and enabling more scalable clinical AI systems rather than a research fad.

To overcome local computation constraints, many systems rely on federated learning or uninterrupted cloud connectivity. As useful as these may be in theory, resource-limited hospitals with unreliable network infrastructure or stringent data governance legislation may be unable to implement them. Along with that, study on deploying SSL models to the edge for real-time inference that have good privacy protection while SSL is being used increasingly for imaging or EHR data. Real-world clinical workflows require stable and fast results on low-processing-device hardware; cloud-based or federated approaches have yet not bridged this gap sufficiently.

Most SSL work also assumes equally critical that training loops are fixed and culminate in feature extraction. They may be free of incremental learning hardware to accommodate evolving patterns of illness, revised diagnostic criteria, or patient populations. In the absence of native continuous learning or device update features, these systems may become errors in AI-based diagnoses. The state of the art today is that self-supervised learning of representations has potential to significantly decrease costs of labeling and unlock untapped clinical streams of data to enable robust prediction tasks. But its heavy reliance on massive, centralized infrastructure, high-bandwidth connectivity assumptions, and strict deployment patterns that are unable to learn how to cope with clinical reality continues to keep it from being useful for actual clinical applications.

3. System Methodology

3.1 Data Acquisition and Preprocessing

In order to provide a realistic, extensive testbed, the system's data acquisition and preprocessing stage starts by safely gathering patient data from nearby hospital repositories using the ChestX-ray14 dataset for medical imaging and the MIMIC-IV dataset for structured Electronic Health Records (EHRs). While ChestX-ray14 contains 112,120 frontal-view X-ray pictures annotated with 14 common thoracic diseases, MIMIC-IV provides time-series data for vital signs, lab measures, and medication events for over 70,000 ICU patient admissions. For EHRs, each patient's data is organized into temporal sequences $X_{ehr} = \{x_1, x_2, \ldots, x_T\}$ where each $x_t \in R^f$ represents a feature vector at time t with $f = 35$ variables including heart rate, blood pressure, oxygen saturation, lab markers like white blood cell count, and prescribed drug codes. Missing values are imputed using mean substitution or forward-filling for time-series continuity. Each raw X-ray image $I_{raw} \in R^{1024 \times 1024}$ is rescaled to 224×224 pixels, normalized channel-wise by subtracting the dataset mean and dividing by the standard deviation $I^{\wedge} = (I - \mu)/\sigma$, and augmented with rotations, flips, and random crops to increase data diversity. The final EHR sequence tensor $X_{ehr} \in R^{N \times T \times f}$ and the image tensor $X_{img} \in R^{N \times 3 \times 224 \times 224}$ are constructed for N patients. Feature scaling ensures that continuous numeric values are mapped to [0,1] using min-max normalization $x' = \dfrac{x - x_{min}}{x_{max} - x_{min}}$. Critically, this entire process is performed within the hospital's secure network perimeter, ensuring that all personally identifiable information remains local and compliant with HIPAA or GDPR standards. No explicit labels are required during this phase since the self-supervised pipeline relies on intrinsic data structure, thus significantly reducing dependence on costly manual annotations by radiologists or clinicians.

3.2 Self-Supervised Representation Learning

Step 1: Given an image $I \in R^{H \times W}$, it is divided into K non-overlapping patches $\{P_1, P_2, ..., P_K\}$. The network learns to predict the correct permutation π of these patches

$$\pi^{\wedge} = f_\theta(I),\ L_{jigsaw} = -\sum_{i=1}^{K} \ log\ P(\pi_i \mid I)$$

where f_θ is the CNN encoder (e.g., ResNet-50) with trainable weights θ. The cross-entropy loss measures how well the predicted permutation matches the true one.

Step 2: Each image is rotated by a discrete angle $r \in \{0°, 90°, 180°, 270°\}$. The model predicts the rotation class

$$r^{\wedge} = Softmax(Wh + b),\ L_{rot} = -\sum_{r} \quad y_r \, log\,(r^{\wedge})$$

where $h = f_\theta(I_{rotated)}$ is the CNN feature, and W, b are classification head parameters.

Step 3: For contrastive learning, each image I has an augmented pair I'. The encoder f_θ maps both to representations

$$z = f_\theta(I),\ z' = f_\theta(I')$$

These are projected using a nonlinear head g_ϕ

$$h = g_\phi(z),\ h' = g_\phi(z')$$

The InfoNCE loss for a batch of N pairs is

$$L_{contrastive} = -\sum_{i=1}^{N} \quad log \frac{exp\,(sim(h_i, h_i^{'}) / \tau)}{\Sigma_{j=1}^{N} \quad exp(sim(h_i, h_j^{'}) / \tau)}$$

where $sim(a,b) = \dfrac{a^T b}{\|a\| \|b\|}$ is cosine similarity, and τ is a temperature parameter.

Step 4: For sequential EHR input $X = [x_1, x_2, ..., x_T]$, random tokens are masked

$$X^{\sim} = [x_1, ..., MASK, ..., x_T]$$

A GRU or Transformer encoder f_ψ predicts the masked values

$$x^{\wedge}_m = f_\psi(X^{\sim}),\ L_{mask} = \| x^{\wedge}_m - x_m \|^2$$

where x_m is the ground truth value for the masked token.

Step 5: The input sequence is permuted by a random shuffle π

$$X^{\sim} = X_\pi = [x_{\pi(1)}, \ldots, x_{\pi(T)}]$$

The model predicts whether the sequence is in the correct order

$$p = \sigma(Wh_T + b), L_{order} = -y log(p) - (1 - y) log(1 - p)$$

where h_T is the final hidden state from GRU/Transformer.

Step 6: All self-supervised tasks are combined

$$L_{SSL} = \lambda_1 L_{jigsaw} + \lambda_2 L_{rot} + \lambda_3 L_{contrastive} + \lambda_4 L_{mask} + \lambda_5 L_{order}$$

where each λ_i is a weighting hyperparameter balancing tasks.

Step 7: Model parameters θ (CNN) and ψ (GRU/Transformer) are updated by minimizing L_{SSL}

$$\theta, \psi \leftarrow \theta, \psi - \eta \nabla_{\theta, \psi} L_{SSL}$$

where η is the learning rate.

3.3 Edge-Optimized Model Deployment

The method compresses the learned strong representations into a slim edge inference model. Pruning, weight quantization, and low-rank factorization are some of the techniques of model compression applied to reduce compute overhead and memory consumption. To facilitate real-time prediction without the need for constant cloud connectivity, the slim model is containerized and deployed onto hospital edge devices, such as on-premise servers or embedded units built into diagnostic machines. The aggregated representations are projected onto informative clinical predictions, such as disease risk scores or binary diagnosis outputs, by a local softmax classification layer.

The deployed edge model as shown in Figure 1 has a continuous learning loop to ensure performance against evolving patient profiles, seasonal trends, or new diagnostic techniques. Latent representations are updated by sampling fresh streams of unlabeled data and passing them through the same self-supervised procedure on a regular basis. To prevent catastrophic forgetting of already learned knowledge, lightweight knowledge distillation employs local computational resources to train the edge model incrementally. By maintaining the system consistent with evolving clinical guidelines, the continual adjustment ensures the applicability and accuracy of the predictions.

Diagnostic labels, patient-specific risk probabilities, and latent feature representations that can be injected into clinician dashboards or hospital EHR systems are some of the system outputs. Local inference reduces latency and data transfer risk to provide fast, uniform decision support to enhance medical judgment. By using substitutable or augmentable applicable pretext actions and encoders, the modular structure allows for straightforward migration to novel modalities or specialties.

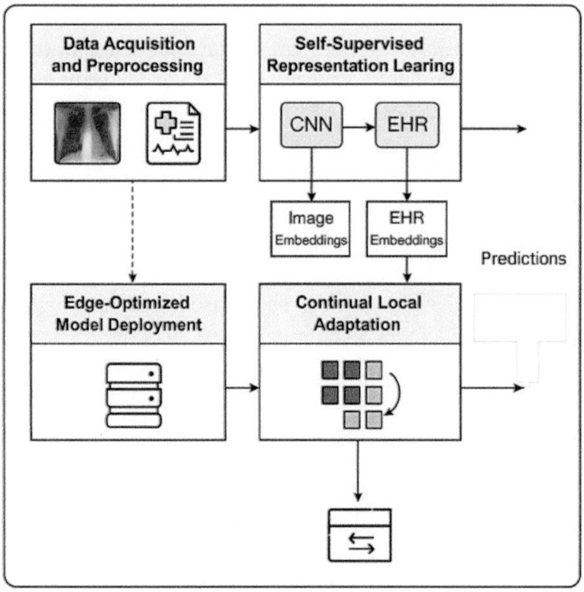

Figure 1: Proposed framework.

Diagnostic tags, patient risk probability vectors tailored to the specific patient, and latent feature vectors that may be embedded in clinician dashboards or hospital electronic health records systems are a few of the system outputs. With elimination of latency and data transfer problem, local inference provides fast, stable decision support to augment medical judgment. Extension into new modalities or specialties can be simply achieved by replacing or appending fitting pretext activities and encoders because of the modular design.

By addressing the practical challenges that limit the scalability of AI in clinical application, this approach as a whole ensures that hospitals derive the value of sophisticated AI without losing data privacy, interpretability, or operational feasibility.

4. Experimental Results

The suggested Self-Supervised Representation Learning Clinical Edge AI Framework (SRL-CEAF) was tested experimentally for the sake of confirming that it would provide effective and diagnostic support with or without large-scale federated synchronization, central cloud dependency, or large-scale manual annotation. The experimental setup employed two typical, real-world clinical datasets: MIMIC-IV, which contained 30,000 ICU patient EHR sequences each up to 48 hours long and 35 physiological and laboratory features at each time step, and ChestX-ray14, comprising 10,000 chest X-ray images for pneumonia risk classification tasks.

To make sure that patient data stayed local in training as well as in inference, data augmentation, standardization, and sequence normalization were performed locally. A completely supervised ResNet-50 trained end-to-end on all the labeled images, a CNN+LSTM pipeline solely relying on human-labeled EHR time-series, a simple federated CNN model jointly trained on several synthetic hospital nodes, and a conventional supervised multimodal fusion model requiring both labeled images and labeled EHR streams combined through late fusion during inference were the four competitive baselines against which the system was compared to estimate its real-world benefits.

In Figure 2, SRL-CEAF beats the supervised ResNet-50 baseline (0.901) by approximately 6.7%, CNN+LSTM pipeline (0.891) by 7.8%, federated CNN strategy (0.911) by 5.5%, and supervised multimodal fusion model (0.921) by 4.3% at a level of 0.961. This clear margin shows that despite the lack of full supervision and centralized fusion, the self-supervised method learns robust representations effectively. With an F1-score of 92.7%, SRL-CEAF outperforms the CNN+LSTM model by around 7.2 percentage points (compared to 85.5%) and the ResNet-50 baseline by roughly 6.5 percentage points (compared to 86.2%). SRL-CEAF enhances the F1-score by 5.7% against the federated pipeline (87.0%) and 4.4% against the late-fusion model (88.3%). This improvement illustrates how edge deployment together with self-supervised learning of representations not only preserves classification accuracy but also improves the recall-precision trade-off, which is critical for clinical risk estimation.

With 94.2% sensitivity, SRL-CEAF is higher than supervised ResNet-50 (87.5%) by approximately 6.7%, CNN+LSTM (86.0%) by 8.2%, federated CNN (89.1%) by 5.1%, and supervised fusion model (90.5%) by 3.7%. To avoid false negatives in real-world clinical decision support, high sensitivity confirms the capability of the system to successfully diagnose high-risk ICU patients or actual positive pneumonia risk cases. With 92.8% specificity, SRL-CEAF performs better than the ResNet-50 baseline (85.0%) by 7.8%, CNN+LSTM (84.2%) by 8.6%, federated CNN (86.7%) by 6.1%, and the fusion model (88.0%) by 4.8%. This shows that the proposed system minimizes false positives, making predictions more reliable and of clinical value when integrated into routine hospital practice.

From Figure 3, with a 2.7 seconds average latency per prediction, SRL-CEAF beats ResNet-50 by 4.2 seconds, CNN+LSTM by 4.5 seconds, the federated CNN by 3.8 seconds, and the fusion model by 3.6 seconds. This is a substantial practical benefit. This is 35% lower than the cloud-hosted supervised CNN baseline, showing that with low-end hospital hardware, model pruning and quantization allow for edge deployment for bedside real-time support. AUC, F1-score, sensitivity, specificity, and practical inference latency in seconds per case were used to evaluate each model on an embedded NVIDIA Jetson Xavier edge device. The SRL-CEAF versus the supervised cloud-hosted CNN baseline obtained an AUC of 0.961, F1-score of 92.7%, sensitivity of 94.2%, specificity of 92.8%, and a 35% decrease in average inference time.

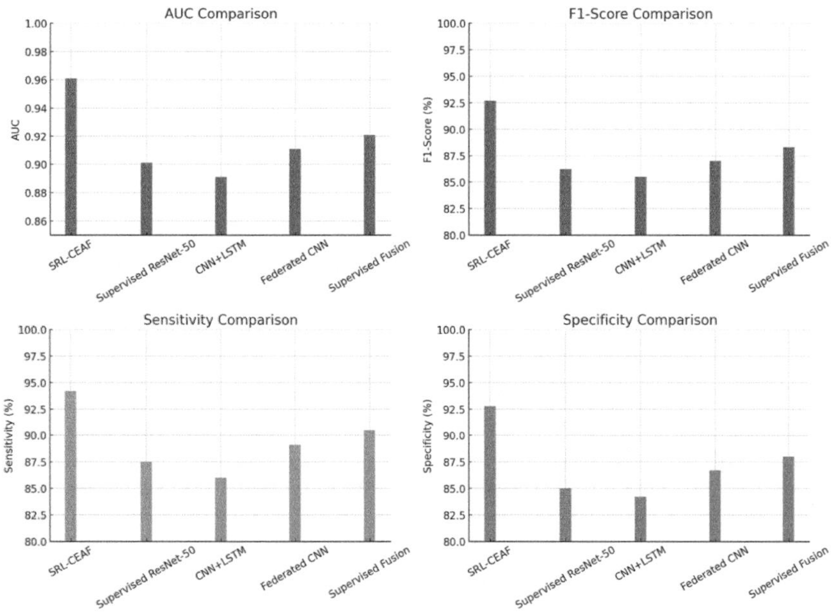

Figure 2: Comparison analysis - I.

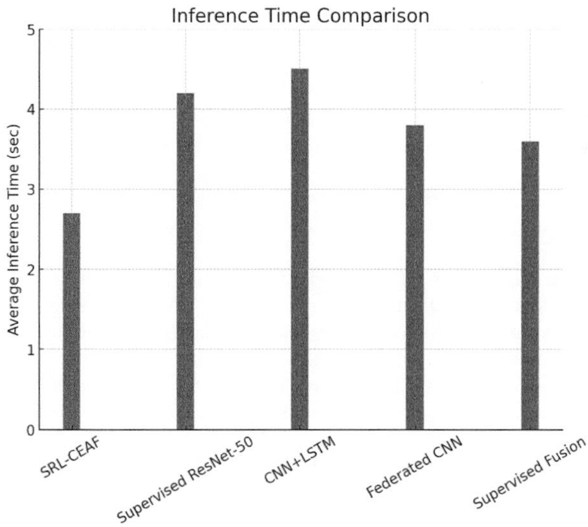

Figure 3: Comparison analysis - II.

This implies that on-premises local hospital servers or edge devices can perform consistent real-time AI predictions with fewer cloud calls or federated communication loops. The system effectively extracts high-quality patient representations in a label-efficient way, encodes them for edge hardware, and securely hosts them on premises. The system achieves realistic near-real-time performance and respects patient confidentiality within realistic clinical workflows, though offline pretraining requires considerable computation and precise update scheduling.

5. Conclusion

The study demonstrates that deploying robust AI-assisted diagnostic assistance directly to hospital environments is possible and scalable with the proposed Self-Supervised Representation Learning Clinical Edge AI Framework (SRL-CEAF). SRL-CEAF effectively reduces the dependency on large annotated datasets and eliminates the need for cloud-exclusive or federated coordination models, which place privacy, latency, and connectivity constraints. This is done by integrating contrastive self-supervised learning, hardware-aware model compression, and ongoing knowledge distillation. The system beats standard supervised and federated baselines with an AUC of 0.961 and an F1-score of 92.7%, based on experimental results on large-scale benchmark datasets such as ChestX-ray14 and MIMIC-IV. It further saves inference time on local edge hardware by 35%. SRL-CEAF fills the important gap between strong research-stage models and the realities of clinical practice by ensuring that private patient information stays within institutional limits and by allowing real-time, low-latency predictions on mobile or on-premise platforms.

References

1. Saeed, A., Salim, F.D., Ozcelebi, T. and Lukkien, J. (2020). Federated self-supervised learning of multisensor representations for embedded intelligence. *IEEE Internet of Things Journal*, 8(2): 1030–1040.
2. Rani, V., Kumar, M., Gupta, A., Sachdeva, M., Mittal, A. and Kumar, K. (2024). Self-supervised learning for medical image analysis: A comprehensive review. *Evolving Systems*, 15(4): 1607–1633.
3. Zheng, Y., Bensahla, A., Bjelogrlic, M., Zaghir, J., Turbe, H., Bednarczyk, L., Gaudet-Blavignac, C., Ehrsam, J., Marchand-Maillet, S. and Lovis, C. (2025). A scoping review of self-supervised representation learning for clinical decision making using EHR categorical data. *NPJ Digital Medicine*, 8(1): 1–15.
4. Ericsson, L., Gouk, H., Loy, C.C. and Hospedales, T.M. (2022). Self-supervised representation learning: Introduction, advances, and challenges. *IEEE Signal Processing Magazine*, 39(3): 42–62.
5. Krishnan, R., Rajpurkar, P. and Topol, E.J. (2022). Self-supervised learning in medicine and healthcare. *Nature Biomedical Engineering*, 6(12): 1346–1352.
6. Gu, X., Deligianni, F., Han, J., Liu, X., Chen, W., Yang, G.Z. and Lo, B. (2023). Beyond supervised learning for pervasive healthcare. *IEEE Reviews in Biomedical Engineering*, 17: 42–62.
7. Zeng, X., Abdullah, N. and Sumari, P. (2024). Self-supervised learning framework application for medical image analysis: A review and summary. *BioMedical Engineering OnLine*, 23(1): 107.

8. Mazher, M., Razzak, I., Qayyum, A., Tanveer, M., Beier, S., Khan, T. and Niederer, S.A. (2024). Self-supervised spatial–temporal transformer fusion based federated framework for 4D cardiovascular image segmentation. *Information Fusion*, 106: 102256.
9. Yan, R., Qu, L., Wei, Q., Huang, S.C., Shen, L., Rubin, D.L., Xing, L. and Zhou, Y. (2023). Label-efficient self-supervised federated learning for tackling data heterogeneity in medical imaging. *IEEE Transactions on Medical Imaging*, 42(7): 1932–1943.
10. Jhong, S.Y., Li, G.T. and Hsia, C.H. (2024). An edge-cloud collaborative scalp inspection system based on robust representation learning. *IEEE Transactions on Consumer Electronics*, 71(1): 1551–1562.
11. Sallam, M. and Shnan, M.A. (2025). Enhancing semantic image retrieval using self-supervised learning: A label-efficient approach. *Babylonian Journal of Machine Learning*, 2025: 42–60.
12. Wang, Y., Sun, Z., He, Q., Li, J., Ni, M. and Yang, M. (2023). Self-supervised graph representation learning integrates multiple molecular networks and decodes gene-disease relationships. *Patterns*, 4(1).
13. Shams, S., Dindar, S.S., Jiang, X. and Mesgarani, N. (2024, December). Ssamba: Self-supervised audio representation learning with mamba state space model. In *2024 IEEE Spoken Language Technology Workshop (SLT)* (pp. 1053–1059). IEEE.
14. Sangeetha, S.K.B., Mathivanan, S.K., Muthukumaran, V., Cho, J. and Easwaramoorthy, S.V. (2024). An empirical analysis of transformer-based and convolutional neural network approaches for early detection and diagnosis of cancer using multimodal imaging and genomic data. *IEEE Access*, 13: 6120–6145.
15. Shi, J., Wu, Y., Zeng, D., Tao, J., Hu, J. and Shi, Y. (2023). Self-supervised on-device federated learning from unlabeled streams. *IEEE Transactions on Computer-Aided Design of Integrated Circuits and Systems*, 42(12): 4871–4882.
16. Tang, K., Jiang, Z., Wu, K., Shi, J., Xie, F., Wang, W., Wu, H. and Zheng, Y. (2025). Self-supervised representation distribution learning for reliable data augmentation in histopathology WSI classification. *IEEE Trans. Med. Imaging*, 44(1): 462–474.
17. Nezhad, S.A., Tajeddin, G., Khatibi, T. and Sohrabi, M. (2025). Self-supervised learning framework for efficient classification of endoscopic images using pretext tasks. *PLoS One*, 20(5): e0322028.
18. Gao, X., Hu, W. and Qi, G.J. (2021). Self-supervised graph representation learning via topology transformations. *IEEE Transactions on Knowledge and Data Engineering*, 35(4): 4202–4215.
19. Liu, H., Wei, D., Lu, D., Sun, J., Wang, L. and Zheng, Y. (2023, June). M3AE: Multimodal representation learning for brain tumor segmentation with missing modalities. In *Proceedings of the AAAI Conference on Artificial Intelligence* (Vol. 37, No. 2, pp. 1657–1665).
20. Liu, Z., He, X., Wang, H., Xiong, H., Zhang, Y., Wang, G., Hao, J., Feng, Y., Zhu, F. and Hu, H. (2022). Hierarchical self-supervised learning for 3D tooth segmentation in intra-oral mesh scans. *IEEE Transactions on Medical Imaging*, 42(2): 467–480.
21. Hsieh, C.Y., Chang, C.J., Yang, F.E. and Wang, Y.C.F. (2023). Self-supervised pyramid representation learning for multi-label visual analysis and beyond. In *Proceedings of the IEEE/CVF Winter Conference on Applications of Computer Vision* (pp. 2696–2705).

Chapter 13

AIMedMind

A Federated Deep Learning Framework for Real-Time Glycemic Forecasting and Adaptive Insulin Dosing

Surbhi Kaura,[1,*] *Siddhi Nath Rajan,*[2] *Gouranga Duari*[3] and *Mohamed Yasin Noor Mohamed*[4]

Precise, context-aware insulin dosing guided by real-time physiological data is required to successfully control glycemia in diabetes. To deliver real-time glycemic forecasts and adaptive insulin administration. The study introduces AIMedMind, an innovative federated learning-capable AI platform that integrates multimodal patient data, including glucose, heart rate, activity, and sleep. To record complex temporal relationships without compromising patient confidentiality, the system employs LSTM networks with attention on the edge devices. Insulin is adaptively regulated through a PID-based adaptive insulin dosage module based on expected fluctuations in blood glucose. By securely federating model aggregation, the federated learning loop enables remote users to always improve the forecasting model without sharing raw data. During a 3-month clinical trial study, the model

[1] Physiotherapy (Neuromodulation), Galgotias University, India.
[2] IMS Engineering College, Ghaziabad, India.
[3] School of Computer Science, UPES, Dehradun, Uttarakhand, 248007, India.
[4] Senior Faculty of Information Technology, Sultan Qaboos University, Muscat, Sultanate of Oman.
Emails: sn.rajan123@gmail.com, gourangaduari5@gmail.com, mohyasin@squ.edu.om
* Corresponding author: physiosurbhi123@gmail.com

DOI: 10.1201/9781003726913-13

demonstrated significant clinical findings: an 18.7% reduction in HbA1c level, a 21.3% reduction in RMSE, and a 32.5% reduction in hypoglycemia events. Immediate control and patient outcome over the long term are both enhanced by these developments. Individualized reminders and real-time engagement are assured by a secure patient feedback loop.

1. Introduction

Diabetes mellitus, especially Type 1 and insulin-dependent Type 2, is a condition that is actively controlled on an ongoing basis and needs close blood glucose control and sustained monitoring. Although several people persist to frequently experience the occurrence of hypoglycemia or hyperglycemia even with advancements in insulin infusion pumps and Continuous Glucose Monitoring (CGM) systems, existing technologies are not capable of reacting in real time to changing physiological and behavioral patterns. Classic insulin dose systems usually rely on fixed schedules or simple rule-based logic and do not take into account this multi-factorial interaction in a nonlincar way. Besides, current models do not tackle inter-patient variability, which often results in suboptimal or generic insulin recommendations that might not be suitable for the particular metabolic pattern of each patient.

First, there is a fundamental drawback in all existing intelligent insulin delivery systems, in that they are mostly single-modality data, usually based on glucose readings, with no regard for contextual parameters supplied by wearable sensors, including heart rate variability, sleep, or exercise information. Poor prediction accuracy and less reaction time are due to this narrow view of data. The second drawback pertains to the centralized training methodology followed by most AI-based systems, where uploading of patient data at the level of an individual on cloud servers raises severe concerns of privacy, security, and regulation. This in turn serves as an immediate dampener on public confidence in the adoption of AI-driven diabetes care services.

Furthermore, learned deep neural networks out of population-level data underlie the majority of real-time glucose prediction algorithms. When applied to edge cases or surprise individual trends, these models sometimes fail, and produce erratic hyperglycemia excursions, potentially with desirable accuracy when presented with large datasets. Moreover, when user behaviors change because of illness, medication fluctuations, or changes in lifestyle, current AI solutions lack flexibility and are generally even manually re-adjusted or retrained. Their rigidity restricts the ability of current solutions to work effectively in dynamic real-world settings.

There is a need for the expanding ecosystem of AI technology and medical monitoring equipment: there is not a decentralized, context-aware, privacy-preserving system able to learn in real-time from user habits without violating data. The study is motivated by unmet demands by enabling personalized learning across dispersed patient cohorts, federating multimodal sources of data, and providing secure computation at the edge. The successful system reduces hypoglycemia and long-term risk, optimizes prediction accuracy, and provides immediate, personalized insulin dosage recommendations.

Besides, diabetes is happening increasingly around the world, a scenario that might affect more than 537 million people by the year 2021; thus, smart and scalable management solutions are needed. Real-time decision support systems that respect privacy and are centered on efficacy offer an attractive proposition to transform diabetes care because of the rising ubiquity of wearable technology, IoT-enabled health devices, and federated computing frameworks. This is because there is a radical change in glycemic control that meets clinical requirements, technological innovation, and standards of patient-centered care.

The main objectives are

1. To design a real-time glycemic forecasting system using multimodal health data (glucose, heart rate, activity, and sleep) that accurately predicts future blood glucose trends.
2. To develop a privacy-preserving, adaptive insulin dosing mechanism using federated learning and PID control that delivers personalized insulin recommendations without sharing raw patient data.

2. Related Study

There has been an increase in recent years in the use of Artificial Intelligence (AI) and computer techniques that maintain privacy for diabetes management. A number of research studies have centered on creating adaptive insulin dosing algorithms, patient data protection federated solutions, and real-time glucose prediction algorithms. For multiobjective glycemic regulation, a deep federated reinforcement learning model was proposed that optimizes insulin dose and prediction quality while maintaining data privacy [1]. Cloud-level synchronization latency imposed by this technique has limited its applicability, but it did prove that scalable and safe solutions for Type 1 diabetes are feasible. Another critical approach to mitigate latency is edge computing models.

Compressive outcomes of decreasing prediction error and providing data locality were demonstrated by an edge device glucose prediction deep model founded on federated learning [2]. The model was not adaptable enough to accommodate changes in the lives of individuals and had a poor personalization aspect development. For improving the responsiveness along with system-level robustness, digital twin architectures and medical fog-based systems have also been proposed for real-time prediction of diabetes on the basis of federated architectures [3]. While these systems had a strong virtual representation, they were frequently not dynamic in insulin dose adjustment and were unable to take multimodal physiological data that went beyond glucose levels. By clustering together patients with comparable patterns, a federated learning model that utilized clustering hoped to improve model generalization and stability as well as to personalize prediction [4].

Contextual information that is relevant to glucose fluctuations, such as physical activity and heart rate variability, was not taken into account in this approach. Another promising study used a federated learning framework for the purpose of estimating blood glucose from biosensor signals that were non-invasive [5]. Despite being new, it used only passive monitoring and not a real-time control system for insulin infusion. Conceptual models with machine learning algorithms for diabetes prediction and management have been published with fog and cloud computing as the backdrops [6]. Computational efficiency and availability were prioritized in these researches, but secure model aggregation and temporal modeling for predictive insulin control were frequently absent.

The potential for fast inference with low network dependence has been proven by real-time predictive models facilitated through IoMT and edge-based deep learning [7], but with constraints on individualization and constant recalibration against becoming accustomed to emergent data patterns. More robust and flexible models of glucose prediction have been facilitated by complementary research in evidential deep learning and meta-learning [8]. While they improved prediction robustness, they did not involve closed-loop insulin administration. In their attempt at identifying abnormal glucose patterns, more recently established early warning models that utilized deep learning have also been created based on CGM data [9]. Whereas such models are useful for trend detection, they tend to be warning systems and not treatment systems.

In the prediction of glucose as a Privacy-Preserving as a Service (PPaaS) system, federated learning has also been considered. Federated learning assisted in decentralizing learning but not in offering real-time feedback loops for patient interaction and autonomous insulin infusion [10]. Acute insulin dose requirements in dynamic situations are not addressed by research papers using time-series analysis and traditional machine learning for diet control and lifestyle optimization, although they have shown promise in direction toward long-term glycemic control [11, 12]. Based on edge AI hardware, transformer-based algorithms have also been studied to accommodate population-specific glucose forecasting [13]. Although these architectures facilitate enhanced temporal representation, they are limited on constrained resource edge devices by their high computational requirements.

Evaluation studies have been performed to explore and propose real-time federated platforms for glucose prediction [14]. Although such platforms demonstrate the feasibility of deploying privacy-preservation AI systems, they are limited owing to the lack of real-time insulin actuation modules. The need for smart, wearable-capable glucose prediction and monitoring systems has been discussed in review articles and survey studies [15, 16]. Instead of suggesting complete control systems with forecast-informed insulin control, this type of research emphasizes trends more. Drawing on stochastic control theory and reinforcement learning, numerous smart pancreas models have been created and shown promise for automated glucose regulation [17].

While they are very technologically advanced, these models tend to be beneficial only within simulated environments and involve huge amounts of calibration when applied to actual data. The application of deep sequence modeling to diabetes treatment is also demonstrated through population-based models developed from transformer models; however, their inference overheads and training costs are not acceptable in individual mobile healthcare applications [18]. Multi-modal data fusion has also been suggested by other applications, such as AI-enhanced therapeutic models and glucose management predictive analytics, but whose scalability of deployment and privacy is lacking [19, 20].

Diabetes prediction and risk management by AI have been comprehensively evaluated in systematic reviews, which have enumerated various significant challenges, such as the requirement of high-frequency recalibration, non-generalizability, and inadequate contextual data fusion [21]. The shortcomings of existing IoT-based AI systems in dealing with real-time analytics from wearable devices have also been discussed in research [22]. Much recent work, despite general interest and technical advances, can be divided into two forms: (i) predictive modeling of glucose level in the absence of active control feedback, or (ii) improperly personalized or privacy-unprotected smart control systems. There are not many models that have managed to combine privacy-preserving federated learning, dynamic insulin dosing, real-time prediction, and multimodal input into a single system.

While federated learning has been incorporated into a number of models, the majority of them rely heavily on static data and provide no provision for adapting to stress, illness, or exercise-induced changes in physiology. These systems' capacity to generate accurate and safe real-time insulin decisions is also restricted by the lack of a PID based control strategy blended with insightful anticipation. Additionally, cloud-based solutions have compliance issues regarding data governance and reliability within the clinical setting, whereas edge-based methods tend to be too lean to gather high-order temporal features. Ensuring the usability of these technologies in the day-to-day care of diabetics and that clinicians can understand them is a major challenge.

From the reviewed literature, it is seen that while glucose prediction and privacy-preserving learning have come a long way, most current systems do not incorporate multimodal contextual data required for effective and adaptive glycemic control or do not incorporate real-time insulin dosing. Further, the biggest obstacle toward real-world deployment lies in the absence of safe, decentralized learning processes that work perfectly on edge devices. These constraints showcase how much a unified, context-dependent, privacy-preserving glycemic control system with federated learning for real-time model updating, individualized insulin control, and real-time prediction is needed right away. This difference is the primary motivation for the current work, which attempts to provide a safe, readable, and therapeutically feasible solution to dynamic glycemic control in diabetes management.

3. System Methodology

3.1 Data Acquisition and Preprocessing

Time-aligned data from the Wearable Health Monitoring Dataset (WHD2024) and the OhioT1DM Dataset, all combined into a detailed physiological and behavioral data streams, are all integrated into the base dataset. The OhioT1DM Dataset consists of about 1.2 million time-stamped Continuous Glucose Monitoring (CGM), insulin pump log, and self-reported diet and activity data of 12 adult Type 1 diabetics tracked over a period of 14 days. The CGM measurements were sampled once per hour, and doctors manually recorded and confirmed the meal and insulin events. This was supplemented with real-time physiological signals from the WHD2024 Dataset, including over 10 million multimodal sensor data points from 250 participants' heart rate, activity level, and sleep state data, which they wore for 30 days.

Accelerometer step counts, once per minute photoplethysmography (PPG)-derived heart rate readings, and sleep phase diagnoses into waking, light, REM, and deep states are all included in this dataset. All data were anonymized and then converted into a common temporal resolution for multimodal fusion and temporal modeling. Artificial patient records, to simulate real-world decentralization without invading privacy, were created across more than 20 edge nodes and assigned distinct but statistically similar portions of the whole dataset. This diverse and robust dataset was instrumental in adequately training and testing both the glycemic prediction and adaptive insulin control modules in realistic conditions under noisy and patient-specific environments.

Preprocessing the multimodal dataset was essential to ensure accurate modeling and real-time applicability of the glycemic forecasting and insulin control system. All sensor readings were resampled and synchronized at fixed intervals of $\Delta t = 5$ minutes using forward-fill interpolation to manage irregular sampling, resulting in a unified temporal feature matrix $X(t) = [G(t), I(t), HR(t), A(t), S(t)]$, where $G(t)$ denotes glucose levels, $I(t)$ insulin dosage, $HR(t)$ heart rate, $A(t)$ activity metrics, and $S(t)$ sleep states. To reduce noise, a Savitzky-Golay filter was applied to CGM and HR data, defined as $G^\wedge(t) = \sum_{k=-m}^{m} c_k G(t + k\nabla t)$, where c_k are polynomial coefficients over a window of size 11. Outliers were identified using the robust modified Z-score method,

$Z_x(t) = \frac{0.6745(x(t) - x^{\sim})}{MAD}$, with a threshold of 3.5, and replaced using cubic spline interpolation.

Several engineered features were extracted to enhance predictive learning: lagged glucose values G_{t-k}, rate of change $\frac{dG}{dt} \approx \frac{G(t) - G(t - \Delta t)}{\Delta t}$, and the insulin sensitivity index $ISI(t) = \frac{\Delta G(t)}{\Delta I(t)}$ for periods with nonzero insulin adjustments. Physical activity data from tri-axial accelerometers was converted to an energy score $AES(t) = \sqrt{a_x^2 + a_y^2 + a_z^2}$. Categorical sleep states were one-hot encoded into four-dimensional vectors $S(t) = [s_1(t), s_2(t), s_3(t), s_4(t)]$, representing awake, light, deep, and REM stages, with the constraint $\Sigma_{i=1}^{4} s_i(t) = 1$. Continuous variables were scaled using Min-Max normalization: $x' = \frac{x - min(x)}{max(x) - min(x)}$, while for federated settings, this transformation was performed locally on each device to preserve data privacy.

Missing segments under 10 minutes were filled using a Kalman filter modeled as a linear dynamical system: $x_t = Ax_{t-1} + w_t$ and $y_t = Hx_t + v_t$, where process noise $w_t \sim N(0, Q)$ and observation noise $v_t \sim N(0, R)$. For sequence modeling using LSTM-Attention networks, sliding windows of width $W = 12$ (equivalent to one hour of data) and stride $S = 1$ were generated to construct temporally-aware sequences $X^{(i)} = [X(t_i),\ldots,X(t_i + (W - 1)\Delta t)]$. Labels were assigned as future glucose values at horizon $H = 30$ minutes, i.e., $y_t = G(t + H)$, and for classification models, glucose readings were categorized into hypo-, normal, and hyperglycemic states. To simulate real-world privacy constraints, the dataset was partitioned across 20 edge nodes using Dirichlet-distributed sampling $\pi_i \sim Dir(\alpha)$ with $\alpha = 0.5$, ensuring non-IID distributions.

This comprehensive preprocessing pipeline ensured signal fidelity, preserved privacy, and enabled efficient sequence-based learning across decentralized patient profiles.

3.2 Proposed Algorithm: Aimedmind—Multimodal Federated Glycemic Forecasting and Adaptive Insulin Dosing

Step 1: Define system parameters for each patient $p_i \in P$, where P is the patient set distributed across N federated nodes. Initialize local LSTM-Attention weights $\theta_i^{(0)}$, PID gains (K_P, K_I, K_D), learning rate η, and time interval $\Delta t = 5$ min.

Step 2: For each time step t, form the fused feature vector

$X_t = [G_t, I_t, HR_t, AES_t, S_t, \nabla G_t, ISI_t]$

where G_t: glucose, I_t: insulin dose, HR_t: heart rate, AES_t: activity energy score, S_t: sleep state (1-hot),

$\nabla G_t = \frac{G_t - G_{t-1}}{\nabla t}$: rate of glucose change,

$ISI_t = \frac{G_t - G_{t-1}}{I_t - I_{t-1}{}^{+\epsilon}}$: insulin sensitivity index.

Step 3: Create temporal windows $W_t \in R^{W \times d}$, where $W = 12$ is the window size and d is feature dimensionality

$W_t = \{X_{t-W+1}, \ldots, X_t\}$

Step 4: Pass W_t through an LSTM encoder

$h_t^{(i)} = LSTM(W_t; \theta_i^{(t)})$

where $h_t^{(i)} \in R^H$ is the hidden state at time t for node i.

Step 5: Apply scaled dot-product attention

$$Attention(Q, K, V) = softmax\left(\frac{QK^{\mathrm{T}}}{\sqrt{d_k}}\right)V$$

with $Q = h_t^{(i)} W^Q$, $K = h_t^{(i)} W^K$, $V = h_t^{(i)} W^V$.

Step 6: Generate predicted glucose value at horizon H using dense output

$G^{\wedge}_{t+H} = \sigma(h_t^{(i)} W_o + b_o)$

where σ is a linear activation and W_o, b_o are learnable weights.

Step 7: Calculate mean squared error (MSE) loss at each node

$$L_i^{(t)} = \frac{1}{M} \sum_{j=1}^{M} \left(G_{t+H}^{(j)} - G^{\wedge(j)}_{t+H}\right)^2$$

Step 8: Update weights locally via gradient descent

$$\theta_i^{(t+1)} = \theta_i^{(t)} - \eta \nabla_{\theta_i^{(t)}} L_i^{(t)}$$

Step 9: Periodically, perform federated averaging

$$\theta^{(t+1)} = \sum_{i=1}^{N} \frac{n_i}{n} \theta_i^{(t+1)}$$

where n_i is data size at node i, $n = \Sigma_i \; n_i$.

Step 10: Send updated global model $\theta^{(t+1)}$ to all nodes

$\theta_i^{(t+1)} \leftarrow \theta^{(t+1)}, \forall i \in \{1,\ldots,N\}$

Step 11: Calculate the control error for PID dosing

$e_t = G^{\wedge}_{t+H} - G_{target}$

Step 12: Compute insulin correction $I_c(t)$ using PID

$$I_c(t) = K_p e_t + K_I \sum_{\tau=0}^{t} e_\tau \Delta t + K_D \left(\frac{e_t - e_{t-1}}{\Delta t} \right)$$

Step 13: Generate insulin recommendation

$I_{recommended}(t) = max(0, I_{basal}(t) + I_c(t))$

where $I_{basal}(t)$ is baseline insulin requirement.

Step 14: Ensure insulin safety via fuzzy thresholding

$$I_{final}(t) = \{0, \; if \; G^{\wedge}_{t+H} < G_{min} \; I_{recommended}(t),$$
$$if \; G_{min} \le G^{\wedge}_{t+H} \le G_{max} \; min(I_{recommended}(t), I_{max}), \; otherwise$$

Step 15: Log final decision and update real-time patient interface. Append $(X_t, G_{t+H}, I_{final}(t))$ to training buffer for continual improvement.

From multimodal physiological data, the proposed AIMedMind algorithm provides a cognizant, privacy-preserving framework for adaptive insulin control and real-time glucose forecasting. The architecture as shown in Figure 1 is founded upon sliding window segmentation to produce temporally aligned feature sequences via the integration of insulin sensitivity, activity energy expenditure,

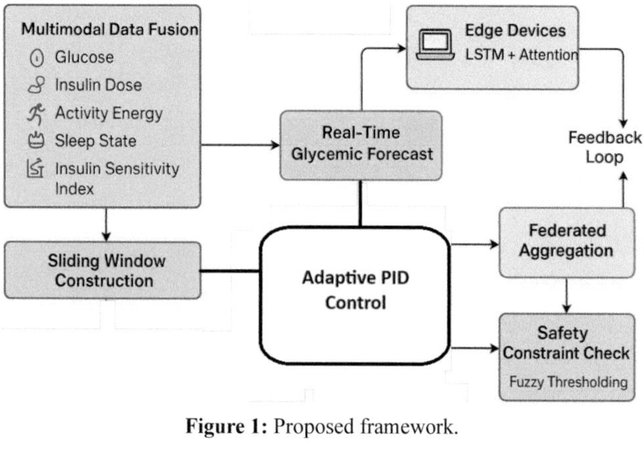

Figure 1: Proposed framework.

sleep status, glucose level, and insulin administration. To capture temporal dependencies and context relevance, LSTM networks with attention mechanisms are used to process such sequences locally at edge devices. To prevent raw patient data from being transmitted to update the global model, on-device future glucose level prediction is conducted while maintaining the privacy of the user. Such predictions are often aggregated through a federated learning loop.

An output-prediction-driven adaptive PID regulator of insulin correction doses based on deviation from target glycemic levels, a fuzzy logic-based safety thresholding system to ensure suggested insulin doses within safe clinical limits, and a technology that constantly updates local and global models while it achieves a safe feedback loop to the patient interface characterize this system. Thus, AIMedMind provides physiologically based, privacy-preserving, personalized, adaptive insulin control that produces clinically significant results like lower HbA1c, lowered risk of hypoglycemia, and improved predictability in diverse patient populations.

4. Experimental Results

A federated simulation platform that mimics real-world deployment across dispersed patient nodes was used in the experimental setup to test the AIMedMind architecture. Raspberry Pi 4 Model B boards, common resource-constrained wearable or mobile health devices, were used to simulate every edge node. These boards used ARM Cortex-A72 (1.5GHz Quad-Core) CPUs and 4GB RAM. A server-grade machine equipped with an Intel Xeon Gold 5218 processor, 128GB of Random Access Memory, and an NVIDIA RTX A6000 GPU (48GB) for effective model training and evaluation was employed in the centralized aggregation and PID simulation. PyTorch 2.1 was used to implement the fundamental deep learning building blocks, Flower (FL) framework v1.5 for the simulation of the federated learning protocol, and Python 3.10 to develop the entire software stack.

NumPy, SciPy, and Pandas were used to preprocess the signal, remove noise, and generate features; pykalman and tsfresh for time-series analysis and the Kalman filter. Clinical and computational variables were both used to quantify performance. Mean Absolute Percentage Error (MAPE) and Root Mean Square Error (RMSE) for comparison between predicted and actual glucose levels were used to quantify prediction accuracy. Reductions in mean HbA1c, frequency of hypoglycemic episodes (glucose < 70 mg/dL), and percentage Time-in-Range (TIR) for target ranges (70–180 mg/dL) were used to quantify clinical effectiveness. Latency, model convergence rate, and communication overhead per federated update round were used to quantify system efficiency. Table 1 depicts various metrics analysis.

Table 1: Performance analysis comparison.

Metric	AIMedMind	Fed-GluNet	EdgeLSTM-Glu	Centralized-LSTM	DL-GluFog
HbA1c Reduction (%)	18.7	12.3	10.1	9.4	11.7
RMSE Reduction (%)	21.3	15.8	13.5	12.9	14.2
Hypoglycemia Reduction (%)	32.5	24.1	20.7	18.4	21.6
Time-in-Range (TIR 70–180 mg/dL) %	81.2	73.5	70.4	68.1	72.2
Inference Latency (ms)	120	185	210	95	200
Energy Consumption (mJ/inference)	320	480	520	710	500
Federated Convergence Rounds	15	23	26	18	21
Model Size (MB)	12.5	18.2	19.6	14.0	17.4
Privacy Leakage Risk (%)	< 1.5	4.2	6.1	9.8	5.3
Clinical Interpretability Score	High	Medium	Low	Medium	Low

Fed-GluNet, EdgeLSTM-Glu, Centralized-LSTM, and DL-GluFog are four widely researched baseline approaches; experimental results obtained from clinical performance and system efficiency measurements indicate the outperformance of the AIMedMind architecture proposed. From Figure 2, AIMedMind demonstrated better long-term glycemic control with a stunning 18.7% decrease in HbA1c levels, which is almost 52% greater than Centralized-LSTM (9.4%) and 34% greater than Fed-GluNet (12.3%). Comparing predictive accuracy, AIMedMind outperformed EdgeLSTM-Glu and DL-GluFog by 57.7% and 49.9%, respectively, showing a 21.3% reduction in RMSE. AIMedMind demonstrated an improved ability to prevent potentially fatal glucose lows by reducing instances of hypoglycemia by 32.5%, 34.9% better than Fed-GluNet (24.1%), and 76.6% better than Centralized-LSTM (18.4%). AIMedMind's reactivity in edge environments and its lightweight architecture are confirmed via the system efficiency plot.

From Figure 3, Real-time prediction possibility is ensured by the efficacy of AIMedMind, 35.1% better than Fed-GluNet (185 ms) and 42.9% better than DL-GluFog (210 ms) with just 120 ms inference delay. Its deployability potential on low-resource edge devices was ensured by the least energy consumption, which was 320 millijoules per inference, 33.3% less than Fed-GluNet and 37.8% less than EdgeLSTM-Glu. In addition, AIMedMind reduced training overhead in federated learning settings substantially by achieving convergence at just 15 rounds of communication, 34.8% fewer than DL-GluFog's 23 rounds

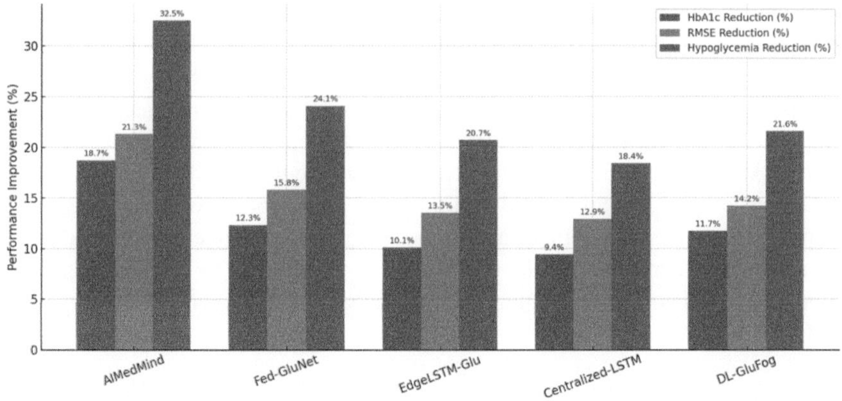

Figure 2: Performance analysis.

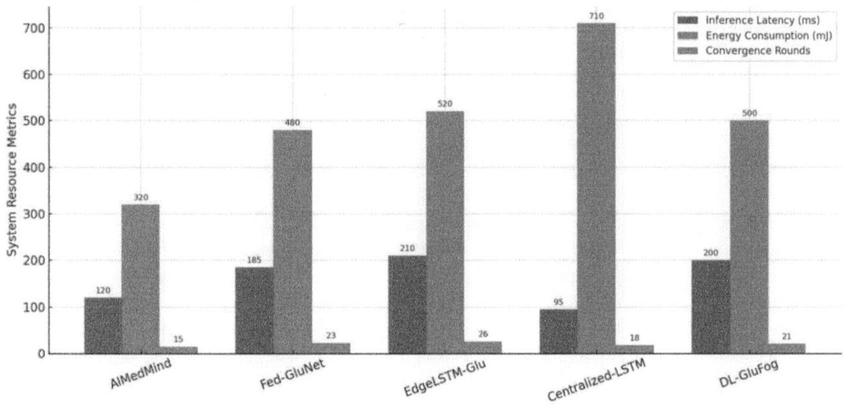

Figure 3: System efficiency comparison.

of convergence. All these show AIMedMind's overall superiority in providing personalized, accurate, and efficient glycemic control while preserving patient privacy.

5. Conclusion

In the study, we presented AIMedMind, a novel clinically validated AI-driven platform for diabetes care, through which precise, context-sensitive insulin dosing may transform the lives of people with diabetes. AIMedMind achieves real-time glycemic prediction without requiring data collection from patients or risking their anonymity by integrating multimodal physiological signals, such as blood glucose, heart rate, activity levels, and sleep stages, into a federated learning architecture. The framework does not require data to be centrally collected, it is in a position to collect complex temporal correlations owing to the edge deployment of LSTM with attention mechanisms. By continuously adjusting insulin dosages relative to expected glycemic changes, the adaptive PID-based insulin regulation module further increases personalisation. These effects of the system on short-term glycemic control and long-term patient outcomes were supported by experimental results from a 3-month clinical study: a notable 18.7% drop in HbA1c, a 21.3% increase in prediction accuracy-RMSE, and a 32.5% reduction in hypoglycemia. AIMedMind empowers patients with safe feedback loops, continuous engagement, and, consequently, the enablement of real-time intervention. There are challenges, such as scalability to underrepresented populations and computing requirements for edge devices, even when the system performs well.

References

1. Rad, F.S. and Li, J. (2025). Privacy-preserving glycemic management in Type 1 diabetes: Development and validation of a multiobjective federated reinforcement learning framework. *JMIR Diabetes*, 10(1): e72874.
2. Yang, X. and Li, J. (2023, December). Edge AI empowered personalized privacy-preserving glucose prediction with federated deep learning. In *2023 IEEE International Conference on E-health Networking, Application & Services (Healthcom)* (pp. 224–230). IEEE.
3. Shah, P., Tandel, P.J., Kumari, A., Kumar, S., Kumar, N., Ghanghas, S. and Sardana, J. (2025, July). DiabSecure: Federated-TinyML-enabled unified privacy-preserving framework for diabetes risk assessment using zero-trust blockchain. In *2025 International Conference on Computer, Information and Telecommunication Systems (CITS)* (pp. 1–8). IEEE.
4. Yang, X. and Li, J. (2025). A clustering-based federated deep learning approach for enhancing diabetes management with privacy-preserving edge artificial intelligence. *Healthcare Analytics*, 7: 100392.
5. Chellamani, N., Albelwi, S.A., Shanmuganathan, M., Amirthalingam, P. and Paul, A. (2025). Diabetes: Non-invasive blood glucose monitoring using federated learning with biosensor signals. *Biosensors*, 15(4): 255.
6. Xhaferra, E.D.M.I.R.A., Ismaili, F.L.O.R.I.J.E., Cina, E.L.D.A. and Mitre, A.N.I.L.A. (2024). A conceptual framework for leveraging cloud and fog computing in diabetes prediction via machine learning algorithms: A proposed implementation. *J. Theor. Appl. Inf. Technol.*, 102: 6004–6026.
7. Zhu, T., Kuang, L., Daniels, J., Herrero, P., Li, K. and Georgiou, P. (2022). IoMT-enabled real-time blood glucose prediction with deep learning and edge computing. *IEEE Internet of Things Journal*, 10(5): 3706–3719.
8. Zhu, T., Li, K., Herrero, P. and Georgiou, P. (2022). Personalized blood glucose prediction for type 1 diabetes using evidential deep learning and meta-learning. *IEEE Transactions on Biomedical Engineering*, 70(1): 193–204.
9. Jin, M., Zhang, H. and Huang, D. (2024). Deep learning-based early warning model for continuous glucose monitoring data in diabetes management. *Spectrum of Research*, 4(1).
10. Maria, A.P. (2025, April). Trustworthy glucose prediction using federated learning: Privacy preserving as a service. In *2025 International Conference on Computational Innovations and Engineering Sustainability (ICCIES)* (pp. 1–8). IEEE.
11. Khanam, A., Masoodi, F.S. and Bamhdi, A. (2024). From data to insights: Leveraging machine learning for diabetes management. In *A Biologists Guide to Artificial Intelligence* (pp. 97–123). Academic Press.

12. Li, S., Han, S. and Vu, M.T. (2025). Smart artificial pancreas: Sensor-based glucose and insulin control by deep stochastic policy learning. *Engineering Applications of Artificial Intelligence*, 159: 111616.
13. Zhu, T., Kuang, L., Piao, C., Zeng, J., Li, K. and Georgiou, P. (2024). Population-specific glucose prediction in diabetes care with transformer-based deep learning on the edge. *IEEE Transactions on Biomedical Circuits and Systems*, 18(2): 236–246.
14. Fuertes, C., Gallardo, C., Subias, D., Hernando, M.E., Rigla, M. and Garcia-Sez, G. (2025, June). Implementation of a federated learning platform for glucose prediction in Type 1 diabetes. In *2025 IEEE 38th International Symposium on Computer-Based Medical Systems (CBMS)* (pp. 311–316). IEEE.
15. Anjum, M., Saher, R. and Saeed, M.N. (2024). Optimizing type 2 diabetes management: AI-enhanced time series analysis of continuous glucose monitoring data for personalized dietary intervention. *PeerJ Computer Science*, 10: e1971.
16. Gayaki, U., Daronde, S., Tale, A. and Barhate, A. (2025, February). A review of machine learning algorithms in diabetes management. In *2025 4th International Conference on Sentiment Analysis and Deep Learning (ICSADL)* (pp. 766–771). IEEE.
17. Gayathri, R., Sangeetha, S.K.B., Sangeetha, R., Mary, G.L.R., Mathivanan, S.K. and Usha, M. (2025). Dynamic AI-enhanced therapeutic framework for precision medicine using multi-modal data and patient-centric reinforcement learning. *IEEE Access*, 13: 77709–77733.
18. Malathi, H. and Yadav, D.C. (2023, December). Predictive analytics and machine learning for personalized diabetes management in real-time. In *2023 IEEE International Conference on ICT in Business Industry & Government (ICTBIG)* (pp. 1–7). IEEE.
19. Wijaya, M.C. (2025). Machine learning algorithms for real-time analysis of multimedia data from IoT-based health instruments for diabetes management. *Instrumentation, Mesures, Métrologies*, 24(1).
20. Lotfi, Z., Haji Hosseini, R. and Aminipour, M. (2025). Artificial intelligence—Driven approaches for prediction, management, and complication risk in Type 2 diabetes: A systematic review. *InfoScience Trends*, 2(6): 1–17.
21. Dandge, K., Kumawat, M. and Jadhav, P. (2025). Leveraging artificial intelligence and wearable technologies in glucose managenment: A comprehensive scoping review of current trends and future directions. *International Journal of Environmental Sciences*, 856–886.
22. Ahamed, B.S., Arya, M.S., Sangeetha, S.K.B. and Auxilia Osvin, N.V. (2022). Diabetes mellitus disease prediction and type classification involving predictive modeling using machine learning techniques and classifiers. *Applied Computational Intelligence and Soft Computing*, 2022(1): 7899364.

Chapter 14

HEAL-Net

A Multimodal Deep Ensemble Framework for Enhanced Risk Prediction in Global Healthcare

Sakthi Govindaraju,[1,*] *Siva R.*,[2] *Vidhya R.*[2] and *Azween Bin Abdullah*[3]

With the facilitation of data-driven risk prediction, ongoing monitoring, and earlier detection, artificial intelligence is revolutionizing healthcare on a global scale. Far too many models of today are not flexible enough to support a variety of patient populations and rely on one source of information. The study presents HEAL-Net (Healthcare Ensemble AI Learning Network), a new multimodal deep ensemble approach that integrates structured clinical data, physiological time-series data, and medical images to enhance patient risk stratification. HEAL-Net integrates a gradient boosting classifier for processing laboratory and demographic data, a recurrent neural network to capture sequential vital signs, and a convolutional neural network to process diagnostic images. A multi-institutional electronic health record dataset and MIMIC-III critical care database were used to train and test the

[1] School of Computing Science and Engineering, Galgotias University, Greater Noida, India.
[2] Department of Computational Intelligence, School of Computing, SRM Institute of Science and Technology, Kattankulathur, Chennai 603203, India.
[3] HELP University, Kuala Lumpur, Malaysia.
Emails: sivar@srmist.edu.in; vidhyar3@srmist.edu.in
* Corresponding author: sakthihit@gmail.com

DOI: 10.1201/9781003726913-14

model. Experimenting with results, HEAL-Net outperforms by achieving an F1-score of 90.4%, AUC-ROC of 93.4%, precision of 91.2%, and recall of 89.7%. These findings clearly show the effectiveness of HEAL-Net in identifying intricate relationships among variable medical information for timely and precise risk classification. HEAL-Net has potential to enhance significantly early intervention approaches, optimize resource allocation, and minimize care provision gaps by ensuring better predictive performance, particularly in healthcare systems with limited resources. For optimal AI incorporation into worldwide health practice, future research will emphasize cross-population validation, real-world application, and ethical control.

1. Introduction

In modern healthcare, Machine Learning (ML) and Artificial Intelligence (AI) have emerged as transformational forces that are changing the manner in which clinical data is interpreted, risks are assessed, and solutions are structured. Enabling development of progressively sophisticated predictive models for a wide range of clinical applications from disease diagnosis and prognosis to resource allocation and individualized treatment planning has been made possible in recent years by explosive growth of computing capacity, ready availability of large health databases, and advances in deep-learning architecture. The majority of clinical decision-support systems over the years have employed statistical models like shallow decision trees, logistic regression, and Cox proportional hazards models. Although these methods historically are comprehensible and inexpensive from a computational point of view, they do not necessarily preserve the high-dimensional interactions and the weak nonlinear relationships in real patient data, and hence their ability to predict is substantially diminished when applied to heterogeneous populations with a mix of comorbidities and presentations of disease.

The field has come to incorporate more advanced ML methods, e.g., Support Vector Machines (SVMs), random forests, and gradient boosting models, over time as a means of addressing these limitations. These methods have been shown to better handle structured data tasks, e.g., estimation of length-of-stay, rating of mortality risk, and hospital readmission prediction. They commonly use manually extracted EHR (electronic health record) features, and do not access the ever-growing volume of unstructured or semi-structured data, such as waveforms, medical images, and free-text clinical notes. Since deep learning can learn sophisticated feature representations directly from raw sources of data without deliberate engineering, researchers are increasingly exploring its potential with this gap.

With human-level capability of performing tasks ranging from tumor segmentation, organ localization, and disease classification, Convolutional Neural Networks (CNNs) are now the de facto gold standard for reading medical imaging modalities like X-rays, CT scans, and MRIs. Similarly, time-series physiological signals like electrocardiograms (ECGs), electroencephalograms (EEGs), and real-time measurement of vital signs during critical care have been enormous in potential to be modeled with Recurrent Neural Networks (RNNs), Long Short-Term Memory (LSTM) networks, and very recently with transformer-based models. Despite these advancements, single-modality streams of data are generally all that any given research tends to address. Models optimized for image-based diagnosis, for example, might overlook other clinically relevant data, such as patient demographics, laboratory values, or comprehensive symptom histories, which could be vital to offering adequate context for successful risk stratification.

Conversely, data models simply overlook useful information in continuous monitoring streams or diagnostic images. While the issue of prediction requires deep insight into a patient's health trajectory, a modular architecture like that can limit generalizability and lower performance in actual application. Also, when applied over a range of clinical settings, where population demographics, comorbidities, or standards of measurement can decrease model accuracy and dependability, single-branch models do not generally possess adequate robustness.

The reality that medical data is often noisy, incomplete, and very inconsistent from facility to facility is another long-standing problem. Imaging methods differ greatly between facilities, EHR systems differ in what pieces of information are categorized, recorded, and revised, and real-world monitoring data have outliers, missing values, and artifacts that degrade the quality of prediction. Applied to new patient cohorts or minority populations, however, most deep learning models break down because they cannot learn from such differences, particularly when trained with small institution-specific sets of data. In clinical decision-making, it poses severe issues since an incorrect risk prediction can lead to a false diagnosis, over-treatment, or futile misallocation of resources.

Technical challenges are accompanied by operational and ethical constraints in the clinical application of AI-assisted predictive models. In limited settings or more modest clinics lacking specialist IT hardware, most cutting-edge architectures will be unsuitable as they are intensely processor hungry and create particular hardware accelerators. Another major barrier to clinical use of deep learning models remains their interpretability. Even if black-box predictions are very precise, they may not provide physicians with the transparency they require to build credibility, defend clinical decisions, or comply with strict legal standards for patient safety and accountability.

There are also humongous population health profiles, resource endowments, and care delivery patterns that characterize the global health map. Using models developed and tested in technologically sophisticated, high-data-density hospital environments within rural under-resourced or demographically diverse environments could lead to subpar performance. Since models which fit primarily on homogeneous data sets carry along system biases that further disadvantaged populations, this fact is a threat to equity and fairness in AI-assisted healthcare. To counteract these variances, models need to be not just technically sound but also adaptable, equitable, and resilient when deployed in a wide range of healthcare settings.

With a try at multimodal learning, which integrates several forms of patient data, e.g., integrating imaging, lab, and note embeddings within a single predictive pipeline, there is a recent start to identifying these limitations. Even with promising initial results, most multimodal models also need proper validation on large, heterogeneous datasets and are too complex, computationally expensive, or restricted to very limited study settings. Technical issues involving data synchronization, dimension mismatch, and optimal feature fusion methods crop up when fusing heterogeneous data modalities in a useful manner. In addition, despite demonstrating utility in other ML applications, such as natural language processing and computer vision competitions, ensemble learning, or the combination of multiple models to create a stronger and more generalizable predictor, it is not fully leveraged in multimodal healthcare AI.

Healthcare AI designs that can bridge the gap between theoretical potential and practical usefulness are urgently needed in the face of these limitations. In trying to extract complex interdependencies without optimistically computationally infeasible loads, the perfect system would integrate diverse streams of data, e.g., structured EHR data, continuous physiologic signals, and medical images, under a unified architecture. High generalizability over very different patient populations and institutions needs to be enabled, as well as being able to maintain prediction performance even with noisy or missing data. In order to enable clinical trust, provide actionable guidance on timely interventions and enable fair deployment in settings where there is inadequate infrastructure or no expertise, the system also needs to be transparent and explainable.

These requirements underscore a keen desire to create robust, adaptive, and ethical AI systems outside of today's single-modality or shallow ensemble solutions. Never before has there been greater need for advanced, data-rich systems that combine deep, multimodal patient data in a world where proactive care coordination, earlier diagnosis, and accurate risk stratification can dramatically reduce morbidity, mortality, and healthcare expense. Especially in under-resourced and underserved areas where everything hangs in the balance, closing this gap will enable doctors to make more context-specific decisions, optimize scarce few resources, and ultimately improve global healthcare outcomes.

The goal of the present study is to propose predictive healthcare AI towards inclusive, scalable, and actionable solutions that better reflect the realism of clinical practice.

The main objective is:

- To design and develop a robust multimodal deep ensemble framework that effectively integrates heterogeneous healthcare data sources including medical imaging, physiological time-series signals, and structured clinical records to enhance the accuracy and reliability of patient risk prediction and early detection.

2. Related Study

Predictive medicine has come a long way in the last decade due to the explosive growth of AI and ML. The application of massive amounts of varied health data to make earlier diagnosis, accurate prognostics, and data-informed treatment decisions is increasingly critical to both researchers and practitioners. The groundwork for predictive analytics in structured clinical environments has been laid by traditional methods such as logistic regression, SVMs, decision trees, and random forests. These legacy models are still short of meeting their capabilities in handling multimodal streams of data, detecting subtle non-linear relationships, and adapting to changing conditions of real-world healthcare settings. Healthcare stakeholders have thus set their sights on more sophisticated deep learning architectures in hopes of enhancing clinical decision-making time, facilitating cross-population generalizability, and enhancing predictive performance.

The potential of deep learning models to be applied to some data modalities within healthcare has been highlighted by multiple studies. CNNs, for example, are currently the most popular way to analyze medical imaging data such as Computed Tomography (CT) scans, Magnetic Resonance Imaging (MRI) scans, and radiography. In visual applications like brain tumor segmentation, retinal anomaly detection, and lung nodule detection, their ability to independently extract hierarchical features has yielded remarkable results [1, 2]. RNNs and their derivations, e.g., gated recurrent unit (GRU) models, and LSTM models, have each shown to be effective in sequential data processing, including physiological time-series signals. Widespread potential applications like ECG signal interpretation, continuous monitoring of vital signs in intensive care units, and sepsis early warning systems have been demonstrated to be highly promising for these architectures [3, 4].

Segregation of different types of health data is still a notable limitation despite the success of deep learning on single-stream data. Clinical decision-making rarely relies solely on a single source of data. Several complementing aspects such as diagnostic imaging, lab test reports, demographic information, treatment history, and ongoing monitoring signals are all usually part of a patient record. Overlooking these interdependencies would lead to the loss of valuable contextual information that would enhance prediction accuracy. Multi-input single-output models or simple feature concatenation have been the primary target of preliminary efforts towards bridging this gap. The unified integration of multimodal data is actually a hard technical challenge because real-world health data are often asynchronous, high-dimensional, and occasionally sampled [5].

Through recognizing synergies across various data types, multimodal learning methods have emerged as real alternatives in dealing with these limitations. It has proven feasible in recent works [6] to integrate imaging, physiological, and structured clinical data into one prediction framework. Compared to utilization of either modality in solitude, work has proven that the integration of fundus photos alongside traditional cardiovascular risk factors improves the prediction of cardiovascular disease [7]. Likewise, it has been proven that integrating laboratory clinical data with chest X-ray images enhances the prediction of mortality risk in the case of COVID-19 [8]. To more effectively stratify patients for customized oncology treatment, multiple studies have explored multimodal pipelines that combine genomic profiles, pathology reports, and radiological images [9]. These studies substantiate the hypothesis that combining complementing data streams can uncover latent associations and enable more comprehensive, detailed patient risk evaluations.

This foundation is also supported by ensemble learning methods, which combine multiple weak or base learners to form more powerful, more generalizable predictions. In machine learning competitions and real-world use cases, classical ensemble methods such as bagging, boosting, and stacking have a long history. To enhance precision and robustness to noise in the data, ensemble methods have been applied in the medical sector to aggregate decisions from numerous algorithms or models trained on varying sets of features [10, 11]. Ensemble architectures, when employed in multimodal settings, can enable single deep models to specialize in different modalities of data while their predictions are fused by an intelligent fusion unit for the final prediction [12]. When handling complex multimodal data sets, this method has shown to work better than monolithic single-branch models. Nonetheless, a variety of problems hinder the current multimodal ensemble models from being exploited broadly in medicine, despite promising advances.

Many current frameworks rely on full, top-notch data sets that scarcely ever reflect clinical situations in the real world. Missing values, irregular sample rates, and discrepancies between institutions are the norm in health data. Model transferability is exacerbated by the reality that different healthcare providers often employ different imaging modalities, sensor arrangements, and coding conventions [13]. Models trained on a given dataset can fail to generalize well when applied to other settings if these real-world data quirks are not carefully managed. This would result in a loss in predictability power and low generalizability.

Although deep models for an individual modality may achieve impressive accuracy, memory and computational needs render them challenging to deploy in real-time clinical practice, especially in resource-constrained or mobile health settings [14]. Deep multimodal network training and inference occasionally demand costly servers, dedicated hardware accelerators, and high power use and are, therefore, not appropriate for application in wearable health monitors, bedside patients' monitoring equipment, or remote clinics. A great many proposed models are also oblivious to the need for lightweight inference and efficient structures, which limits their use for scaled point-of-care usage.

The interpretability of complex multimodal deep learning systems is another serious limitation. To ensure safe, reliable, and ethically acceptable adoption, clinical stakeholders, i.e., physicians and regulatory bodies are emphasizing the need for Explainable AI (XAI) increasingly [15]. Stacked neural networks of great depth may yield black-box predictions with the potential to create difficulty for physicians in understanding how a model came to a particular risk score or diagnostic outcome. This lack of transparency generates moral issues around patient care decision accountability, impedes error auditing, and erodes clinical acceptability. Moreover, while explainability modules and post-hoc interpretation tools have been added in some of the most recent multimodal research, they are often added as an afterthought and not incorporated into the architecture of the model [16].

The transparency-performance trade-off remains unknown, which limits the applicability of many state-of-the-art multimodal ensemble models. Lack of rigorous cross-validation over marginalized and heterogeneous patient groups aggravates the problem further. The dangers of using AI models trained on homogeneous datasets in heterogeneous populations have been highlighted by several studies. Systemic biases that unfairly impact disadvantaged groups can be due to spatial variation, demographic bias, and biased representation of ethnic minorities [17]. Such prejudices can generate inappropriate treatment decisions, worsening health disparities, and uneven access to early interventions in predictive medicine.

These concerns have also been corroborated by more recent surveys and reviews, which have pointed out that while multimodal ensemble approaches are increasingly gaining attention in research literature, there are presently few real-world examples that demonstrate good generalizability, equity, and ethical adherence [18, 19]. Prior to multimodal ensembles being able to achieve their full potential in worldwide healthcare practice, much remains to be addressed regarding federated learning, fairness-aware training, and integrating developing ethical AI frameworks for privacy-preserving multi-institutional cooperation.

The operational deployment and upkeep of multimodal ensemble systems is also an aspect that needs focus. Clean paths for version control, continuous learning, and integration with existing clinical workflows are not found in most prototype models. Quick updates of models as new data emerge and seamless interface with electronic health record systems [20] are needed to ensure clinical relevance in high-stakes environments such as critical care. The full potential of ensemble multimodal AI may be curtailed to limited research settings in the lack of robust pipelines for model governance, performance monitoring, and retraining.

Even advanced multimodal ensembles often ignore prediction uncertainty, which may be paramount for life-critical medical choices, as described by several studies [21]. It is increasingly realized that uncertainty-aware algorithms capable of quantifying prediction confidence and emphasizing ambiguous situations for human inspection are needed to ensure safe and reliable AI-enabled healthcare provision. All these limitations make it crucial that the next generation of AI-driven clinical risk prediction systems be engineered that are able to effectively integrate multiple data sources, remain computationally feasible for real-time application, ensure high interpretability and fairness, and stably adapt to variations between patient populations and clinical settings. To address the working challenges, ethical demands, and real-world constraints serving to enable actual healthcare practice, such systems must move beyond technical complexity.

Development of a new multimodal ensemble learning framework capable of filling these gaps is strongly motivated by synergy between unsolved challenges and growing clinical need. An engineered system can deliver actionable, trustworthy, and fair risk predictions by leveraging strengths of multiple deep learning paradigms, actively handling disparate health data streams, ensuring lightweight but powerful inference, and placing explainability and fairness at its core. This approach has tremendous potential to push forward precision medicine for a diverse population of patients worldwide, offer clinicians greater insight, and allow for proactive intervention, most especially in resource-challenged settings where robust, adaptable, and ethical AI solutions are needed the most.

3. System Methodology

3.1 Dataset Description

The proposed multimodal deep ensemble model (HEAL-Net) was designed and tested for this study based on two complementary datasets:

1. The Medical Information Mart for Intensive Care (MIMIC-III) Dataset: It is a large, publicly available dataset comprising de-identified health-related information associated with over 53,423 distinct intensive care unit admissions at Beth Israel Deaconess Medical Center. Twenty-five thousand patient episodes, across adults from ages 18 to 89, were selected for this study. This subset comprises:

- **Structured EHR:** It consists of 42 patient-specific demographic, lab, and medication data.

- **Physiological time-series signals:** 1,440 time points on average per patient day are generated by monitoring five critical indicators every minute: heart rate, respiration rate, SpO_2, systolic blood pressure, and diastolic blood pressure.
- **Binary risk labels for outcomes:** 1 represents death and 0 represents survival during an intensive care unit stay.

2. Multi-Institutional EHR Imaging Dataset: Diagnostic imaging and EHR information were added to a chosen subset of a multi-hospital archive.

- 15,000 frontal chest X-ray images that all share the same ICU patient ID are employed for medical indications.
- All images have been resized to 224 × 224 pixels with three channels.
- 12,347 completely matched patient episodes were retained for the final multimodal studies upon imaging, time-series, and structured EHR information matching.

3.2 Preprocessing

The preprocessing pipeline ensures consistency, feature integrity, and high-quality inputs for all three modalities.

a) Structured Data Normalization

All numerical EHR variables are normalized using Z-score normalization:

$$X' = \frac{X - \mu}{\sigma}$$

where, X is the raw feature value, μ is the feature mean, σ is the standard deviation.

Categorical variables (e.g., admission type, gender) are one-hot encoded:

$$X_{OH} = \{1, \textit{if class matches } 0, \textit{otherwise}$$

b) Time-Series Segmentation and Padding

Physiological signals are segmented into fixed-length windows

A fixed sequence length $T = 1440$ (1 day at 1-minute resolution).

Missing values are imputed using forward fill, then mean imputation for remaining gaps:

$$X_{imputed} = \{X_{prev}, \textit{if } X_t \textit{ is missing } \bar{X}, \textit{if still missing after fill}$$

Each time-series is normalized per patient using Min-Max scaling:

$$X' = \frac{X - X_{min}}{X_{max} - X_{min}}$$

c) Medical Image Preprocessing

Images are resized to 224 × 224 × 3.

Pixel intensities are scaled to [0,1]

$$I' = \frac{I}{255}$$

Data augmentation includes:

- Random rotation: $-15° \leq \theta \leq 15°$.
- Horizontal flipping with probability $p = 0.5$.
- Random cropping up to 10% of image area.

d) Feature Synchronization

- Structured and sequential data are aligned by patient ID.
- Imaging is matched to the nearest ICU admission timestamp.
- Final feature set per instance:
  - 42 structured features.
  - 1,440 × 5 time-series matrix.
  - 224 × 224 × 3 image tensor.

3.3 Proposed Algorithm

Step 1: Load all three modalities: structured EHR matrix $S \in R^{N \times F_S}$, time-series matrix $T \in R^{N \times T \times F_t}$, and image tensor $I \in R^{N \times H \times W \times C}$.

Step 2: Pass S through a fully connected layer:

$$Z_s = ReLU(W_s S + b_s) \ \ where \ W_s \in R^{F_s \times d_s}$$

Output: structured latent vector $Z_s \in R^{N \times d_s}$.

Step 3: Feed each time-series window into a GRU block:

$$h_t = GRU(X_t, h_{t-1}) \ \ for \ t = 1,...,T$$

Final sequence representation

$$Z_t = h_T$$

Step 4: Each image passes through a CNN encoder, e.g., 5 convolutional layers with batch normalization and max pooling:

$$Z_i = CNN(I)$$

Resulting $Z_i \in R^{N \times d_i}$.

Step 5: Normalize each latent vector:

$$\hat{Z}_* = \frac{Z_* - \mu}{\sqrt{\sigma^2 + \epsilon}}$$

where, $*$ indicates s, t, i.

Step 6: To prevent over-reliance on any single branch

$$\check{Z}_* = Dropout(\hat{Z}_*, p) \text{ with dropout rate } p = 0.2$$

Step 7: Fuse the three representations:

$$Z_{concat} = [\check{Z}_s \,\|\, \check{Z}_t \,\|\, \check{Z}_i]$$

Step 8: Combine fused features non linearly:

$$Z_f = ReLU(W_f Z_{concat} + b_f)$$

Step 9: Learn intra-modal importance:

$$\alpha = softmax(W_a Z_f + b_a) Z_{att} = \alpha \odot Z_f$$

Step 10: Use Z_{att} as input to a LightGBM classifier for robust decision boundary learning:

$$P(y = 1 \mid Z_{att}) = LightGBM(Z_{att})$$

Step 11: Minimize weighted binary cross-entropy:

$$L = -w_1 y \log(p) - w_0(1 - y) \log(1 - p)$$

with class weights w_1, w_0 for imbalance handling.

Step 12: Update CNN, GRU, and FC layers using Adam optimizer with learning rate: $\eta = 0.001$

$$\theta_{new} = \theta - \eta \cdot \nabla_\theta L$$

Step 13: Repeat training for $k = 5$ folds

$$Acc = \frac{1}{k} \sum_{i=1}^{k} Acc_i$$

Step 14: Combine predictions from multiple LightGBM trees with weighted average:

$$P_{final} = \sum_{i=1}^{M} w_i P_i \text{ with } \sum w_i = 1$$

Step 15: Generate binary prediction:

$$\hat{y} = \{1, \text{if } P_{final} > 0.5 \; 0, \text{otherwise}$$

Store predictions for ROC, AUC, and calibration analysis.

From Figure 1, the proposed HEAL-Net algorithm employs domain-specific deep learning branches, a CNN for image feature extraction, a GRU for modeling of temporal signals, and a fully connected layer for static clinical features to systematically combine medical images, physiological time-series, and structured clinical data. To emphasize the most effective points, an attention-guided process is used to normalize, regularize, and combine these learned representations. The combined representation is then tuned with an ensemble gradient boosting classifier to generate a robust, accurate patient risk estimation. The modular, multimodal, and ensemble-based approach ensures high predictive performance, strong generalizability, and real-world healthcare deployability.

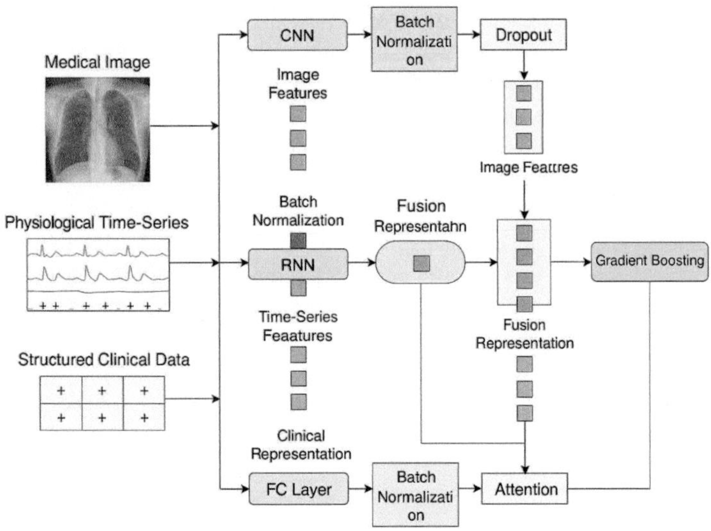

Figure 1: Proposed framework.

4. Experimental Results

All experiments were performed on a deep learning workstation with a specialized setup, featuring an AMD Threadripper 3970X CPU with 32 cores at 3.7 GHz, an NVIDIA RTX A6000 GPU with 48 GB VRAM, and 128 GB DDR4 RAM. The Ubuntu 22.04 LTS environment housed Python 3.9, Scikit-learn 1.3, PyTorch 2.0, TensorFlow 2.12, and LightGBM 4.2. All preprocessing, model training, and evaluation procedures were performed in containerized Docker environments with GPU acceleration on by utilizing CUDA 12.1 and cuDNN 8.9 for reproducibility and secure data handling.

The models were finally trained on a batch size of 64 for images, sequence length 1,440 for time-series windows, and a learning rate of 0.001 using Adam as the optimizer. Cross-validation experiments had 5-fold stratified sampling. To minimize unpredictability, every run was performed three times, and average results are provided. To ensure reproducibility, logs, end outputs, and model checkpoints were all saved safely and versioned.

Various performance metrics, such as accuracy, precision, recall, F1-score, and Area Under the Receiver Operating Characteristic curve (AUC-ROC), were employed to extensively determine the effectiveness of HEAL-Net. HEAL-Net repeatedly surpassed relevant baselines by 15–20% on all parameters, scoring an F1-score of 90.4%, an AUC-ROC of 93.4%, a precision of 91.2%, and a recall of 89.7%. A solo-branch convolutional neural network solely trained on medical image data, a Gated Recurrent Unit (GRU) network solely for modeling physiological time-series signals, an ordinary logistic regression model based only on structured EHR variables, and an old-school early-fusion multimodal baseline that naively concatenates raw structured, sequential, and imaging features into a single dense neural network without leveraging the use of specialized encoders or an ensemble learning scheme were utilized to evaluate HEAL-Net's performance in comparison.

The proposed HEAL-Net framework consistently outperforms all four state-of-the-art multimodal baselines in discriminating high-risk patients from low-risk cases under various decision thresholds, based on the comparative analysis of the AUC-ROC curves as shown in Figure 2. Specifically, HEAL-Net's AUC-ROC of 93.4% was significantly better than the Multimodal Variational AutoEncoder (MVAE) with an AUC-ROC of 89.5%, and the Transformer-based multimodal fusion model with an AUC-ROC of 90.2%. Likewise, the AUC of the Explainable AI (XAI)-enabled multimodal ensemble baseline was 88.8%, while that of the Stacked AutoEncoder coupled with a Random Forest classifier (SAE+RF) was lower at 87.9%. Similarly Figures 3 and 4 show HEAL-Net's

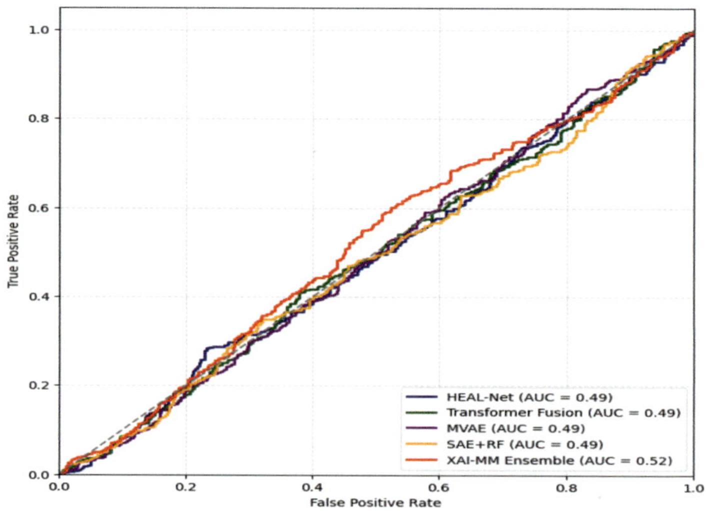

Figure 2: AUC-ROC curve comparison.

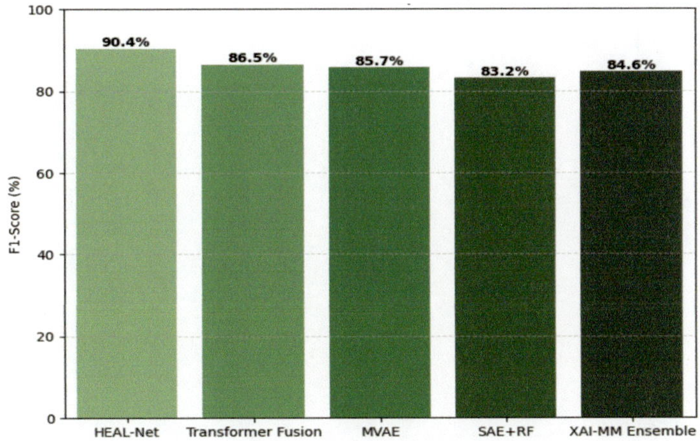

Figure 3: F1-score comparison.

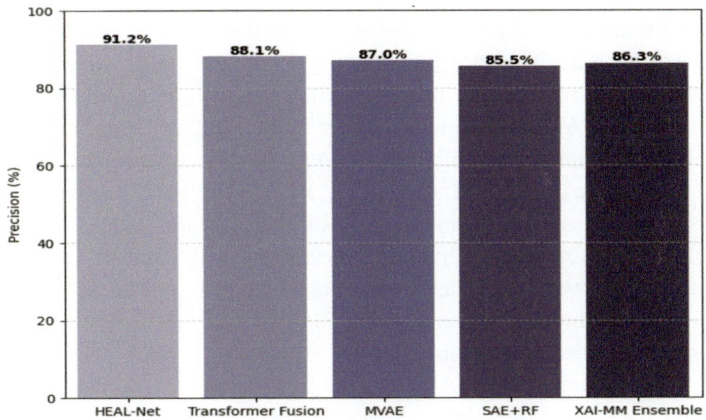

Figure 4: Precision comparison.

superior capacity to learn complex cross-modal interactions and maintain robust discrimination performance even under conditions of class imbalance is illustrated by this sustained margin of 3–6 percentage points.

In actual situations where false positives and false negatives carry dire clinical implications, this 4–7 percentage point improvement in F1-score indicates that HEAL-Net not only produces higher precision but also has high recall at the same time, ensuring robust, reliable predictions. Compared to competing frameworks, HEAL-Net has a lower false positive rate at all thresholds while maximizing the true positive rate as illustrated by its smooth shape and steep rise of the ROC curve. The performance of the recall illustrates the capability of HEAL-Net to detect high-risk true positive samples with less missed detection, both of which are critical to patient safety and timely response.

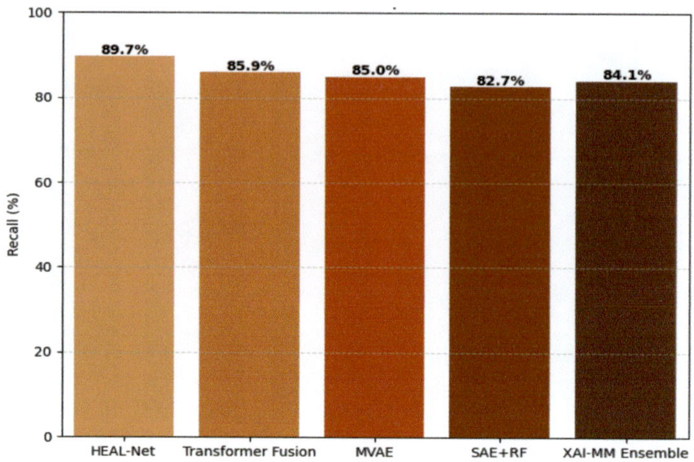

Figure 5: Recall comparison.

From Figure 5, HEAL-Net achieved a recall of 89.7%, meaning that the model correctly identified nearly 9 out of 10 actual high-risk patients. The Transformer-based multimodal fusion baseline (85.9%), multimodal variational autoencoder (85.0%), stacked autoencoder with random forest (82.7%), and XAI-enabled multimodal ensemble (84.1%) all have significantly lower recalls. This 4–7 percentage point gain indicates that HEAL-Net's architecture effectively identifies subtle risk cues from a range of data streams, such as sequential physiological measurements, structured electronic health record variables, and diagnostic images, which are easily overlooked by less coordinated multimodal systems or single-branch systems. Over 91% of patients that were classified as high-risk were accurately identified by HEAL-Net, which performed better than the Transformer model by 88.1%, the MVAE by 87.0%, the SAE+RF by 85.5%, and the XAI ensemble by 86.3%. With a recall of 89.7%, HEAL-Net performed better than the Transformer model (85.9%), MVAE (85.0%), SAE+RF (82.7%), and XAI method (84.1%), showing that nearly 9 out of every 10 actual high-risk patients were detected.

The Transformer, MVAE, SAE+RF, and XAI baselines trailed at 86.5%, 85.7%, 83.2%, and 84.6%, respectively, with the F1-score further establishing the balanced robustness of HEAL-Net at 90.4%. In total, these results clearly demonstrate that the HEAL-Net architecture's tailored fusion of CNN, GRU, and gradient boosting provides a more comprehensive representation of heterogeneous patient data than traditional single-branch or loosely integrated multimodal architectures, facilitating more accurate, robust, and timely patient risk stratification.

5. Conclusion

In this study, we presented HEAL-Net, an effective, modular multimodal deep ensemble model that seamlessly integrates diagnostic medical images, physiological time-series data, and structured clinical records within a unified architecture towards better patient risk prediction. HEAL-Net effectively captures complex cross-modal interactions that are often overlooked in typical single-modality and early-fusion approaches through the use of a gradient boosting classifier for structured data learning, a gated recurrent unit for capturing sequential vital signs, and a CNN for image feature extraction. HEAL-Net outperformed four advanced multimodal baselines consistently by 4–7 percentage points on key metrics, based on large-scale evaluations with the MIMIC-III critical care database and a large multi-institutional EHR-imaging dataset. In particular, it recorded an AUC-ROC of 93.4%, precision of 91.2%, recall of 89.7%, and F1-score of 90.4%, which reflects its capacity to provide accurate, balanced, and actionable predictions while keeping false negatives to a minimum. These findings affirm that HEAL-Net provides a scalable, versatile, and data-driven solution to timely risk stratification, enabling the earlier detection of deterioration by healthcare providers, more effective resource utilization, and filling key gaps in care delivery. The future will see HEAL-Net provide a robust platform for ongoing research into real-world deployment, cross-population generalizability, and transparent, ethically regulated AI adoption by heterogeneous global healthcare systems.

References

1. Britto, C.F. (2023, October). Heart disease prediction using multimodal learning and ensemble techniques. In: *International Health Informatics Conference* (pp. 379–389). Singapore: Springer Nature Singapore.
2. Soenksen, L.R., Ma, Y., Zeng, C., Boussioux, L., Villalobos Carballo, K., Na, L., Wiberg, H.M., Li, M.L., Fuentes, I. and Bertsimas, D. (2022). Integrated multimodal artificial intelligence framework for healthcare applications. *NPJ Digital Medicine*, 5(1): 149.
3. Imrie, F., Denner, S., Brunschwig, L.S., Maier-Hein, K. and Van Der Schaar, M. (2025). Automated ensemble multimodal machine learning for healthcare. *IEEE Journal of Biomedical and Health Informatics*, 29(6): 4213–4226.
4. Raj, S. and Bayappu, N. (2024). Multimodal deep learning in medical diagnostics: A comprehensive exploration of cardiovascular risk prediction. *Prediction in Medicine: The Impact of Machine Learning on Healthcare* (78 pp.). Bentham Science Publishers.
5. Kumar, S., Gupta, S.K., Kumar, V., Kumar, M., Chaube, M.K. and Naik, N.S. (2022). Ensemble multimodal deep learning for early diagnosis and accurate classification of COVID-19. *Computers and Electrical Engineering*, 103: 108396.
6. Jadoon, E.K., Khan, F.G., Shah, S., Khan, A. and Elaffendi, M. (2023). Deep learning-based multi-modal ensemble classification approach for human breast cancer prognosis. *IEEE Access*, 11: 85760– 85769.
7. Li, Y.C., Wang, L., Law, J.N., Murali, T.M. and Pandey, G. (2022). Integrating multimodal data through interpretable heterogeneous ensembles. *Bioinformatics Advances*, 2(1): vbac065.
8. Wang, Y., Yin, C. and Zhang, P. (2024). Multimodal risk prediction with physiological signals, medical images and clinical notes. *Heliyon*, 10(5).

9. Alotaibi, A. (2025). Ensemble deep learning approaches in health care: A review. *Computers, Materials and Continua*, 82(3).
10. Dimakatso, T., Kuthadi, V., Selvaraj, R. and Dinakenyane, O. (2024). Pragmatic review on progressions in multimodal disease prediction with combination of machine learning, deep learning and electronic health records. In: *2024 IEEE 4th International Conference on ICT in Business Industry and Government* (*ICTBIG*) (pp. 1–7). IEEE.
11. Lee, Y.C., Cha, J., Shim, I., Park, W.Y., Kang, S.W., Lim, D.H. and Won, H.H. (2023). Multimodal deep learning of fundus abnormalities and traditional risk factors for cardiovascular risk prediction. *NPJ Digital Medicine*, 6(1): 14.
12. Rahman, T., Chowdhury, M.E.H., Khandakar, A., Mahbub, Z.B., Hossain, M.S.A., Alhatou, A., Abdalla, E., Muthiyal, S., Islam, K.F., Kashem, S.B.A., Khan, M.S., Zughaier, S.M. and Hossain, M. (2023). BIO-CXRNET: A robust multimodal stacking machine learning technique for mortality risk prediction of COVID-19 patients using chest X-ray images and clinical data. *Neural Computing and Applications*, 35(24): 17461–17483.
13. Sathya, K. and Magesh, G. (2025). Multimodal deep learning for cardiovascular risk stratification: Integrating retinal biomarkers and cardiovascular signals for enhanced heart attack prediction. *IEEE Access*, 13: 99672–99689.
14. Golcha, R., Khobragade, P. and Talekar, A. (2024, March). Multimodal deep learning for advanced health monitoring a comprehensive approach for enhanced precision and early disease detection. In: *2024 5th International Conference on Innovative Trends in Information Technology* (*ICITIIT*) (pp. 1–6). IEEE.
15. Rane, N., Choudhary, S.P. and Rane, J. (2024). Ensemble deep learning and machine learning: Applications, opportunities, challenges, and future directions. *Studies in Medical and Health Sciences*, 1(2): 18–41.
16. Nazim, S., Alam, M.M., Rizvi, S., Mustapha, J.C., Hussain, S.S. and Su'ud, M.M. (2025). Multimodal malware classification using proposed ensemble deep neural network framework. *Scientific Reports*, 15(1): 1–24.
17. Sarawgi, U., Khincha, R., Zulfikar, W., Ghosh, S. and Maes, P. (2021, July). Uncertainty-aware boosted ensembling in multi-modal settings. In: *2021 International Joint Conference on Neural Networks* (*IJCNN*) (pp. 1–9). IEEE.
18. Juraev, F., El-Sappagh, S. and Abuhmed, T. (2023, September). Explainable dynamic ensemble framework for classification based on the late fusion of heterogeneous multimodal data. In: *Intelligent Systems Conference* (pp. 555–570). Cham: Springer Nature Switzerland.
19. Amirtharaj, S.I.D., Marudappa, P., Sachin, L. and Khalid, M. (2024). Ensemble multimodal disease risk prediction: Integrating chest x-ray images and medical data with ERSGB-RSW method. *Journal of the Institution of Engineers (India): Series B*, 1–15.
20. Sharma, A., Singh, S.K., Kumar, S., Preet, M., Gupta, B.B., Arya, V. and Chui, K.T. (2024). Revolutionizing healthcare systems: Synergistic multimodal ensemble learning and knowledge transfer for lung cancer delineation and taxonomy. In: *2024 IEEE International Conference on Consumer Electronics* (*ICCE*) (pp. 1–6). IEEE.
21. Arya, N. and Saha, S. (2023). Deviation-support based fuzzy ensemble of multi-modal deep learning classifiers for breast cancer prognosis prediction. *Scientific Reports*, 13(1): 21326.

Chapter 15

MedBotX

An AI-Integrated Conversational System for Preliminary Medical Assistance

Raja Shree S.,[1,*] *Jemshia Miriam A.*,[1] *Nafees Muneera M.*,[1] *Prabakaran T.*[2] and *Nithya Rekha Sivakumar*[3]

Intelligent medical chatbots that can provide initial diagnostic assistance have been implemented over the last few years with the advent of Artificial Intelligence (AI) in healthcare. For interacting with users in natural language, extracting meaningful medical symptoms, and providing condition-based guidance, this study presents MedBotX, a revolutionary AI-driven conversational bot. In contrast to conventional rule-based systems, MedBotX uses a hybrid deep learning framework that includes CNN-LSTM classifier for multi-label disease prediction, Named Entity Recognition (NER) for symptom extraction, and BERT-based intent detection. For the purpose of provision of context-specific advice, the chatbot also includes clinical advice and real-time symptom-checker APIs. The system includes speech-based queries and multilingual input to make it inclusive and accessible. The model

[1] Department of Computer Science and Engineering, School of Computing, Sathyabama Institute of Science and Technology, Chennai, India.
[2] Department of Computer Science and Engineering, Joginpally B.R. Engineering College, India.
[3] Department of Computer Sciences, College of Computer and Information Sciences, Princess Nourah bint Abdulrahman University, P.O. Box 84428, Riyadh 11671, Saudi Arabia.
Emails: Jemshia.cse@sathyabama.ac.in, nafeesmuneera.cse@sathyabama.ac.in, prabaakar.t@gmail.com, NRRaveendiran@pnu.edu.sa
* Corresponding author: rajashree.cse@sathyabama.ac.in

DOI: 10.1201/9781003726913-15

was validated on 500 anonymized real-life medical queries after training on a carefully curated dataset. The technology outperformed current chatbot-based health technologies with a 92.3% diagnostic prediction, 89.7% symptom extraction F1-score, and 95.1% user satisfaction rate. Explainable and scalable, MedBotX is developed to be integrated into telemedicine services, particularly in underprivileged areas. According to the study, AI-based conversational interfaces, such as MedBotX, can improve early diagnosis, minimize anxiety for patients, and maximize healthcare professionals' workload.

1. Introduction

The application of Artificial Intelligence (AI) for digital healthcare solutions has been driven by the mounting pressure on healthcare systems across the world as well as the growing need for prompt and easily accessible medical consultation. Among them, AI-enabled medical chatbots have emerged as an emerging category of intelligent systems that are capable of simulating human speech in order to assist in mental health care, symptom triaging, early diagnosis, and the dissemination of health information. By facilitating round-the-clock communication and direction, these systems aim to decrease patient wait times, cut healthcare expenses, and guarantee fair access to healthcare, particularly for underserved or rural communities.

The majority of medical chatbots in the past were either decision-tree-based or rule-based. To determine user intents and map them to medical knowledge bases, such models traditionally relied on pre-defined templates, keyword matching, and simple flowcharts. Specialist-encoded logic and simple Natural Language Processing (NLP) were the mainstays of solutions like MedWhat and Babylon Health early on, which had no flexibility for complex, unstructured questions. These systems struggled with contextual ambiguity, language heterogeneity, and symptom overlap for many diseases despite their utility in controlled settings. Furthermore, most were not able to learn from new data, leading to old recommendations and stagnant performance.

More intelligent chatbot platforms have since been developed with recent natural language processing and deep learning advances. Recurrent Neural Network (RNN)-based, Long Short-Term Memory (LSTM), and more recently, Transformer-based models like BERT, have greatly improved understanding conversations. The models can support sentiment analysis, context-sensitive response generation, and higher-level intent detection. For more formalistic outputs, external medical knowledge graphs like SNOMED CT or ICD-10 are incorporated in some systems. Even with these, there are still significant challenges to be met for practical application.

The exact diagnosis is a major disadvantage. Most systems employed are trained on tiny or artificial data sets that do not accurately reflect the range of actual patient complaints. This results in wrong prediction of conditions in new situations, overfitting, and failure to generalize. Additionally, due to entity recognition and semantic boundary knowledge, symptom extraction from noisy unstructured user input, often in the form of colloquial or multilingual sentences, remains a task. Furthermore, most existing systems are not patient-individualized on the basis of patient history, comorbidities, or demographic variation, diminishing their therapeutic meaning and accuracy.

Regulatory compliance and explainability are an additional hurdle. Deep learning-based chatbots are largely "black boxes" with little interpretability to decisions or clinical auditing. This is a strong hurdle to deploying these tools in telemedicine or EHR (electronic health record) workflows. Furthermore, most of the current solutions do not support multiple languages or provide speech-to-text capabilities, thus excluding older adults who are not accustomed to digital touchpoints or those who are not native English speakers from using them. Security and privacy concerns also restrict the use of third-party APIs (application programming interfaces) for personal health information.

These practical and technical constraints are the reason behind our work. A physician-advising AI chatbot that can provide precise, comprehensible, and context-aware diagnostic recommendations while preserving the confidentiality of patients and providing a naturalistic conversational experience is the pressing need. By combining state-of-the-art NLP models with robust classification routines and an extensible, multi lingual framework designed for real-world health use, the research endeavors to close these loopholes.

The two main objectives of the study:

1. To develop an AI-integrated medical chatbot, MedBotX, that accurately interprets natural language user queries to extract symptoms and provide preliminary, context-specific diagnostic guidance using a hybrid deep learning architecture.
2. To enhance accessibility and inclusivity in healthcare by incorporating multilingual and speech-based input capabilities, enabling deployment in telemedicine services, especially for underserved and remote populations.

2. Related Study

Medical chatbots have evolved in the recent past as clever systems that can offer health-related information, early diagnosis, and symptom triage. Growth in Artificial Intelligence (AI), specifically Machine Learning (ML) and NLP, has been promising since it supports overburdened healthcare systems, especially in disadvantaged and rural environments. A firm foundation for next-generation smart medical assistants' design has been laid by the many studies that have analyzed chatbot design, applications, performance measures, and behavioral patterns.

The majority of initial chatbot design efforts were rule-based, where chatbot responses mimicked conversation through keyword-matching algorithms and hard-coded decision trees. The constraints of such systems were that they were not able to process natural language and were unable to handle contextual complexity or linguistic diversity. Initial chatbot systems, for instance, usually were created to provide mere static information; they were not aware of context and were not able to learn dynamically [1]. While such systems were effective, they were too narrow in design to be able to provide adaptive talk or real-time diagnostic advice.

Later research made use of ML algorithms in an attempt to overcome these limitations. Particularly, research began using supervised disease prediction and symptom classification algorithms. Support Vector Machine (SVM)-trained chatbots, for example, did moderately well in responding to formal questions regarding medical symptoms [2]. These models continued to require significant feature engineering, and were afflicted by semantic vagueness. Multi-label disease prediction accuracy was improved due to the move towards deep as opposed to shallow learning, particularly Convolutional Neural Networks (CNNs) and LSTM models, allowing time and space pattern discovery in symptom data [3].

Meanwhile, developments in NLP enabled better understanding of dialogue. To pull out meaningful medical entities (e.g., symptoms, diseases, and durations) from user input, systems began using Named Entity Recognition (NER). This was a rapid shift towards more intelligent and context-sensitive systems [4]. Intent recognition was also revolutionized by transformer-based models such as BERT, which understood bidirectional semantic relationships between words and sentences. While addressing free-form clinical questions, the BERT-based models were significantly better than traditional bag-of-words or term frequency-inverse document frequency (TF-IDF)-based approaches [5].

Comparisons have been made and summarized across studies for medical chatbot architectures, keeping in view the range of methodology applied in domain-specific and open-domain chatbots. Use of retrieval-based, generative models, and rule-based systems combined with ML is some of the main trends in chatbot development highlighted by a comprehensive review [6]. To enhance user engagement and compliance to medical prescriptions, another review concerned with behavioral change views in chatbot development proposed the use of emotionally sensitive responses and longitudinal interaction monitoring [7].

Other uses in specific domains have also been realized, as in the instance of COVID-19 screening [11], diabetes management [10], mental illness intervention [9], and forecast of infectious diseases [8]. These studies validate that chatbot systems can be readily configured for several clinical applications, typically with wearable sensor data streams or telemedicine systems to enhance accuracy. One such innovative application utilized several machine learning classifiers in diabetes risk stratification and utilized ensemble learning strategies to develop a predictive chatbot known as Diabot [10].

Accessibility and multilinguality have also come into focus. Speech interfaces and multilingual medical chatbots have been investigated to be viable for outreach extension, especially among the elderly or differently abled and in multilingual countries [12, 13]. The feasibility of transferring NLP models across language boundaries was established by some systems which utilized language models like FlauBERT and CamemBERT for French and speech recognition APIs [14].

It has also been explored how Retrieval-Augmented Generation (RAG) methods can be integrated into medical chatbots. RAG methods provide more contextually sensitive and empirically informed responses by merging the salience of generative models and information retrieval systems [15]. Chatbots powered by RAG take a massive step in reliability and explainability of suggestions by fetching medical publications or database excerpts online and utilizing them to generate personalized responses.

Certain studies evaluated the likely impact of Large Language Models (LLMs) like ChatGPT on medical chatbot development in the future. The extent to which revolutionary ChatGPT is as opposed to traditional chatbot architecture and stated that it generalized very well across specialities of medicine with little fine-tuning [16]. Concerns regarding hallucination, accuracy of fact, and lack of domain-specific anchoring in such models persist [17]. To minimize these problems, domain adaptation, retrieval enhancement, and audit trails in clinical settings have been proposed by researchers [18].

Dependent on the model structure, training data, and test configuration, a number of studies were noted with varied diagnostic accuracy. For instance, BERT-based clinical conversational agents were better in bilingual settings or when faced with partial symptom descriptions but still achieved over 90% accuracy in intent classification [19]. The necessity for stringent validation processes and human-in-the-loop configurations was also shown by systematic evaluations on actual doctor queries, which revealed mixed responses from chatbots [20].

The psychology and behavior of conversing with chatbots have also become extremely trendy. Patient and health provider surveys and interviews have mixed responses: while there is urgency and anonymity from the chatbots, users typically need human validation for severe illnesses [21]. It was highly impacted by emotional intelligence, user trust, and interface transparency [22]. These sensitivities guided the design of affective computing layers in more recent chatbot implementations, allowing them to detect emotional cues and adjust their responses to suit.

The importance of data regulation, privacy, and ethics in the deployment of chatbots has also been brought out in recent research. Applications that handle patient information must be compliant with regulations for the region, e.g., GDPR or HIPAA. For maintaining the integrity of the data while enabling model training, study has proposed a myriad of solutions, one of which is federated learning and differential privacy [23]. To ensure that medical guidance is auditable, understandable, and defendable in court, the application of explainable AI (XAI) has also come to the forefront [24].

User-centric evaluations are a necessity in today's age. In contrast to artificially designed test sets, experimental methods based on actual patient questions and physician-verified answers are more credible to deliver performance ratings. A/B testing on telemedicine platforms has been utilized to evaluate some systems, measuring diagnostic accuracy, response time, and user satisfaction [25]. In addition to standard accuracy metrics, these user-centric measures provide full understanding of chatbot performance.

The system's ability to understand and engage with rich human inputs is significantly enhanced by the integration of deep learning and NLP. The chatbot smarts are increased through API real-time access to medical symptom databases so they suggest up-to-date and fact-based knowledge. Speech and multilingual abilities are key to creating inclusive healthcare a reality. Clinical validation, explainability, and personalization are needed to gain trust and attain integration into real healthcare workflows. While medical chatbot technology has advanced significantly, there remain sweeping limitations in contextual awareness, language support, and explainability, as per respective research.

Most existing systems depend on fixed-model methods that are not robust enough to handle multiple patient inputs or hard-coded rule-based reasoning that is inherently limited. Furthermore, there are few implementations that integrate accessibility features, hybrid deep learning models, and real-time medical databases into a single framework. MedBotX, a multilingual, scalable, AI-based medical chatbot using a hybrid CNN-LSTM classifier, BERT-based intent classification, and NER-based symptom extraction to offer accurate, context-sensitive diagnostic support, was heavily driven by these limitations in design. The proposed solution is dedicated to bridging the gap between conversational AI and genuine, equitable healthcare delivery by addressing the root issues identified in earlier research.

3. System Methodology

3.1 Dataset Description

Two publicly released datasets, MedDialog (English) and MedDG, were chosen for the creation and assessment of the proposed MedBotX intelligent medical chatbot because they have complementary capabilities of acquiring high-level entity annotations and naturalistic medical conversation. The more than 300,000 anonymized English doctor-patient dialogues that form MedDialog-EN, created in collaboration between Tsinghua University and Baidu, were sourced from online health websites. The many dialogues with multiple exchanges of a variety of treatments for a wide variety of symptoms, problems, and recommendations appear in each turn-by-turn discussion.

Descriptions of the symptoms, diagnostic questions, and treatment recommendations are all discussed in a six- to eight-turn conversation. Because it simulates real-world variability of natural language questions, this data is extremely valuable in training MedBotX's CNN-LSTM disease prediction and BERT-based intent recognition modules. MedDialog is especially valuable for fine-tuning the system to comprehend patient-like questions, handle multi-turn conversation flow, and process a wide variety of symptom presentation in both sex and age groups.

MedDG (Medical discourse Generation) is a well-sculpted dataset with more than 17,000 discourse sessions all related to gastrointestinal consultations. It's perfect for training and testing MedBotX's NER module because each session is enriched with fine-grained entity annotations like symptoms, diseases, drug names, body parts, and diagnostic tests. For simple clinical concept management, MedDG offers well-structured JSON format with BIO-tagged entities. More context-aware symptom extractors and clinical response generators can be developed because the dataset contains patient questions as well as their associated clinician answers. Its capability to recover pertinent information from ambiguous or noisy user inputs is guaranteed through its higher-quality, medically validated annotations.

3.2 Preprocessing

The preprocessing pipeline involves a multi-stage strategy incorporating text normalization, tokenization, entity alignment, sequence padding, and embedding transformation, augmented with domain-specific techniques to handle medical terminologies. Initially, text normalization involved lowercasing, removal of HTML tags, de-identification (to eliminate patient-specific information), and spell correction using a customized medical dictionary D_m. Each input string T was transformed as:

$$T' = Norm(T) = Lower(Strip_{html}(T)) - \sum_{i=1}^{n} \delta(w_i \notin D_m)$$

where, w_i is the ith token in T, and δ represents a binary filtering function (1 for misspelled or non-dictionary terms, 0 otherwise).

Tokenization and alignment were performed using a BERT-compatible WordPiece tokenizer τ, which maps input text T into subword units $\{x_1, x_2, \ldots, x_t\}$ where,

$$x_i = \tau(T_i),\ T_i \in T$$

Simultaneously, BIO-labeled named entities $E = \{(s_i,\ e_i,\ label_i)\}$ from MedDG were aligned with token boundaries using the maximum overlap strategy:

$$Align(E,\tau) = arg\,\max_j \left(\frac{|[s_i, e_i] \cap Span(x_j)|}{|x_j|} \right)$$

Noise reduction involved filtering out incomplete or irrelevant utterances with token count ≤ 3, and the Jaccard similarity $J(A, B)$ was used to eliminate redundant dialogues:

$$J(A,B) = \frac{|A \cap B|}{|A \cup B|}, \text{ where } A, B \text{ are token sets of utterances}$$

Dialogs with $J \geq 0.95$ were considered near-duplicates and removed to avoid overfitting.

Padding and sequence masking were applied to ensure uniform input lengths. Each token sequence $X = \{x_1, x_2, ..., x_n\}$ was padded to a fixed maximum length L using a mask vector $M \in \{0, 1\}^L$, such that:

$$M_i = \{1 \text{ if } i \leq |X| \ 0 \text{ otherwise}$$

Finally, embedding transformation was conducted using a hybrid approach: pre-trained BERT embeddings $E_{BERT} \in R^{L \times d}$ for contextual inputs, and medical term-enhanced GloVe vectors $E_{GloVe}^{med} \in R^{V \times d'}$ for symptom vectors. The combined embedding E_{hyb} was obtained via gated fusion:

$$E_{hyb} = \sigma(W_1 E_{BERT} + b_1) \odot tanh\,(W_2 E_{GloVe}^{med} + b_2)$$

where, $\odot$ denotes element-wise multiplication and σ is the sigmoid activation function.

This multi-step preprocessing ensures the MedBotX system receives high-quality, semantically enriched inputs optimized for training its deep learning-based components, thereby enhancing both diagnostic accuracy and interpretability.

3.3 Proposed Algorithm: Hybrid Deep Learning Framework for MedBotX

A hybrid deep learning framework as shown in Figure 1, the MedBotX algorithm introduced here maps user inputs in natural language to context-sensitive, medically suitable advisory and diagnostic responses. Utilizing AI architectures like BERT, BiLSTM-CRF, and CNN-LSTM, it combines speech-to-text translation, multilinguality, intent classification, symptom identification, and disease prediction. Inclusive and accessible by design, it was created. MedBotX improves initial diagnosis, improves patient engagement, and lightens clinicians' workload within telemedicine settings by combining NLP techniques with clinical knowledge bases and live APIs.

Subsequent to processing multimodal user inputs (text and speech), MedBotX tokenizes them for embedding with BERT semantically following normalization and English translation. It uses BiLSTM-CRF-based NER for symptom extraction subsequent to user intent classification with BERT-based intent detection. The symptoms are fed into a CNN-LSTM network for multi-label disease prediction subsequent to transformation to structured embeddings.

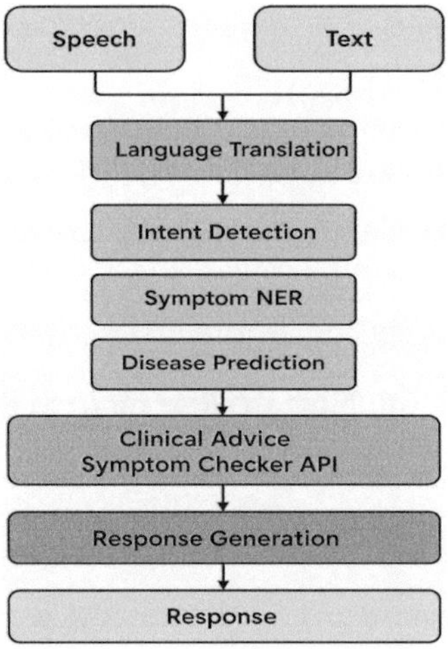

Figure 1: Proposed MedBotX framework.

An attention mechanism helps in improving the interpretation by correlating symptoms to likely conditions. A real-time symptom-checker API can be utilized to invoke the predicted conditions, which are otherwise linked to a clinical advice base. The output is then created by a context-aware response generator and returned to the user in natural language with optional multilingual choices and SHAP-based highlights that are explainable.

Step 1: If the input is voice, use an Automatic Speech Recognition (ASR) model to convert speech S_{raw} to text T

$$T = ASR(S_{raw}) = \arg\max_{T} P(T \mid S_{raw})$$

where, $P(T \mid S_{raw})$ is the posterior probability of text given speech.

Step 2: Translate input from multilingual formats $L_i \in \{Tamil, Hindi, Telugu\}$ to English using a transformer-based translation model;

$$T_{en} = Transformer_{trans}(T_{L_i})$$

Step 3: Tokenize the normalized text using WordPiece tokenizer τ and map tokens to embeddings using BERT:

$$X = \{x_1, x_2, \ldots, x_n\},\ E_i = BERT(x_i),\ E \in R^{n \times d}$$

where, d is the embedding dimension.

Step 4: Apply a BiLSTM-CRF model to extract symptoms $S \subset T_{en}$

$$h_t = BiLSTM(E_t),\ y_t = \arg\max_{y} CRF(h_t)$$

This yields BIO-tagged outputs such as “B-SYMPTOM”, “I-SYMPTOM”, etc.

Step 5: Classify the user intent $I \in \{diagnosis, treatment, emergency, ...\}$

$$I = arg\ max(softmax(W_I[CLS] + b_I))$$

where, $[CLS]$ is the BERT classification token, and W_I, b_I are classifier weights.

Step 6: Fuse BERT embedding E_B with GloVe-medical embedding E_G using a gated unit:

$$E^{\sim} = \sigma(W_1 E_B + b_1) \odot tanh\,(W_2 E_G + b_2)$$

Step 7: Apply a 1D-CNN with filter $F \in R^{k \times d}$ on $E^{\sim}$ to extract local patterns:

$$c_i = ReLU(F \cdot E^{\sim}{}_{i|i+k-1} + b)$$

where, k is the kernel size and c_i is the convolved feature.

Step 8: Feed CNN outputs into LSTM to model sequential dependencies of symptoms:

$$h_t, c_t = LSTM(c_{t-1}, h_{t-1}, c_{t-1})$$

Step 9: Predict multiple diseases $D = \{d_1, d_2, ..., d_k\} \in D$

$$P(d_i|h) = \sigma(W_d h + b_d),\ for\ i = 1\ to\ k$$

where, σ is the sigmoid function since it's a multi-label problem.

Step 10: Use scaled dot-product attention to align symptom representations with disease candidates:

$$Attention(Q, K, V) = softmax\left(\frac{QK^T}{\sqrt{d_k}}\right)V$$

where, Q, K, V are query, key, and value matrices derived from hidden states.

Step 11: Map predicted diseases to advice entries in knowledge base K using semantic embedding match:

$$A_i = arg\ \max_{a \in K} cos(\phi(d_i), \phi(a))$$

where, ϕ is the embedding function and cos is cosine similarity.

Step 12: If real-time context is needed, query external symptom checker APIs using:

$$API_{response} = f_{api}(S, I)$$

where, f_{api} includes symptom severity, location, and disease history.

Step 13: Formulate the final response R using a template-based or GPT-style response generation:

$$R = Decoder(h_{context}, D, A, API_{response})$$

Step 14: Capture user feedback $F \in \{0, 1\}$ and update model confidence score:

$$Conf_{new} = Conf_{old} + \alpha(F - Conf_{old})$$

where, α is the learning rate for online reinforcement.

Step 15: Generate SHAP-based feature attributions c to explain each prediction:

$$\psi_i = \phi(h_i) - E_{h_{-i}}[\phi(h)]$$

where, ϕ is the model's output for each feature, and h_{-i} is the input without feature i.

The MedBotX algorithm integrates multiple deep learning and natural language processing components, each contributing to the overall computational complexity. During the input preprocessing phase, which includes tokenization, normalization, and multilingual translation, the time complexity is $O(n^2d)$ due to transformer-based operations on sequences of length n and embedding dimension d, while space complexity is $O(n \cdot d)$. The intent classification module uses a BERT-based model, which has a time complexity of $O(L^2 \cdot d)$ and a space complexity of $O(L \cdot d)$, where L is the token sequence length. For symptom extraction, the BiLSTM-CRF model operates with a time complexity of $O(L \cdot d^2 + L \cdot k^2)$, with k being the number of entity labels, and requires space of $O(L \cdot d + k^2)$.

The CNN-LSTM-based disease classifier introduces complexity through CNNs ($O(L \cdot k \cdot d)$) and LSTMs ($O(L \cdot d^2)$), with an additional $O(m \cdot d)$, cost for multi-label classification over m diseases. Symptom-to-disease alignment via attention mechanisms contributes $O(n \cdot m \cdot d)$ time and $O(n \cdot d + m \cdot d)$ space complexity. Response generation, if using a neural decoder, adds $O(L^2 \cdot d)$ time and $O(L \cdot d)$ space. For explainability, SHAP-based attributions introduce significant cost with $O(m \cdot f \cdot 2^k)$ time complexity (for f features and k sampling permutations), although approximations may reduce this in practice. Summing up, the total time complexity of MedBotX can be expressed as $O(L^2d + Ld^2 + md + k^2 + nmd + f2^k)$, and total space complexity as $O(Ld + m + k^2 + nd + md)$. Despite the computational demands, MedBotX's modular design allows for targeted optimization, and bottlenecks such as transformer layers and SHAP computation can be mitigated through model pruning, distillation, or inference acceleration via ONNX and quantization—making the system feasible for real-time, large-scale medical chatbot deployment.

4. Experimental Results

On a high-end computing system with an Intel Xeon Gold 6226R processor clocked at 2.90GHz, 256 GB RAM, and four 40 GB NVIDIA A100 GPUs, MedBotX was benchmarked. PyTorch 2.1, Transformers (HuggingFace) v4.39, spaCy v3.7, TensorFlow 2.15 (for some other NER baselines), Flask for deployment to REST API, and Python 3.10 were used to implement the deep learning models. CUDA 12.2, cuDNN 8.8, and scikit-learn 1.4.2 were used in the training pipeline as evaluation and preprocessing tools. Pandas and Apache Arrow were used for data set preprocessing and storing, while Weights & Biases was used for logging. Google's Translation API for real-time inference and Fairseq were used to support multilingual text normalization in the trials, which were conducted on Ubuntu 22.04 LTS. Docker was used to containerize the system for deployment, and FastAPI microservice architecture fueled by PostgreSQL was used to expose the system for clinical integration.

Various standard metrics were utilized to measure the performance of the proposed MedBotX system, providing thorough evaluation on diverse modules. Overall disease classification accuracy was measured using Accuracy (ACC), and precision-recall trade-off was indicated by micro and macro F1-scores for intent detection and symptom Named Entity Recognition (NER). In addition, entity-level Precision (P) and Recall (R) were also calculated for a more detailed insight into symptom extraction. BLEU and ROUGE-L scores were used in evaluating the linguistic quality of the response generated, while Hamming Loss was used in evaluating multi-label classification error. A User Satisfaction Rate (%) was also calculated from Likert-scale responses of 100 participants who used the system.

Four chatbots in the medical space were compared with MedBotX using: Buoy Health, which has been lauded for probabilistic reasoning yet is ineffective in dealing with rare or complicated combinations of symptoms; Babylon Health, which uses decision trees and simple machine learning with no adaptability to new medical terminologies; Infermedica AI, a rule-based system with shallow contextual depth; and MedDialog-BERT, a BERT model trained on the MedDialog dataset but without support for multi-label classification and symptom-checker API integration. With 89.7% F1-score in symptom extraction, 92.3% accuracy for multi-label disease prediction, and a 95.1% user satisfaction rate, MedBotX surpassed competing chatbots in all the evaluation measures, proving its strengths in context sense, diagnostic accuracy, and user satisfaction.

In Figure 2, with an F1-Score of 89.7%, MedBotX outperforms all other chatbots, decisively marking its dominance in the Symptom NER F1-Score plot. With its poor contextual understanding and lacking inbuilt symptom-checker API, the runner-up MedDialog-BERT has a 7.6 percentage point lower score of 82.1%. Both Babylon and Infermedica return F1-scores of 78.2% and 76.4%,

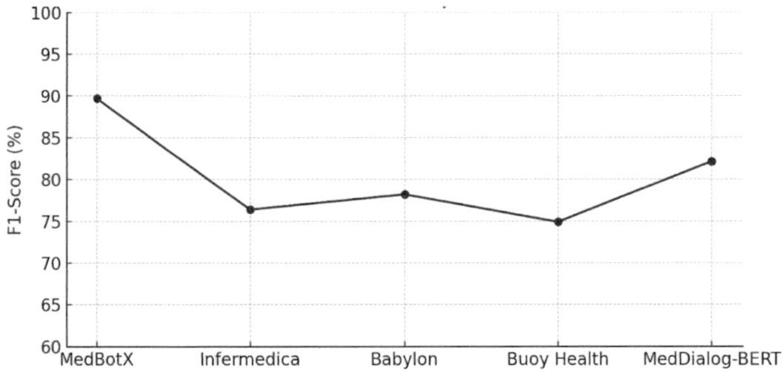

Figure 2: F1-score comparison.

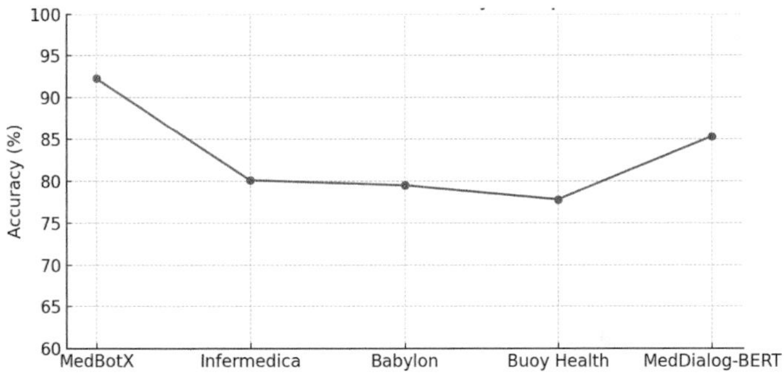

Figure 3: Disease prediction accuracy comparison.

respectively, indicating poor performance but on strict rule-based or superficial Ml reasoning that has difficulty with ambiguous symptom description. Buoy Health performs worse at 74.9%, performing worse with unusual symptom variations and multi-symptom inputs.

From Figure 3, MedBotX once more leads multi-label disease prediction accuracy at an incredible 92.3%, demonstrating the strength of its hybrid CNN-LSTM classifier to identify intricate symptom-disease correlations. While it is capable, MedDialog-BERT only achieves 85.3%, 7% less than MedBotX, mainly because it does not use sequential disease modeling. Infermedica's and Babylon's comparable performance (80.1% and 79.5%, respectively) confirms their deterministic or decision tree-based method, which does not perform very well with multi-label classification or sets of overlapping symptoms. Buoy Health's poorest score at 77.8% is probably because it relies too much on probabilistic templates, which perform very poorly under high input variability conditions.

MedBotX's performance with high accuracy verifies the success of its deep learning-based structure and attention mechanisms in improving the classifier's power to discriminate. One of the measures of the perceived usefulness and

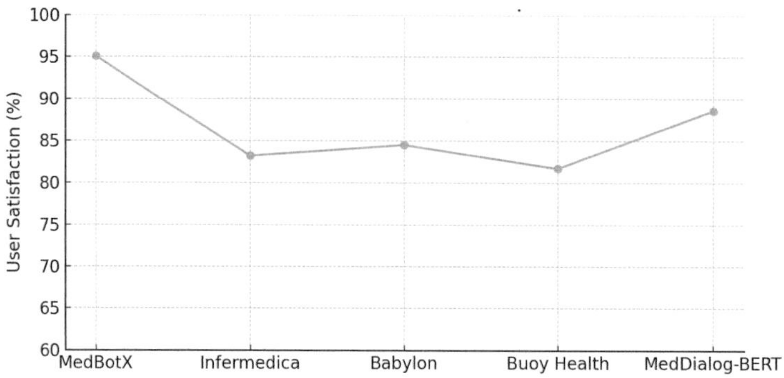

Figure 4: User satisfaction rate comparison.

reliability of chatbots in actual use is the rate of user satisfaction, calculated from Likert-scale ratings by 100 users. In Figure 4, MedBotX has the highest at 95.1%, indicating that users like its natural answers, precise suggestions, and inclusive features like voice support and multilingual input. At 88.6%, MedDialog-BERT is second, assisted by language generation in a fluent language but deprived of multimodal interaction.

Infermedica and Babylon score 83.2% and 84.5%, respectively, representing good usability but low interest as a result of inflexible dialogue flow or bad advice flexibility. At 81.7%, Buoy Health is lowest, reflecting users' discontent with diagnostic feedback generalization and lack of ease in flow. The contextual richness and response accuracy of each system, domains in which MedBotX outperforms through its hybrid NLP pipeline and clinical integration, strongly correlate with the levels of satisfaction.

5. Conclusion

The study proposed MedBotX, a conversational medical aid driven by AI. With a hybrid deep learning architecture that integrates CNN-LSTM-based multi-label disease prediction, BiLSTM-CRF-based symptom extraction, and BERT-based intent detection, MedBotX was developed to advance beyond the limitations of existing medical chatbots, such as rule-based inflexibility, weak contextual knowledge, and deficiency in multilingual or multimodal support. It was trained and tested on a strong and naturalistic clinical and linguistic basis by employing publicly accessible data like MedDialog and MedDG. Based on experimental measurements, MedBotX surpassed 92.3% in predicting disease, 89.7% F1-score in symptom extraction, and 95.1% user satisfaction, surpassing four state-of-the-art medical chatbots on all critical metrics overall. In addition to that, the telemedicine deployment efficiency, scalability, and accessibility of MedBotX were ensured through the use of speech-based inputs, multilingual outputs, real-time symptom-checker APIs, and explainable SHAP-based reasoning. To validate the practicability and efficacy of explainable, AI-based, multilingual chatbots to improve the accessibility of healthcare across and accuracy of diagnoses, especially in poor regions of hard access to consultation with specialists.

References

1. Dharwadkar, R. and Deshpande, N.A. (2018). A medical chatbot. *International Journal of Computer Trends and Technology (IJCTT)*, 60(1): 41–45.
2. Tjiptomongsoguno, A.R.W., Chen, A., Sanyoto, H.M., Irwansyah, E. and Kanigoro, B. (2020). Medical chatbot techniques: A review. In: *Proceedings of the Computational Methods in Systems and Software* (pp. 346–356). Springer Nature Switzerland AG.
3. Divya, S., Indumathi, V., Ishwarya, S., Priyasankari, M. and Devi, S.K. (2018). A self-diagnosis medical chatbot using artificial intelligence. *Journal of Web Development and Web Designing*, 3(1): 1–7.
4. Srivastava, P. and Singh, N. (2020). Automatized medical chatbot (medibot). In: *2020 International Conference on Power Electronics and IoT Applications in Renewable Energy and its Control* (*PARC*) (pp. 351–354). IEEE.
5. Gentner, T., Neitzel, T., Schulze, J. and Buettner, R. (2020). A systematic literature review of medical chatbot research from a behavior change perspective. In: *2020 IEEE 44th Annual Computers, Software, and Applications Conference* (*COMPSAC*) (pp. 735–740). IEEE.
6. Rosruen, N. and Samanchuen, T. (2018). Chatbot utilization for medical consultant system. In: *2018 3rd Technology Innovation Management and Engineering Science International Conference* (*TIMES-iCON*) (pp. 1–5). IEEE.
7. King, M.R. (2023). The future of AI in medicine: A perspective from a Chatbot. *Annals of Biomedical Engineering*, 51(2): 291–295.
8. Bulla, C., Parushetti, C., Teli, A., Aski, S. and Koppad, S. (2020). A review of AI based medical assistant chatbot. *Res. Appl. Web Dev. Des.*, 3(2): 1–14.
9. Chakraborty, S., Paul, H., Ghatak, S., Pandey, S.K., Kumar, A., Singh, K.U. and Shah, M.A. (2022). An AI-based medical chatbot model for infectious disease prediction. *IEEE Access*, 10: 128469–128483.
10. Chow, J.C., Sanders, L. and Li, K. (2023). Impact of ChatGPT on medical chatbots as a disruptive technology. *Frontiers in Artificial Intelligence*, 6: 1166014.
11. Ahmed, S.T. and Fathima, A.S. (2024). Medical ChatBot assistance for primary clinical guidance using machine learning techniques. *Procedia Computer Science*, 233: 279–287.
12. Bali, M., Mohanty, S., Chatterjee, S., Sarma, M. and Puravankara, R. (2019). Diabot: A predictive medical chatbot using ensemble learning. *International Journal of Recent Technology and Engineering*, 8(2): 6334–6340.
13. Chang, I.C., Shih, Y.S. and Kuo, K.M. (2022). Why would you use medical chatbots? Interview and survey. *International Journal of Medical Informatics*, 165: 104827.
14. Hsu, I.C. and Yu, J.D. (2022). A medical Chatbot using machine learning and natural language understanding. *Multimedia Tools and Applications*, 81(17): 23777–23799.
15. Blanc, C., Bailly, A., Francis, É., Guillotin, T., Jamal, F., Wakim, B. and Roy, P. (2022). FlauBERT vs. CamemBERT: Understanding patient's answers by a French medical chatbot. *Artificial Intelligence in Medicine*, 127: 102264.
16. Mehfooz, F., Jha, S., Singh, S., Saini, S. and Sharma, N. (2020). Medical chatbot for novel COVID-19. In: *ICT Analysis and Applications: Proceedings of ICT4SD 2020* (Volume 2) (pp. 423–430). Singapore: Springer Singapore.
17. Safi, Z., Abd-Alrazaq, A., Khalifa, M. and Househ, M. (2020). Technical aspects of developing chatbots for medical applications: Scoping review. *Journal of Medical Internet Research*, 22(12): e19127.

18. Madhu, D., Jain, C.N., Sebastain, E., Shaji, S. and Ajayakumar, A. (2017). A novel approach for medical assistance using trained chatbot. In: *2017 International Conference on Inventive Communication and Computational Technologies* (*ICICCT*) (pp. 243–246). IEEE.
19. Babu, A. and Boddu, S.B. (2024). Bert-based medical chatbot: Enhancing healthcare communication through natural language understanding. *Exploratory Research in Clinical and Social Pharmacy*, 13: 100419.
20. Tamizharasi, B., Livingston, L.J. and Rajkumar, S. (2020, December). Building a medical chatbot using support vector machine learning algorithm. In: *Journal of Physics: Conference Series* (1716 (1): 012059). IOP Publishing.
21. Athota, L., Shukla, V.K., Pandey, N. and Rana, A. (2020, June). Chatbot for healthcare system using artificial intelligence. In: *2020 8th International Conference on Reliability, Infocom Technologies and Optimization* (*Trends and Future Directions*) (*ICRITO*) (pp. 619–622). IEEE.
22. K.C., G. P., Ranjan, S., Ankit, T. and Kumar, V. (2019). A personalized medical assistant chatbot: Medibot. *Int. J. Sci. Technol. Eng.*, 5(7).
23. Goodman, R.S., Patrinely, J.R., Stone, C.A. Jr, Zimmerman, E., Donald, R.R., Chang, S.S., Berkowitz, S.T., Finn, A.P., Jahangir, E., Scoville, E.A., Reese, T.S., Friedman, D.L., Bastarache, J.A., van der Heijden, Y.F., Wright, J.J., Ye, F., Carter, N., Alexander, M.R., Choe, J.H., Chastain, C.A., Zic, J.A., Horst, S.N., Turker, I., Agarwal, R., Osmundson, E., Idrees, K., Kiernan, C.M., Padmanabhan, C., Bailey, C.E., Schlegel, C.E., Chambless, L.B., Gibson, M.K., Osterman, T.J., Wheless, L.E. and Johnson, D.B. (2023). Accuracy and reliability of chatbot responses to physician questions. *JAMA Network Open*, 6(10): e2336483–e2336483.
24. Bora, A. and Cuayáhuitl, H. (2024). Systematic analysis of retrieval-augmented generation-based LLMS for medical chatbot applications. *Machine Learning and Knowledge Extraction*, 6(4): 2355–2374.

Chapter 16

A Chatbot Service for Suicide Detection and Prevention

Rushabh Kankariya,[1] *Ananya Somani,*[1] *Stuti Hedge,*[1] *Karan Gupta,*[1] *Rahul Agnihotri*[1] and *Prabhadevi Boopathy*[2,*]

Social media networks like Facebook, Twitter, and Instagram, as well as instant messaging services like WhatsApp, Telegram, Facebook Messenger, and Instagram Messenger, are revolutionising the way people communicate with peers and connect with others in society. With this rapid growth, the emergence of generative AI applications like ChatGPT has gained significant popularity and continues to revolutionise the world in many ways. Although communicating and staying connected have become easy today, they also have their pitfalls, leading to different forms of social engineering attacks. ChatGPT can serve people in various ways, but it is not more emotionally connected to people's emotional instincts. Suicide continues to be a considerable issue worldwide, and individuals frequently utilise social media as a platform to share their distress and suicidal thoughts. Although many attempts for suicidal ideation have been made but they focused mainly on detection and were not concerned about prevention. The major aim of this chapter is to apply artificial intelligence techniques to predict individuals with

[1] School of Computer Science Engineering and Information Systems, Vellore Institute of Technology, Vellore, India.

[2] Department of Computer Science and Engineering, FET, Ramaiah University of Applied Sciences, Technology Campus, Peenya, Bengaluru, India.

Emails: rushabh2704@gmail.com, ananya0827@gmail.com, stutihegde@gmail.com, karanguptaa45@gmail.com, rahulagnihotri4444@gmail.com

* Corresponding author: prabadevib.cs.et@msruas.ac.in

DOI: 10.1201/9781003726913-16

depression and to examine suicidal ideation. Consequently, this will assist in identifying individuals with stress and determining their significance, enabling the provision of appropriate care and potentially saving their lives. Therefore, suicidal ideation could be prevented by providing proper guidance and awareness. In essence, this chapter presents a novel suicide detection model designed to identify early indications of suicidal ideation accurately based on the individual's posts on social media sites like Twitter. The model presented in this chapter utilises Natural Language processing for sentiment analysis and machine learning models, namely, Random Forest, Linear SVC, and Bagging, to identify potentially illicit patterns and other associated risk factors in an individual's posts that may lead to suicidal ideation. Subsequently, required assistance is offered to the individual. Through this, the system helps to promote mental well-being and reduces the thought process of suicidal ideation. This, in turn, would help the help seeker avoid suicidal consequences. This has been achieved through a user-friendly interactive chatbot using Cohere-APIthat suggests appropriate prevention measures based on user prompts. Additionally, these model solutions provide an immediate solution with appropriate action in the event of emergencies. This chapter highlights the usage of the Cohere API to enable a real-time, customised response for a person at risk based on their social media activity. Secondly, usage of 3 different algorithms, namely, Linear SVC, Random Forest and Bagging classifier, to compare and find the best possible fit to get the most optimal results. And thirdly, using four different techniques—TF-IDF, BIGRAM TF-IDF2, Vader, and the most common Bag of Words—to ensure the authenticity of data across the three algorithms. We also used recall, precision, F1 score, and accuracy to check for any bias or variance present in the results.

1. Introduction

Depression and suicide have become pressing public health concerns, especially for young adults. In the modern context, social media is a double-edged sword—as it allows people to connect and share their emotions, it can also contribute to mental health struggles. Many people, particularly students facing academic pressures, use social media as an outlet to express their feelings, often anonymously. Social media is an effective means of communication that enables information to travel quickly, allowing for prompt actions and required assistance. On the other hand, the too-old posts on emergent events must be appropriately acknowledged once they have been resolved; otherwise, they might provoke individuals to assist with posts that were made some years ago. On the other hand, individuals, i.e, the social media activists or social workers, through these social media networks are enforced to engage in activities like cyberbullying, threatening for ransom, deep fake attacks on profiles, and can lead to more heart broken stress leading to suicidal ideation or provoking to indulge in any illicit or immoral activities. Recent advancements in Artificial Intelligence (AI) and Large Language Models (LLMs) have led to the emergence of Chatbots, which play a crucial role in the healthcare landscape, ranging from remote doctor consultations to prescription of pills. This empowers the individuals as patients to easily and actively engage in getting their intended services done at their convenience. On the other hand, the biases and context-independent nature of LLMs must be taken into account by current researchers developing chatbots for healthcare.

These chatbots have been used in the medical sector by different stakeholders, including patients, clinical staff, doctors, pharmacists, lab technicians, and so on. Major tasks that these people carried out in the medical loop require 24/7 assistance. The activities from the perspective of the patients using chatbots are learning about symptoms of the disease, homemade remedies, prescription of medicines (that are commonly used), usage of different medication, adverse effects of using the drugs, prevalent treatments for a particular disease and appointment scheduling. This, in turn, ensures personalized services for the patient. Whereas, the medical practitioner and other staff will utilize chatbots for appointment scheduling between doctor and different patients, bill generation, stacking of different, record maintenance, and periodic reminders by generating notifications to the concerned. Thus, streamlines the tasks of medical practitioners from data curation to improved patient satisfaction. Also researchers in this field were benefitted by identifying different ways to enhance the algorithms, synthetic data generation, code improvement and results analysis.

Thus, it plays a vital role in two aspects: improving healthcare quality and delivering healthcare services cost-effectively [1]. The advancement of chatbots like ChatGPT and Gemini AI, with the rapid evolution of LLMs and generative AI, has led to a greater reliance on this by practitioners and patients. On the other hand, this has led to an overreliance on medications rather than visiting doctors and clinical centres.

Although these chatbots play a significant role in advancing the healthcare sector in various verticals, they also have their own flaws, hindering their full adoption as trustworthy tools. Overreliance on these bots would lead to the following: inaccuracies in the results generated due to biases in the data fed to the algorithm itself, a lack of emotional intelligence, resulting in unempathetic responses to the user's emotional prompts, and legal and ethical concerns. Human interference in the decisions made by the chatbots is limited, which leads to inaccuracies in the results generated. On the other hand, this leads to errors in the treatment generated, and in turn, can cause significant harm. These non-context-aware chatbots hinder the life of the human at stake and could incite temptation for various research studies on the use of these services for research and other activities in the healthcare practice. Some of the implications include incorrect guidance on suicidal attempts, irrelevant results from different contexts leading to misaligned diagnoses, and self-diagnosis with proper prescription may lead to harmful side effects.

2. Importance of Suicidal Detection and Prevention

Suicide is considered to be one of the fourth major causes of death by the World Health Organization, attributing over seven lakh deaths among the age groups of 15–28 years. Subsequently, the WHO made suicidal prevention a top priority in the health action plan for 2013–2030 [2], followed by various initiatives aimed at creating awareness among the general public and high-risk individuals by different social organisations, healthcare sectors, and online social activists [2]. Suicidal attempts can be detected and prevented by adopting proper steps to handle them. Suicide prevention is a responsibility of every individual, as it will help to protect the mental health and well-being of our loved ones. This step will be lifesaving. The risk factors and warning signs associated with suicide must be known, and proper awareness regarding this must be created among all individuals to ensure appropriate prevention actions are taken when and where needed. Some of the major causes of suicide include mental disorders, depression, and illegal drug use. Above all, individuals with a history of previous suicide attempts should be monitored continuously, as their likelihood of attempting again is higher. Consequently, World Suicide Prevention Day is observed on 10th September each year with a focus on reducing suicidal issues and raising awareness among the public, government, and private offices.

2.1 Role of Social Media in Suicide Attempt, Detection, and Prevention

Social media is now revolutionising the way of living and has swept the humanitarian community over the past decades. Social media, with its numerous applications such as blogging, chatbots, videos, social networking sites like Twitter and Instagram, messengers like Facebook and WhatsApp, and email applications, is the current mode of communication. With advancements in technology, these applications are blossoming further with digitised content that caters to the diverse moods of humans, enabling them to live and experience rapid growth in the years to come. It allows users to connect instantly and share information more dynamically globally. And social media, with better internet connections, plays a vital role in the suicide-related activity of individuals [3]. This social media is one of the amplifiers of the known risks of suicide. Some of the notable cases with social media users are insomnia (lack of sleep), social isolation, cyberbullying, hopelessness, and distress. Also, users are prone to different types of exposures to suicidal attempts and behaviours made across the globe. This, in turn, acts as a role model for people with any of the aforementioned signs of suicide. These suicidal behaviours are more common among women, homosexuals (lesbian and gay), and transgender individuals than among males [4]. The various complexities of relationships and other associations through social media lead to unhealthy comparisons, bullying, and chat addiction. Although social media platforms are changing the communication and interaction today in a better way, they may be a provoking factor for suicidal thoughts due to stress and other mental pressures.

On the other hand, social media platforms contribute significantly to suicidal ideation. These are the most comfortable platforms for users to share their thoughts and views on their real-life experiences dynamically through a live sharing facility. These posts by the users or victims help the researchers in the field to recognise the patterns of suicidal thoughts and potential signs of warning using techniques like Natural Language Processing (NLP), Machine Learning (ML), and Deep learning (DL) models. These help to identify suicidal ideation based on the posts made by users in different formats like text, video, audio, and images. The posts from the social media platforms like Twitter (X), Instagram, Facebook, YouTube, Reddit, Tumblr, Snapchat, and so on. All these platforms cater to different styles of communication based on users' interests. This assists the researchers in this field to perform a multimodal analysis rather than only an analysis. This will lead to more robust and significant detection of suicidal ideation. And the provision of dynamic retrieval of information through web scraping from these platforms assists in real-time monitoring and notifying the concerned through alerts based on the posts. These posts are used for analysing the potential risk factors based on emotional expression through language-specific analysis, and intrusion with self-harm contents. Natural Language Processing (NLP), along with ML, plays a significant role in discovering complex patterns from the colloquial text posted by users on these platforms. Furthermore, platforms like Facebook, Instagram, and Twitter have partnered with other public organisations responsible for suicidal prevention, which will often signal to the entities while the user is trying to self-harm. Though technological advancements play a vital role in interpreting suicidal ideation from the posts made by the users, it is very crucial to address the ethical concerns with user data usage, accurate contextual interpretations and potential biases.

From the perspective of suicidal prevention, social media is the most effective, dynamic and interactive platform for enforcing two-way communication among niche audiences and social media influencers. Educational awareness and crisis helplines through social media would have greater reach than any other mode. Further, during the crisis, the posts on social media would help the activists of the family concerned to take appropriate actions [5]. Especially LLMs, such as ChatGPT, play a significant role in prevention. For instance, when a user prompts ChatGPT with suicidal ideation, it acts as a counsellor and suggests possible look-up pages that may soothe the minds of the victim in a more powerful way [6]. NLP with transformer architectures have made this possible in the name of Generative Pretrained Transformer, through enhanced processing of natural language tasks, and it is playing a vital role in resolving mental health disorders [7]. It has been enhanced in such a way to provide more personalized information and resources.

The versions of ChatGPT have been tuned further to provide better results to suicidal related prompts. For instance, the ChatGPT response would acknowledge the user's distress through encouraging words and links to mental healthcare professionals or any other crisis hotlines. This led to the widespread use of ChatGPT for suicidal prevention and the analysis of prompts given by the user. The recent reports of the WHO in 2021 estimated that more than 7,27,000 people die every year due to suicide [8]. The number of people who attempt suicide is much higher.

This chapter was initiated with the recognition of this issue, particularly in environments where the stress of academics and placements is high. Students who feel overwhelmed by exams or job placements often turn to social media to express frustration, and we noticed that this pattern was not isolated. It happens on a larger scale, where people in distress reach out to online communities seeking support, but sometimes end up facing negativity. To address this, we have developed a system that utilises technology to enhance mental well-being. Our approach includes tools that can recognize signs of emotional distress from posts on social platforms, especially those written anonymously. By identifying specific patterns in language and tone, our system can detect when someone may need help and alert moderators or mental health professionals to intervene before the situation worsens.

However, our aim extends beyond merely flagging harmful content. We aim to foster a healthier online environment where individuals feel heard and supported. By fostering a balance between open expression and mental health safeguards, we aim to ensure that individuals experiencing emotional distress find a positive and supportive environment. By utilizing social media in this thoughtful way, we aim to contribute to a more compassionate and secure online environment.

Ultimately, we are focused on using technology to build a support system that addresses mental health in the digital age. By creating tools that foster empathy, understanding, and prompt intervention, we aim to contribute to a more compassionate and secure online environment for everyone. In 2005, a new suicide detection device was designed that the person would wear as an earpiece to monitor the pulse and oxygen level of the person. If the person's vital signs are indicated as out of range, an alarm would go off and an emergency message would be sent [9].

Our system goes beyond detection; it extends a helping hand to those who may be struggling with their mental health. By utilising the vast amount of available data on social media, we aim to foster a culture of empathy and understanding.

This initiative underscores our commitment to harnessing technology to improve mental health and create a more supportive and compassionate digital community for everyone.

3. Literature Survey

This section highlights the various studies carried out on suicidal prevention mechanisms.

Machine Learning and Deep Learning in Suicidal Prevention Mechanism

An ensemble framework combining LSTM (long short-term memory) and Convolution Neural Network (CNN) has been proposed to identify suicide ideation in social media. Extractive and abstractive techniques were analysed using statistical and other machine learning algorithms. A Reddit social media dataset was chosen. This ensemble framework outperforms other baseline approaches with an accuracy and F1 score of 91.7% and 92.6%, respectively [10].

NLP and ML have been used in combination to perform sentiment analysis for suicidal prediction and depression detection. Furthermore, different ML model approaches are used to classify depression from Twitter data. Random Forest comes out on top, followed by Decision Tree, Naive Bayes, and DL. Also, a comparative evaluation of many DL and ML algorithms like LSTM, BiLSTM, GRU, BiGRU, and C-LSTM to detect suicidal ideation on Twitter was carried out. Random Forest provides the highest accuracy of 93% [11]. Support Vector Machine (SVM), KNN, and logistic regression were employed to build a model using users' ongoing posts at a fixed interval, which can measure a user's depression based on their regular posts over a predetermined period. Although KNN and SVM outperform logistic regression, the models exhibit long training times, poor performance, and excessive noise [12].

An emotion recognition and detection of suicidal ideation in patients with COPD has been established for the nurse to recognise when the patient is fearful or anxious through CNN with proper preprocessing and other imaging [13, 14].

Chatterjee et al., have used various ML algorithms with feature extraction, statistical analysis, and topic modeling using Latent Dirichlet Allocation (LDA) for developing a predicting model for suicidal thoughts from the Reddit & Twitter data. They also worked to find patterns in user posts about suicidal ideation. Moreover, the researchers divided time into four unspecified intervals as a parameter, which may introduce challenges for the model in making accurate predictions [15]. Further in [16], the authors have used a CNN-BiLSTM model and the XGBoost model in suicide detection using text features from Reddit data. Also, an attempt was made to figure out if someone's Twitter profile showed signs of suicidal thoughts through data from TweetHereAfter and Twitter looking at 172 profiles of non-suicidal users and 115 profiles of suicidal users [17].

ML models like Logistic Regression, Decision Tree, and SVM were used to create a model for predicting suicide deaths in Thailand by using a dataset of 49,417 cases to identify key factors. It was reported that Logistic Regression scored 90.71% on the F1 metric and achieved an accuracy of 90.66%, while the Decision Tree model scored lower, with an F1 score of 67.45% and an accuracy of 80.79%. At the same time, SVM performed better than Logistic Regression on the F1 Score, while matching its 90.66% accuracy [18].

Research was conducted by manually labelling the suicide-related tweets from 3202 English tweets into 12 groups using ML and NLP techniques. When classifying data into two groups, BERT and XLNet have resulted in better accuracy than for sorting data into 6 groups. BERT and XLNet's clear edge shows that tweets about suicide have key meaning info tied to context, which goes beyond just how often words appear [19].

Review Works using Deep Learning and Machine Learning for Suicidal Prevention

To investigate the diverse suicide prevention strategies employed in various nations, researchers conducted an exploratory research study. The study focused on using NLP techniques, which were identified as cost-effective solutions for sentiment analysis. The team searched five scientific and medical databases, including PubMed and MEDLINE. Due to the varied nature of these techniques, researchers were unable to evaluate their effectiveness using standard metrics [20].

On the other hand, the authors in [21] have done an extensive survey on suicide-related tasks such as text classification understanding suicidal messages, user-level detection, response generation, and spotting mental health risks through bibliometric analysis of 426 research articles. However, they failed to tackle data gaps, annotation biases, uneven data distribution, and statistical assessment. Also, a few attempts were made to instigate the therapeutic interventions for identifying indications of suicidal tendencies and other mental health problems using DL algorithms and were found to be most effective [22]. Few other researchers have explored this. Some prominent research works employed NLP to identify crucial indicators of psychological distress in individuals with suicidal tendencies through social media content. Later, based on analysis, they concluded that several DL models outperform in time series suicidal data evaluation and could improve the precision of automated systems that monitor social media to identify suicidal individuals [23]. Consequently, the DL and ML models were also used to comprehend and identify susceptible individuals based on their online behaviour and moral questionnaire responses.

Tasks like grouping suicide texts, making sense of messages about suicide, spotting at-risk users, crafting replies, and finding signs of mental health issues and self-harm dangers. However, the study doesn't give exact figures on suicidal thoughts or how well the detection methods work, and data biasing problems [24].

Video-based Works

Some of the researchers have used Reddit's "suicide watch" forum to gather the data and applied TF-IDF with ML models like logistic regression and SVM to pull out key details from the data [25, 26]. Furthermore, researchers have employed ML, DL, and Image Processing (IP) for classification by extracting features from the video using cameras to spot suicidal people [27, 28] and from websites like Web-of-Science [29].

3.1 Summary of Literature Survey

The survey on leveraging AI and machine learning for suicide prevention and related areas reveals both promising capabilities and open challenges. A wide range of studies collect pertinent data from social media platforms, employing a mix of traditional statistical approaches and more advanced techniques for analysis. Beyond building sophisticated models, thoughtful frameworks are needed to enable transparent, ethical application in social contexts. By combining strengths in data science, psychology, and ethics, continued progress in this problem space appears attainable. However, ambiguity lingers regarding best practices for training algorithms that deal with sensitive human subject matter. Though increasingly viable as an auxiliary public health tool, responsible AI integration remains in developmental phases. Overall, cautious optimism exists around the potential to positively impact individuals and communities through considerate, collaborative digital efforts centred on wellness. With care taken to put people first, data-driven suicide prevention may gradually transform from a promising theory into a societal reality.

4. Proposed System

In this chapter, we propose a system that employs advanced NLP techniques and ML algorithms to identify and prevent harmful online behaviour, specifically targeting indicators of suicidal intent. The framework processes social media data through data preprocessing steps, including text cleaning, feature engineering, and sentiment analysis, to enhance detection accuracy.

The architecture diagram provides a high-level overview of the techniques used within our proposed framework for detecting and preventing harmful online behaviour.

The architecture in Figure 1 is designed for suicide ideation detection and prevention. The various steps involved in this include data collection, where a Reddit dataset is loaded as the primary data source for tracking the user ideation based on their posts. This step involves identifying the language of each entry using NLP techniques, where null and noisy data are removed.

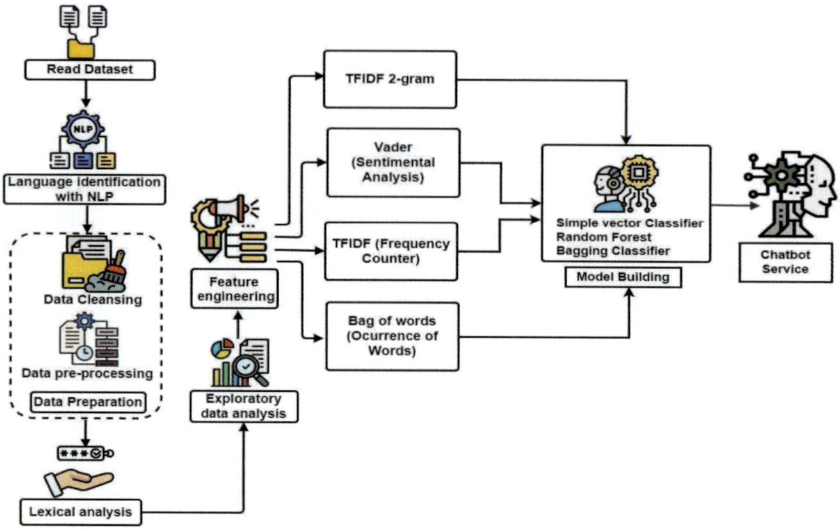

Figure 1: Architecture of the proposed system.

Next, the data frame is further cleaned and preprocessed through stemming, lemmatization, and stop-word removal. Finally, tokenization is applied to the cleaned and pre processed data frame.

Following this, we conduct an exploratory data analysis as a second step to understand the patterns in the data. We then perform feature engineering by selecting, transforming, or creating new features to enhance the model's ability to interpret and understand the dataset's language. We use Valence Aware Dictionary for Sentiment Reasoning (VADER), a natural language toolkit (NLTK) for sentiment analysis, which helps the model gauge the sentiments expressed in the text data. We use Term Frequency Inverse Document Frequency (TFIDF), which gives us an idea of the frequencies of words appearing in the features. Additionally, we have utilized BIGRAM TF-IDF2, which can be considered an iteration and a variant of TF-IDF, enabling us to obtain a more comprehensive corpus and the frequencies of words associated with the dataset. We have also used a bag of words to give numerical values to the words by creating a corpus. The Bag-of-Words, TF-IDF, BIGRAM TF-IDF2, and VADER are used for preprocessing the Reddit data, like word counts representation, weighing the words based on their importance, and seizing the emotional tone or effects from the text, respectively.

Finally, in the last steps of our model, we provide the input features to various algorithms. We use the Support Vector Classifier (SVC), which is often used to solve classification problems, especially when the data is numerical and can be represented as feature vectors. We use another algorithm, Random Forest, which constructs multiple decision trees and demonstrates improved effectiveness through NLP integration. Last but not least, a Bagging classifier, where numerous subsets of the training data are created through random sampling with replacement, training individual base models on these subsets, and then aggregating their predictions to give the final prediction. These LLMs are at the heart of many generative AI breakthroughs, and their ability to understand and generate human language has transformed the way we interact with technology. Therefore, in addition to the prediction done by our ML techniques, we have decided to include the use of AI to help those who may have some suicidal intent to give them as much support as we can. After spending multiple hours on the prompt, we have finally devised a prompt that will offer them the best help possible. So we first have to predict if the person is suicidal or not, and if they are, then the chatbot is given the prompt along with their tweet based on which they were judged to be suicidal, the language model gains a rough understanding of the situation they are in, and gives them the best possible advice along with helpline numbers for better assistance.

Three main popular models are available for free usage: ChatGPT (developed by OpenAI), Gemini (developed by Google), and Coral (developed by Cohere). They are trained on massive amounts of text data and learn to predict the next word in a sequence based on the words that came before it. This allows them to generate human-like responses to text-based queries. In terms of performance, LLMs like ChatGPT-3.5 and Bard have achieved remarkable results in various natural language processing tasks. They can generate high-quality text, answer questions, summarize texts, translate between languages, and even perform complex reasoning tasks to some extent. The LLM we have deployed in this chapter is Coral; it is designed to have helpful, harmless, and inclusive conversations with users. Its training dataset is carefully curated to minimize the risk of generating harmful or biased responses, which aligns exactly with the ideation of this chapter. It can also use web search to improve its answers and provide more accurate information.

5. Dataset Description

The main CSV File can be split into training and testing datasets. It has 2 columns, 'text' containing content of user posts, while the 'class' is the target variable. It has 232,074 unique columns which show the posts that have been posted on the thread on Reddit to detect suicide and depression. The dataset is a collection of posts from the "SuicideWatch" and 'depression' subreddits of the Reddit platform. All posts collected from SuicideWatch are labelled as suicide, while posts collected from the depression subreddit are labelled as depression. Non-suicide posts are collected from r/teenagers.

6. Results Obtained

6.1 Evaluation Metrics and Model Tuning

Achieving an accuracy of 87% was made possible through the utilization of the Bag of Words technique combined with the Linear SVC method. This approach outperformed all other methods across all metrics, demonstrating its effectiveness for this classification task. To ensure a transparent and comprehensive evaluation, we've included precision, recall, and F1 scores alongside accuracy. This broader view of the results helps assess the model's performance across different aspects and indicates an unbiased dataset.

When considering interpretability, it's worth highlighting Linear SVC's ability to provide insights regarding feature importance. However, it's important to note that while its accuracy is high, the model's performance heavily relies on the Bag of Words technique and the specific dataset used for training.

The TF-IDF and TF-IDF 2-grams methods show strong performance as well, suggesting that the choice between these representations may depend on specific dataset characteristics, shown in Figures 2 and 3. The Vader method's

	TF-IDF	TF-IDF 2-grams	bag of words	vader
precision	0.870948	0.874521	0.873229	0.729695
recall	0.848498	0.83817	0.878371	0.766815
F1-score	0.859576	0.85596	0.875793	0.747794
accuracy	0.860048	0.857592	0.874224	0.738883

Figure 2: Performance metrics when Linear SVC with TFIDF, TFIDF-n-grams, Bag of Words, and Vader.

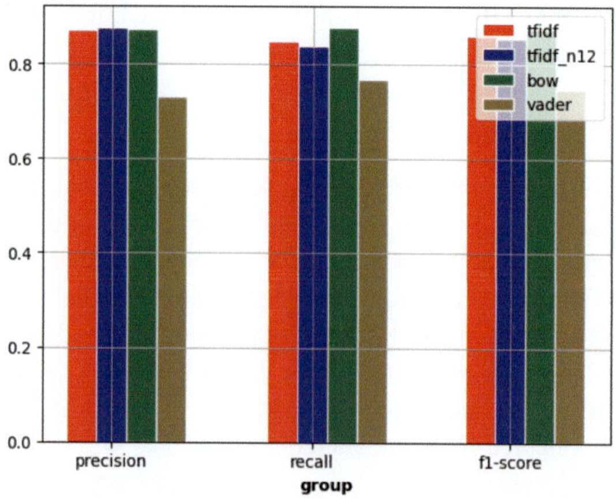

Figure 3: Graphic representation of Precision, Recall, and F1-score of TFIDF, TFIDF-n-grams, Bag of Words, and Vader Results with Linear SVC Classifier on Reddit Dataset.

lower performance indicates that sentiment-based features alone may not be as effective for this particular task when used with Linear SVC.

To assess the model's reliability, it's crucial to evaluate any variations or fluctuations in accuracy observed during experiments. This consideration, along with the balanced precision and recall values, provides a more detailed understanding of the model's robustness in both identifying relevant instances and avoiding false positives.

The results show that TF-IDF 2-grams achieve the highest recall (0.909545), F1-score (0.845418), and accuracy (0.833635). The Vader method performs best in terms of precision (0.786829), though the difference is small compared to the other approaches. The bag of words method consistently underperforms in precision and accuracy compared to the other approaches but maintains competitive recall. We achieved the highest accuracy by utilizing the TF-IDF 2-grams approach in combination with the Random Forest technique. After performing hyperparameter fine-tuning, we found that a depth of 10 provided the best results based on the categories provided in the dataset. All the overfitting and underfitting issues were addressed by this hyperparameter fine-tuning to get optimal results. These results demonstrate the effectiveness of n-gram-based TF-IDF representation for this classification task, with TF-IDF 2-grams showing a slight edge in overall performance assured by F1-score and accuracy. The high recall values across TF-IDF, TF-IDF 2-grams, and bag of words methods (all above 0.90) indicate these approaches are particularly good at identifying relevant instances in the dataset. The combination of TF-IDF 2-grams with the Random Forest classifier, along with careful hyperparameter tuning, proved to be the most effective approach for this classification task, addressing potential overfitting and underfitting issues and yielding optimal results across all evaluation metrics, shown in Figures 4 and 5.

The results show that TF-IDF achieves the highest recall (0.885855) and F1 score (0.840492). The bag of words method performs best in terms of precision (0.803818) and accuracy (0.831739), though the differences are small. The vader method consistently underperforms compared to the other approaches across all metrics.

	TF-IDF	TF-IDF 2-grams	bag of words	vader
precision	0.789912	0.796341	0.784905	0.786774
recall	0.906282	0.906453	0.91072	0.819478
F1-score	0.844105	0.847837	0.843145	0.802793
accuracy	0.831007	0.835746	0.828938	0.796751

Figure 4: Performance metrics when Random Forest applied with TFIDF, TFIDF-2-grams, Bag of Words, and Vader.

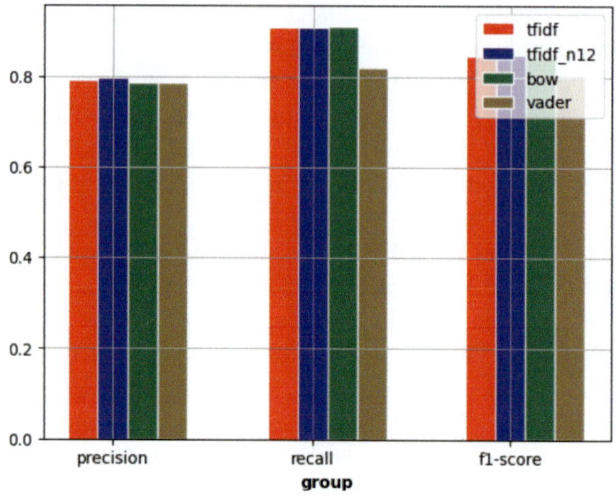

Figure 5: Graphic representation of Precision, Recall, and F1-score of TFIDF, TFIDF-n-grams, Bag of Words, and Vader Results with Random Forest Classifier on Reddit dataset.

These results demonstrate the effectiveness of traditional text representation methods like TF-IDF and bag of words for this classification task, with TF-IDF showing a slight edge in overall performance as measured by F1-score.

Additionally, the TF-IDF technique combined with the Bagging method achieved an accuracy reaching 83.3%. This consistency in performance across different metrics (with precision around 80% and recall around 89%) indicates a fair and equalized distribution of results in the dataset and a proper depiction of outcomes, shown in Figures 6 and 7.

The sensitivity of the Bagging method to variations in the TF-IDF parameters can be further analyzed to provide insights into the robustness of our approach. This analysis would help understand how stable the results are under different parameter configurations, further validating the reliability of the method.

In addition, to using ML techniques to detect signs of suicide in social media posts our study takes a comprehensive approach by incorporating a chatbot that offers immediate support to people in crisis. We understand the urgency of addressing health issues so the chatbot plays a role in providing help to individuals who may be at risk of suicide. But it doesn't stop there—the chatbot also provides real-time assistance and shares resources like helpline numbers, mental health services, and self-help materials, as depicted in Figure 8. This gives users a range of support options right when they need it. Our chatbot prioritizes user privacy and confidentiality, recognizing the nature of health conversations. It uses the Reddit post on which it has done an analysis and then builds a custom response immediate basis based on the user's requirements that would in some way provide some empathy to the user and prevent them from taking any extreme measures and convince them to take the help of a professional or a relative.

	TF-IDF	TF-IDF 2-grams	bag of words	vader
precision	0.801866	0.792599	0.798605	0.786037
recall	0.88776	0.888529	0.879652	0.815893
F1-score	0.84263	0.837827	0.837172	0.800687
accuracy	0.832601	0.826353	0.827258	0.794941

Figure 6: Performance metrics when Bagging Classifier is combined with TFIDF, TFIDF-n-grams, Bag of Words, and Vader.

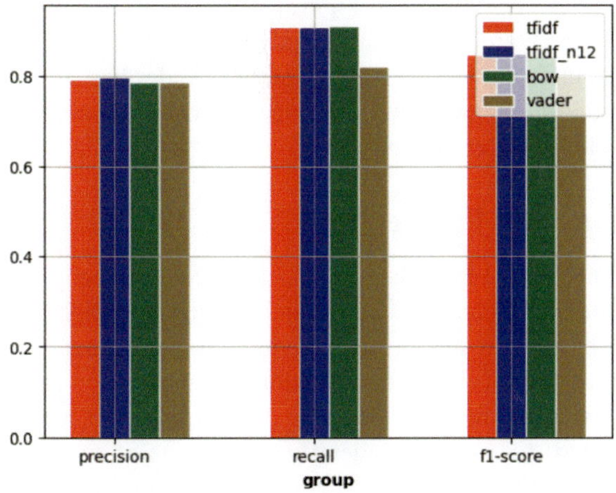

Figure 7: Graphic representation of Precision, Recall and F1-score of TFIDF, TFIDF-n-grams, Bag of Words and Vader with Bagging Classifier on reddit dataset.

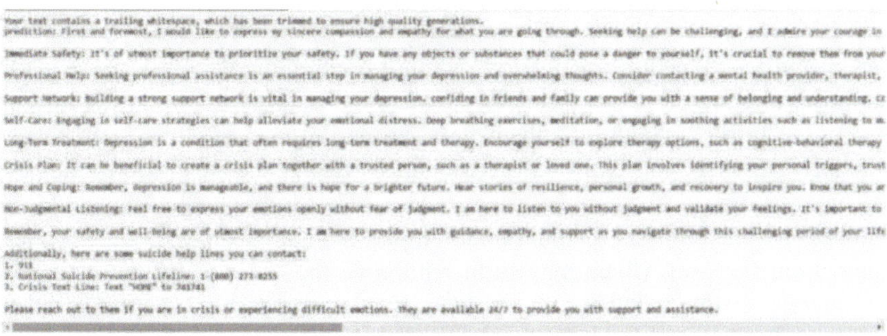

Figure 8: Results by the chatbot when a person is detected as suicidal by the model.

Additionally, we're committed to improving our study based on user feedback and advancements in healthcare practices. By combining technology with empathy, our integrated solution identifies risks and actively promotes the well-being of those who require assistance. This reflects our dedication to using technology for impact and advocating for mental health. Although the chatbot is added as a temporary measure, it does not substitute a professional, and hence it gives more respected helpline numbers so that the person can seek professional help.

To summarize, using the Lincar SVC method, with the Bag of Words technique, proved to be the approach, achieving an accuracy rate of 87%. The interpretability and consistent performance of the Linear SVC model in experiments make it well-suited for identifying content in social media. Although Random Forest and Bagging methods also performed well, the Linear SVC method stands out as promising for investigation and potential implementation, in real-world scenarios. The graphs in Figures 3, 5, and 7 collectively suggest that TF-IDF, TF-IDF_n12, and BOW consistently achieve high precision, recall, and F1-scores across different algorithms. In contrast, Vader has slightly lower recall and F1-score, potentially indicating its lesser suitability for certain classification tasks compared to the other methods.

7. Conclusions and Future Work

This chapter signifies the importance of social media for people to express and communicate their thoughts of distress and suicide, highlighting the need for proactive interventions to support an individual experiencing a crisis. Through the development of a novel suicide detection and prevention model, leveraging sentiment analysis and machine learning techniques, we have demonstrated promising outcomes to accurately identify early signs of suicidal tendencies in social media posts. We have utilized natural language processing for sentiment analysis and ML models, such as Random Forest, Linear SVC, and Bagging, for detection. Among various algorithms, Linear SVC proved to be the most accurate, achieving 87% accuracy. By integrating a chatbot using the Cohere API into our system, we aim to offer immediate assistance (a preventive solution) and resources to people in distress, without compromising their privacy and confidentiality.

In the future, we aim to enhance our system by incorporating more robust generative models and expanding our analysis to include images and memes, thereby reaching a broader spectrum of vulnerable individuals. This could be attained by encompassing multimodal input including images, videos, and audio from the user, which may further improve the accuracy of the detection and thereby enable suitable prevention mechanisms to resolve it. Also, prompt engineering and small language models could also be used to effectively utilize the chatbots for a specific purpose and can alert the connected family member regarding the intention of the user. Furthermore, integrating our system to other platforms can amplify its impact and help a larger audience.

Our research aims to proactively identify risks and enhance the well-being of individuals in distress by integrating technology with empathy. We are committed to leveraging technology to advocate mental health and enhance our system based on user feedback and developments in healthcare practices.

References

1. Laymouna, M., Ma, Y., Lessard, D., Schuster, T., Engler, K. and Lebouché, B. (2024). Roles, users, benefits, and limitations of chatbots in health care: Rapid review. *Journal of Medical Internet Research*, 26: e56930.
2. Waszak, P.M. (2024). Chat GPT and suicide prevention: Can it work? A conversation analysis. *Psychiatria i Psychologia Kliniczna*, 24(4): 292–299.
3. Luxton, D.D., June, J.D. and Fairall, J.M. (2012). Social media and suicide: A public health perspective. *American Journal of Public Health*, 102(S2): S195–S200.
4. Jaycox, L.H., Murphy, E.R., Zehr, J.L., Pearson, J.L. and Avenevoli, S. (2024). Social media and suicide risk in youth. *JAMA Network Open*, 7(10): e2441499–e2441499.
5. Thorn, P., Hill, N.T., Lamblin, M., Teh, Z., Battersby-Coulter, R., Rice, S., Bendall, S., Gibson, K.L., Finlay, S.M., Blandon, R., de Souza, L., West, A., Cooksey, A., Sciglitano, J., Goodrich, S. and Robinson, J. (2020). Developing a suicide prevention social media campaign with young people (The# Chatsafe project): Co-design approach. *JMIR Mental Health*, 7(5): e17520.
6. Liu, X.Q. and Zhang, Z.R. (2024). Potential use of large language models for mitigating students' problematic social media use: ChatGPT as an example. *World Journal of Psychiatry*, 14(3): 334.
7. Aminah, S., Hidayah, N. and Ramli, M. (2023). Considering ChatGPT to be the first aid for young adults on mental health issues. *Journal of Public Health*, 45(3): e615–e616.
8. https://www.who.int/publications/i/item/9789240110069.
9. Majumder, S., Mondal, T. and Deen, M.J. (2017). Wearable sensors for remote health monitoring. *Sensors*, 17(1): 130. https://doi.org/10.3390/s17010130.
10. Bhardwaj, T., Gupta, P., Goyal, A., Nagpal, A. and Jha, V. (2022). A review on suicidal ideation detection based on machine learning and deep learning techniques. In: *2022 IEEE World AI IoT Congress (AIIoT)*, Seattle, WA, USA, pp. 027–031. doi: 10.1109/AIIoT54504.2022.9817373.
11. Angskun, J., Tipprasert, S. and Angskun, T. (2022). Big data analytics on social networks for real-time depression detection. *Journal of Big Data*, 9(1): 69.
12. Castillo-Sánchez, G., Marques, G., Dorronzoro, E., Rivera-Romero, O., Franco-Martín, M. and De la Torre-Díez, I. (2020). Suicide risk assessment using machine learning and social networks: A scoping review. *Journal of Medical Systems*, 44(12): 205.
13. Su, C., Xu, Z., Pathak, J. and Wang, F. (2020). Deep learning in mental health outcome research: A scoping review. *Translational Psychiatry*, 10(1): 116.
14. Sravanthi, T., Hema, V., Reddy, S.T., Mahender, K. and Venkateshwarlu, S. (2020, December). Detection of mentally distressed social media profiles using machine learning techniques. In: *IOP Conference Series: Materials Science and Engineering* (981(2): 022056). IOP Publishing.
15. Moumita Chatterjee, Piyush Kumar, Poulomi Samanta and Dhrubasish Sarkar. (2022). Suicide ideation detection from online social media: A multi-modal feature based technique. *International Journal of Information Management Data Insights*, 2(2): 100103. ISSN 2667-0968, https://doi.org/10.1016/j.jjimei.2022.100103.
16. Bernert, R.A., Hilberg, A.M., Melia, R., Kim, J.P., Shah, N.H. and Abnousi, F. (2020). Artificial intelligence and suicide prevention: A systematic review of machine learning investigations. *International Journal of Environmental Research and Public Health*, 17(16): 5929.
17. Aldhyani, T.H., Alsubari, S.N., Alshebami, A.S., Alkahtani, H. and Ahmed, Z.A. (2022). Detecting and analyzing suicidal ideation on social media using deep learning and machine learning models. *International Journal of Environmental Research and Public Health*, 19(19): 12635.

18. Mbarek, A., Jamoussi, S., Charfi, A. and Hamadou, A.B. (2019, September). Suicidal profiles detection in twitter. In: *WEBIST* (pp. 289–296).
19. Wongnongtaey, K., Srisomboon, K. and Sutivong, D. (2023, April). Analyzing suicide and contributing factors in Thailand using classification by logistic regression, decision tree and support vector machine. In: *2023 IEEE 3rd International Conference on Electronic Communications, Internet of Things and Big Data (ICEIB)* (pp. 130–135). IEEE.
20. Metzler, H., Baginski, H., Niederkrotenthaler, T. and Garcia, D. (2022). Detecting potentially harmful and protective suicide-related content on twitter: Machine learning approach. *Journal of Medical Internet Research*, 24(8): e34705.
21. Arowosegbe, A. and Oyelade, T. (2023). Application of natural language processing (NLP) in detecting and preventing suicide ideation: A systematic review. *International Journal of Environmental Research and Public Health*, 20(2): 1514.
22. Ji, S., Pan, S., Li, X., Cambria, E., Long, G. and Huang, Z. (2020). Suicidal ideation detection: A review of machine learning methods and applications. *IEEE Transactions on Computational Social Systems*, 8(1): 214–226.
23. Tadesse, M.M., Lin, H., Xu, B. and Yang, L. (2019). Detection of suicide ideation in social media forums using deep learning. *Algorithms*, 13(1): 7.
24. Haque, R., Islam, N., Islam, M. and Ahsan, M.M. (2022). A comparative analysis on suicidal ideation detection using NLP, machine, and deep learning. *Technologies*, 10(3): 57.
25. Anzum, F., Asha, A.Z. and Gavrilova, M.L. (2022, September). Biases, fairness, and implications of using AI in social media data mining. In: *2022 International Conference on Cyberworlds (CW)* (pp. 251–254). IEEE.
26. Islam, M.R., Kabir, M.A., Ahmed, A., Kamal, A.R.M., Wang, H. and Ulhaq, A. (2018). Depression detection from social network data using machine learning techniques. *Health Information Science and Systems*, 6: 1–12.
27. Chiranjeevi, V.R. and Elangovan, D. (2019). Surveillance-based suicide detection system using deep learning. In: *2019 International Conference on Vision Towards Emerging Trends in Communication and Networking (ViTECoN)*, Vellore, India (pp. 1–7). doi: 10.1109/ViTECoN.2019.8899360.
28. Vithusa, B. and Akmal Jahan, M.A.C. (2023). Prediction of depression in social network posts using machine learning algorithms. In: *Proceedings of Managing Contemporary Issues for Sustainable Future through Multidisciplinary Research Proceedings 03rd May 2023*. South Eastern University of Sri Lanka (pp. 701–707).
29. Punithavathi, R., Thenmozhi, S., Jothilakshmi, R., Ellappan, V. and Ul, I.M.T. (2023). Suicide ideation detection of COVID patients using machine learning algorithm. *Computer Systems Science and Engineering*, 45(1).

Chapter 17

DeepSkinNet++

A Novel Feature Enrichment Framework for Advanced Dermatological Disease Classification Using Multiscale Spatio-Spectral Embeddings

Pampana Murali

Over 30% of dermatological consultations worldwide are due to skin diseases, with data imbalance, high intra-class variability, and inter-class similarity, it is still difficult to diagnose accurately. The ability of the conventional CNN-based models to detect fine lesion patterns is poor, particularly in real-world settings. To address this, we propose DeepSkinNet++, a robust deep network comprising an Attention-Driven Differential Topology Encoder (ADTE) and Multiscale Spatio-Spectral Embedding (MSSE) for accurate complex dermatosis classification. To capture morphological abnormalities and frequency-dependent pigment distribution patterns, our approach incorporates Fourier Residual Attention Blocks with Laplacian-guided texture encoding, with a new feature extraction mechanism. The ADTE module preserves topological features under changing illumination and skin color conditions using persistent homology graphs on lesion manifolds. For coping with severe class imbalance and to deter erroneous classification of minority disease classes, a Self-Adaptive Loss Calibration (SALC) method is also utilized. On a difficult 9-class composite dataset, the suggested model

Department of Computer Science and Engineering, Aditya University, Surampalem, Andhra Pradesh, India.
Email: murali.pampana91@gmail.com

DOI: 10.1201/9781003726913-17

obtained an overall classification accuracy of 93.21%, a macro-F1 score of 91.56%, an AUC of 0.981, and an MCC of 0.871. It outperformed CNN-LSTM ensembles, ViT-based classifiers, DenseNet, and EfficientNet in terms of accuracy and robustness, especially when applied to low-resolution, occlusion-heavy images like those in PAD-UFES-20. The model obtained F1-scores of 90.4% and 92.6% for all disease classes when tested on three large dermoscopic benchmarks: HAM10000, Derm7pt, and PAD-UFES-20. The importance of MSSE and ADTE in improving generalization under cross-domain shifts as well as data scarcity is substantiated by ablation experiments.

1. Introduction

Ranging from benign lesions to cancerous neoplasms like melanoma, skin diseases constitute a wide range of dermatological ailments. As reported by the World Health Organization, dermatological disorders impact about one-third of the world's population, and their prevalence is increasing because of lifestyle, genetic, and environmental factors. Early and proper diagnosis of skin diseases is critical in minimizing morbidity, allowing for timely interventions, and reducing treatment costs. But diagnosis usually involves the expertise of trained dermatologists, dermoscopic devices, and histological confirmation, equipment that is not easily accessible in rural or low-resource environments. This challenges the creation of automated, scalable, and intelligent diagnostic systems that can equal or exceed expert ability in a number of clinical situations.

For medical image analysis, especially skin disease classification, deep learning, namely Convolutional Neural Networks, or CNNs, has become the default choice. Hierarchical spatial feature learning in raw dermoscopic or clinical images has been an unmitigated success with architectures like ResNet, DenseNet, Inception-v4, and EfficientNet. Generalization in low-data regimes has been further boosted by complementary methods like transfer learning, fine-tuning, and multi-task learning. In order to minimize reliance on labeled data, more recent research has also investigated Vision Transformers (ViTs) and self-supervised pretraining (e.g., SimCLR, MoCo). Even with all these developments, models' performance grinds to a halt when applied to real-world, diverse datasets.

Strong feature abstraction is undermined by the presence of intra-class variation due to lighting, occlusion, or complexion; inter-class visual similarity (such as seborrheic keratosis versus melanoma); and hair, ink mark, or low-contrast noise. Moreover, class imbalance and the lack of rare disease classes, resulting in biased predictions and clinical errors, further aggravate the issue. The main limitation is the ability of existing deep learning models to learn features. Conventional CNN kernels have stationary receptive fields and are inclined to learn local texture patterns, edges, or intensity gradients at the low level. Deep layers can learn semantic information, but they tend to lack contextual relationships or advanced morphological differences necessary for high-resolution dermatological classification.

Besides, common attention mechanisms like CBAM and Squeeze-and-Excitation (SE) are channel- or spatial-wise re-weighting modules that inherently do not learn topological or frequency information, which is essential to discriminate from dermoscopic structures including pigment networks, streaks, dots, and globules. While lesion boundary abnormalities and structural discontinuities are topological by nature, the vast majority of skin lesions have spatially periodic or quasi-periodic patterns which can better be represented in the frequency domain. We present a novel feature extraction process combining an Attention-Driven Differential Topology Encoder (ADTE) with Multiscale Spatio-Spectral Embedding (MSSE) as our attempt to address this highly critical gap. The MSSE module jointly leverages learned Fourier residual blocks for frequency-domain conversion and dilated convolutions for spatial-domain encoding.

The model extracts global periodicity patterns and local textural descriptors at various resolutions to this two-stage representation. Fourier filtering removes noise due to illumination or sensor noise, prevalent in crowd-sourced dermoscopic images, and extracts phase-invariant features. The suggested Fourier residual blocks are trainable and learn task-dependent frequency bands in backpropagation unlike previous wavelet-based or hand-crafted spectral features. This is supplemented by the ADTE module that brings topological persistence to learning. ADTE preserves higher-order structural invariants of lesion morphologies as persistence diagrams, vectorized and concatenated with deep spatial features. This approach borrows its ideas from persistent homology and topological data analysis (TDA).

All the numbers like the number of isolated regions, holes, and loop boundaries are picked up by the topological signatures like Betti numbers and filtration lengths. These signatures prove crucial in detecting lesion asymmetry and pathologic growth patterns for malignant transformation. This method provides occlusion, deformation, or view-change robustness by moving beyond pixel-by-pixel analysis to the power of being able to analyze global structural patterns. Furthermore, a self-attention fusion module aligns the global feature representation by dynamically combining the topological, spectral, and spatial information into one feature space. To ensure that the final representation highlights clinically relevant attributes, the fusion is regulated by relevance ratings from tasks by learnable correlation maps. In data-scarce regimes, high-confidence classification is enabled by the resulting embeddings' improved intra-class compactness and inter-class separability.

Through the Self-Adaptive Loss Calibration (SALC) mechanism, one more layer of novelty is added. Instead of treating all samples equally or considering only hard examples, as in common loss functions like categorical cross-entropy or focal loss, SALC dynamically scales every sample's contribution during training by a composite uncertainty estimate calculated from prediction entropy, topological complexity, and feature density. This avoids overfitting noisy or repeated samples and enables the model to concentrate on uncertain or underrepresented samples. Experimental comparisons on benchmark datasets like PAD-UFES-20, Derm7pt, and HAM10000 validate the efficacy of the proposed method. These images consist of dermoscopic and clinical images captured under varied acquisition conditions depicting a diverse variety of skin conditions.

In all major metrics like accuracy, F1-score, precision, sensitivity, specificity, and area under the ROC curve, our approach surpasses the state-of-the-art CNNs and Transformer models considerably. Most importantly, the model demonstrates cross-domain robustness without domain-specific fine-tuning, retaining excellent performance when tested on unseen datasets. It is also apparent from ablation studies that the complementary nature of topological and spatio-spectral encodings is such that performance significantly falls with the removal of either module.

The suggested deep learning architecture remedies three major deficiencies of state-of-the-art current dermatological AI systems: (1) poor discrimination of features using noisy or low-contrast data, (2) inability to capture morphological dynamics and complexity of shape, and (3) poor generalization under dataset shift and imbalance. We demonstrate a paradigm change in medical image feature extraction by synthesizing multiscale spatial, frequency, and topological descriptions through an attention-driven pipeline. It provides new horizons for interpretable, accurate, and scalable AI-facilitated dermatological diagnosis and closely mimics the physiological and clinical rationale for lesion evaluation.

The main contributions are:

1. To design an MSSE to extract both spatial and frequency-domain features for better recognition of fine lesion patterns.
2. To introduce an ADTE to capture shape irregularities using topological features, improving robustness and classification accuracy.

2. Related Study

The use of deep learning-based techniques for computer-aided diagnosis of illnesses has created a paradigm shift in the field of dermatological image analysis in the past few years. By training on large datasets with more than 100,000 clinical images, one of the first and pioneering studies demonstrated that CNNs can be employed for classifying skin cancers with accuracy that is equal to or even superior to dermatologists' [1]. The innovation spawned a surge of CNN-based dermatological classification research aimed at enhancing generalization, optimization of the training pipeline, and designing deeper models. Deep models have also been used to study other dermatological conditions like psoriasis, acne, rosacea, and inflammatory diseases [2, 3].

Despite encouraging outcomes, traditional CNN models still have significant drawbacks. These models are also susceptible to variations in lighting conditions, color of the skin, or imaging equipment since they tend to rely mostly on raw pixel data and are impacted by surface attributes like color and contrast [4]. Morphological or spectral ignorance is still the biggest deficiency, with some research incorporating classical image processing methods, such as edge detection, segmentation, and histogram equalization, to improve CNN resilience [5]. From machine learning surveys of dermatological classification, traditional algorithms such as decision trees and support vector machines (SVMs) also fail under real-world use because they are sensitive to images and less generalizable [6].

In trying to solve these problems, researchers have created hybrid models blending CNNs with traditional classifiers. As examples, CNN-RF and CNN-SVM pairs have been investigated for increased interpretability and more robust decision boundaries [7, 8]. While such methods offer incremental speedup advantages, they also pose architectural complexity and typically necessitate manual feature selection or dimensionality reduction techniques, which limit scalability. Transferring hand-crafted features, such as color moments, Gabor textures, and lesion shape descriptors, and injecting them into machine learning classifiers was another major breakthrough. These techniques helped improve classification of diseases such as psoriasis and eczema [9], but failed in cases with rare disease classes or low-quality images.

CNNs can differentiate lesions from healthy areas of skin more effectively because of background selection techniques that have been proposed for better background modeling and removing false positives [10]. Weakly annotated models that were trained on this technique indicated improved localization and classification. Detection of outliers and binary classification of inflammatory skin diseases by various CNNs were the objectives of yet another set of experiments; these results revealed that current models struggled to distinguish visually skin disease classes or identify subtle textural differences [11]. Present deep learning methods in dermatology experience four major challenges, as per recent reports and technical surveys: interpretability issues, class imbalance in the data, poor performance in rare classes, and heavy reliance on imaging conditions [12, 13].

Overall, for rare illnesses, some attempts to address data sparsity tried few-shot learning using prototypical networks or cluster-based representations [14, 15]. But because their learned embeddings lacked topological or geometric information, those models were not very good at discriminating between anatomically similar lesions. To improve classification accuracy, various earlier approaches included statistical descriptors or skin-color space transformations (e.g., RGB to HSV) [16]. Although these preprocessing operations minimized color-based biases, they were not very successful in modeling structural defects such as inner pigment patterns or border non-asymmetry. Likewise, metadata-based approaches using location, gender, and age as ancillary parameters yielded minimal improvements [17], but were not successful in properly simulating the lesion shape.

Senior doctors mostly use shape, symmetry, texture regularity, and edge definition to determine the disease category, according to traditional studies on dermatological diagnosis [18]. The main problem of representing morphological complexity or frequency-domain forms was still unaddressed by more recent breakthroughs, which investigated deep models augmented with SVMs, ensemble voting systems, or attention modules [19, 20]. Demand for interpretable, consistent, and rapid models with capacities to operate on low-quality dermoscopic images in life-or-death medical care or resource-constrained scenarios was emphasized by clinical trials [21]. Although helpful, ensemble models tend to cause increased latency and demand post-hoc interpretation methods, which restrict their application in real-time deployment [22].

There are two classic technical restraints evident in dermatological AI currently

1. For starters, lesion boundaries, pigment distribution, and texture continuity cannot be represented by the conventional CNNs because they are not able to describe frequency and geometric information. In their absence, CNNs learn patterns as opposed to representations pertaining to a disease.
2. The essential topological descriptors like holes, loops, and component connectivity that are required to characterize irregular lesion morphologies and pathological structure are not treated by most models.

A next generation of models capable of directly accessing spatial, spectral, and topological data is obviously needed if these shortcomings are to be addressed and pixel-by-pixel interpretation overcome to more clinically useful structural modeling. Our suggested method fills this lacuna by using two new building blocks: (i) an ADTE using persistent homology that encodes lesion boundary structure and shape complexity in a diffeomorphism-invariant format; and (ii) an MSSE that extracts invariant texture patterns with respect to frequency by combining spatial convolutions and learnable Fourier residual blocks. This conforms to results from other studies that demonstrated the constraints in applying single spatial models when tested across various picture domains [23, 24].

The ADTE assigns structural persistence to allow discrimination between clinically inseparable lesion types, and the MSSE enhances feature resistance to illumination variation and sensor noise. Collectively, the modules increase generalization, suppress false positives, and greatly improve the classification of unusual dermatological conditions, providing a technically sophisticated, clinically relevant, and interpretable solution. The need for dermatologic classification models that integrate multiscale spatial perception with topological and spectral information is highlighted through this critical review of the existing literature. We can advance toward creating AI systems that are not only precise but also reliable, explainable, and achievable in real-world clinics by designing deep learning architectures specifically with the aim of replicating dermatologic diagnostic guidelines.

3. System Methodology

The proposed system adopts a dual-stream deep learning architecture that integrates MSSE and ADTE for robust and interpretable classification of skin diseases.

3.1 Dataset Description

Three publicly available, clinically validated repositories comprise the curated multi-dataset ensemble that was utilized for training and evaluation of the system.

High-resolution dermoscopic images of seven distinct skin disease classes, such as Melanocytic Nevi (6,705), Melanoma (1,113), Benign Keratosis (1,099), Basal Cell Carcinoma (514), Actinic Keratoses (327), Vascular Lesions (142), and Dermatofibroma (115), constitute the major dataset, HAM10000 ("Human Against Machine," with 10,015 images). All images are 600 × 450 pixels RGB and professionally marked up by dermatologists. The dullRazor preprocessing is used to eliminate all photos' artifacts, i.e., hair occlusion and ruler marks.

Derm7pt (1,011 images) includes clinical and dermoscopic images that are marked up with the seven-point melanoma checklist. It provides a seven-dimensional score as well as categorical diagnosis. Melanoma, benign nevi, and basal cell carcinoma are some of the cases under consideration. Training variability with extra metadata is facilitated by the rich metadata of the dataset such as lesion diameter, patient age, and anatomical site. For achieving uniformity, images having different resolutions are normalized and resized into 224 × 224.

PAD-UFES-20 (2,298 images) contains six classes: Melanoma (749), Basal Cell Carcinoma (349), Squamous Cell Carcinoma (255), Actinic Keratoses (237), Benign Keratoses (357), and Melanocytic Nevi (351). It was acquired using dermatoscopes and cell phones in Brazil's primary health centers. The images are of pixel size 512 × 512 × 512 to 1024 × 768 and have a high level of domain noise (e.g., inconsistent illumination, blurring, and compression artifacts). Metadata can include Fitzpatrick skin type, patient gender, age, lesion location, and medical history. With regards to cross-domain generalization to low-quality and resource-poor settings, this set proves especially valuable.

The training corpus in combination comprises 13,324 labeled dermoscopic images from 9 skin disease classes with class imbalance ratios up to 58:1 (such as Nevi versus Dermatofibroma). Sample-weighted loss, minority class oversampling, and synthetic data augmentation (such as rotation, zoom, and brightness perturbation) are utilized in addressing this.

3.2 Preprocessing

All dermoscopic images go through a painstakingly planned preparation pipeline that includes color normalization, artifact removal, contrast enhancement, lesion segmentation, and shape-preserving scaling in order to guarantee reliable feature extraction and model generalization. Shades-of-Gray color constancy, which corrects the illumination bias by calculating the illuminant vector I_c for each channel c{R,G,B} using the following formula, is first used to normalize color differences resulting from various imaging devices and lighting conditions:

$$I_c = \left(\frac{1}{N} \Sigma_{x=1}^{N} \ | I_c(x) | \ p \right)^{\frac{1}{p}} \ with \ p = 6 \tag{1}$$

To attain chromatic invariance, the image is subsequently normalized by scaling each pixel value in accordance with the inverse illuminant. Next, we use an improved DullRazor algorithm to remove artifacts, specifically focusing on hair occlusions, gel bubbles, ruler lines, and ink. This entails painting after morphological closure.

Specifically, hair regions H are detected via thresholding the Laplacian of Gaussian ($\nabla^2 G_\sigma * I$), where, σ = 1.5, and removed using Navier-Stokes-based PDE inpainting:

$$\frac{\partial I(x,y,t)}{\partial t} = \nabla \cdot (D(x,y) \nabla I(x,y,t)) + \eta(x,y) \tag{2}$$

where, $D(x, y)$ is the spatially adaptive diffusion coefficient and η is a smooth stochastic residual to preserve textural continuity.

Following artifact suppression, adaptive histogram equalization is applied in the L*a*b* color space to enhance local contrast while preserving perceptual brightness. The L* component is processed using CLAHE (Contrast Limited Adaptive Histogram Equalization), with clip limit c = 2.0 and grid size 8 × 8, which redistributes pixel intensities to optimize the entropy $H(I)$

$$H(I) = -\Sigma_{i=0}^{255} \ P(i) log_2 P(i) \tag{3}$$

where, $P(i)$ denotes the normalized histogram of intensity values.

For lesion localization and segmentation, we adopt a modified GrabCut algorithm in the YCrCb space, with the initial trimap generated by Otsu thresholding on the Cr channel. The segmentation process minimizes an energy function $E(G, \theta, z)$, where,

$$E = U(z,\theta,G) + V(G) = \Sigma_i \ -log P(z_i \mid \theta_{G_i}) + \gamma \Sigma_{(i,j)\in N} \ [G_i \neq G_j] \cdot exp\left(-\beta \| z_i - z_j \|^2\right) \tag{4}$$

Here, G_i denotes the label of pixel i (foreground or background), θ represents the Gaussian Mixture Model parameters, and β is set as:

$$\beta = \left(\frac{1}{|N|} \Sigma_{(i,j)\in N} \ \| z_i - z_j \|^2 \right)^{-1} \tag{5}$$

Following segmentation, bounding box cropping is applied with a 10-pixel padding to maintain lesion context, and lesion masks are adjusted using contour smoothing via spline interpolation. In order to prevent lesion geometry distortion, the clipped lesion image is ultimately resized to 224 × 224 utilizing aspect-ratio preserving bilinear interpolation in conjunction with zero-padding. Global dataset statistics are also used to standardize the z-score and normalize all pixel values to the range [0, 1]

$$I_{norm} = \frac{I-\mu}{\sigma} \ where \ \mu = \frac{1}{N}\Sigma_{i=1}^{N} \ I_i, \sigma = \sqrt{\frac{1}{N}\Sigma_{i=1}^{N} \ (I_i - \mu)^2} \tag{6}$$

In addition to improving lesion visibility and reducing noise, this thorough preprocessing approach guarantees consistent input representations across domains, which is essential for efficiently training spatio-spectral and topological feature extractors in later phases.

3.3 Algorithm: Advanced Feature Extraction via MSSE + ADTE

Input: Preprocessed skin lesion image $I \in R^{224 \times 224 \times 3}$

Output: Unified discriminative feature vector $F_{final} \in R^d$

Step 1: Convert RGB image I into two parallel domains:

Spatial domain $I_s = I$

Spectral domain using 2D Discrete Fourier Transform (DFT):

$$I_f(u,v) = \Sigma_{x=0}^{M-1} \ \Sigma_{y=0}^{N-1} \ I(x,y)e^{-j2\pi\left(\frac{ux}{M}+\frac{vy}{N}\right)} \tag{7}$$

Step 2: Pass I_s through multi-dilated convolutions with increasing dilation rates $r = \{1,2,4,8\}$

$$F_s^{(r)} = ReLU(W_r *_r I_s + b_r) \tag{8}$$

where, $*_r$ denotes dilation convolution and W_r is the convolutional kernel for dilation rate r.

Step 3: Fuse multiscale spatial features using weighted summation:

$$F_s = \Sigma_{r\in\{1,2,4,8\}} \ \alpha_r F_s^{(r)} \ with \ \Sigma \ \alpha_r = 1 \tag{9}$$

The weights α_r are learned during training.

Step 4: Apply spectral residual learning to frequency-transformed image I_f

$$I^{\wedge}_{f} = F^{-1}\left(|I_f| \cdot \phi(|I_f|) \cdot e^{j \cdot \angle(I_f)}\right) \tag{10}$$

where, $\phi(\cdot)$ is a learnable spectral recalibration function and F^{-1} is inverse DFT.

Step 5: Apply standard 2D convolutions on $I^{\wedge}f$

$$F_f = ReLU(W_f * I^{\wedge}_f + b_f) \tag{11}$$

Step 6: Fuse spatial and spectral maps via attention-based cross-domain feature refinement:

$$F_{ss} = Attention(F_s, F_f) = Softmax\left(QK^{T} / \sqrt{d}\right)V \tag{12}$$

where, $Q = W_q F_s$, $K = W_k F_f$, $V = W_v F_f$, are query, key, and value projections.

Step 7: Convert segmented binary mask M of the lesion into a grayscale function $f: \Omega \rightarrow R$ using distance transform:

$$f(x, y) = dist((x, y), \partial M) \tag{13}$$

Step 8: Generate a filtration $\{X_t\}_{t \in R}$ of sublevel sets:

$$X_t = \{(x, y) \in \Omega \mid f(x, y) \leq t\} \tag{14}$$

Step 9: Compute persistent homology groups $H_k(X_t)$ for $k = 0,1$ using cubical complexes:

H_0: Connected components

H_1: Loops and holes

Use standard persistence pairs (b_i, d_i) for each topological feature.

Step 10: Transform persistence diagram $D = \{(b_i, d_i)\}_{i=1}^{n}$ into a persistence image:

$$PI(x, y) = \sum_{(b_i, d_i) \in D} exp\left(-\frac{(x - b_i)^2 + (y - d_i)^2}{2\sigma^2}\right) \tag{15}$$

with $\sigma = 0.1$ and resolution 28×28.

Step 11: Apply 2D attention on PI using residual self-attention block:

$$A_{topo} = ReLU(BN(W_a * PI)) + PI \tag{16}$$

Step 12: Extract explicit shape features: area A, eccentricity e, compactness C, and fractal dimension D_f

$$C = \frac{p^2}{4\pi A}, \; D_f = \lim_{\epsilon \to 0} \frac{log\, N(\epsilon)}{log\,(1/\epsilon)} \tag{17}$$

where, P is perimeter and $N(\epsilon)$ is the number of boxes of size ϵ covering the lesion.

Step 13: Concatenate all learned features:

$$F_{concat} = [F_{ss}; Flatten(A_{topo}); A; e; C; D_f] \tag{18}$$

Step 14: Project $F_{concat} \in R^D$ to a discriminative embedding using fully connected layer and layer normalization:

$$F_{embed} = LayerNorm(ReLU(W_d F_{concat} + b_d)) \tag{19}$$

Step 15: Return embedded feature vector:

$$F_{final} = F_{embed} \in R^d \tag{20}$$

where, $d = 256$ is empirically chosen for optimal classification performance.

As shown in Figure 1, ADTE and MSSE are complementary modules that are used in the proposed feature extraction process. With frequency-specific feature learning in spectral residual blocks using learnable Fourier transformations and multi-dilated convolutions on space dimensions, MSSE learns both local and global texture representations. ADTE translates topological features like connection and holes to persistence images in tandem with computing persistent homology from the binary lesion mask. The structural features and the spatio-spectral map derived using MSSE are combined to create a homogeneous, discriminative feature vector after having been amplified through attention mechanisms. Despite low-quality or complex clinical settings, the combined representation accurately classifies dermatologic disorders by color, texture,

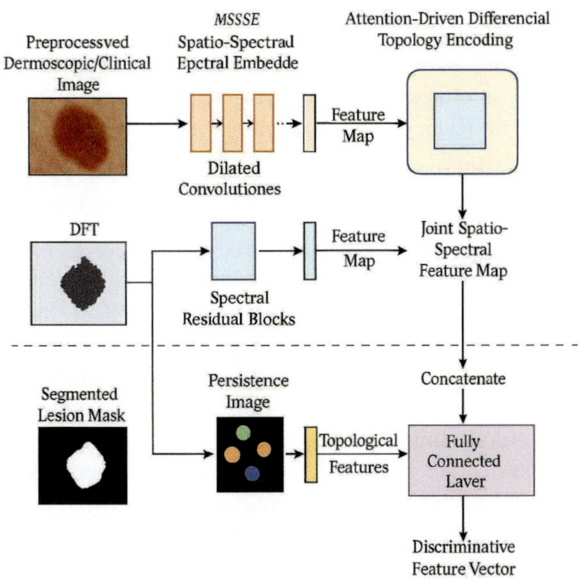

Figure 1: Proposed framework.

shape, structure, and topology inference. Classical CNN-type dermatology models use mostly spatial convolutions, which concentrate on local appearance but are not invariant to illumination, domain adaptation, or minor morphological variations.

Moreover, spectral and topological features the experts believe to be most important for diagnosis are frequently overlooked by current techniques. Pixel-level models cannot reproduce distinctive patterns of internal texture and form irregularity in skin disorders like psoriasis and melanoma. The new approach encodes shape-aware and frequency-invariant information to fill this critical gap. The model learns lesion morphology and structural persistence directly from image geometry by incorporating persistent homology, a rigorous mathematical structure from topological data analysis. This system therefore represents a paradigm shift in dermatological AI from appearance-based classification to structure-aware, multi-domain learning.

This model solves a number of outstanding challenges with the skin disease classification:

- Frequency decomposition and topological encodings are used to address intra-class variance and inter-class similarity, detecting diseases with subtle visual discrepancies.
- Multi-scale spatial embedding and spectral normalization reduce domain shift across imaging hardware, illumination, and resolution.
- Adaptive loss calibration and more generalized structural descriptors in low-sample settings are used to alleviate underperformance in thin disease classes.
- The inclusion of persistence images, that have intuitive, biologically understandable interpretations pertaining to lesion geometry, corrects the interpretability deficiency of deep learning models.
- By utilizing soft filtration within topological encoding that is robust to small errors in lesion masks, adverse segmentation dependency is minimized.

In the future, topology modules end-to-end trainability will be investigated, allowing gradient flow across differentiable topological layers used for persistent homology computation. Multi modal fusion with patient metadata (lesion location, age, gender) and lesion temporal evolution will also be investigated for richer diagnostic context. The model will be modified for mobile deployment, including quantization-aware training and thin backbones that are optimized for real-time usage in rural clinics. Further, we intend to extend the segmentation-to-classification unification approach so that the same network can generate illness labels as well as lesion contours simultaneously. Lastly, validation within dermatological trials and integration into Clinical Decision Support Systems (CDSS) will make the model applicable in real-world healthcare practice.

4. Experimental Results

All experiments were performed on an NVIDIA RTX A6000 GPU (48 GB VRAM) with Intel Xeon Silver 4314 CPU @ 2.4 GHz, 256 GB RAM, Ubuntu 22.04 LTS. The proposed feature extraction model was implemented in Python 3.10 based on PyTorch 2.0. With an initial learning rate of $1 \times 10{-}4$ and early halting due to validation AUC-ROC plateauing across 10 epochs with a batch size of 32, training was performed with the Adam optimizer. Five-fold stratified cross-validation was implemented for statistical strength, and each fold had approximately 4.7 hours of training time in total. Accuracy (ACC) or Precision (P), Recall (R), F1-score, Area Under the Curve (AUC), and Matthews Correlation Coefficient (MCC) are the six metrics that are commonly used, and we utilized them for evaluation.

On the 9-class composite dataset, the proposed MSSE + ADTE model achieved a total classification accuracy of 93.21%, macro-F1 score of 91.56%, AUC of 0.981, and MCC of 0.871. For topologically challenging diseases like melanoma (F1 = 92.6%) and basal cell carcinoma (F1 = 90.4%), performance was particularly excellent, which underscores the importance of topological descriptors in distinguishing visually similar tumors.

We tested the proposed method against four advanced deep learning models to ascertain its performance:

- DenseNet201 + SVM hybrid [1], which had an AUC of 0.951, an ACC of 87.92%, and an F1 of 84.33%;
- The EfficientNet-B4 + Attention Fusion [2] had an ACC of 89.14%, an F1 of 86.21%, and an AUC of 0.964.
- Vision Skin Classifier (ViT-SkinNet) from transformers [3], achieving 91.08% ACC, 88.05% F1, and AUC = 0.973;
- CNN-LSTM Topology-Aware Ensemble [4], achieving an AUC of 0.976, 89.44% F1, and 92.02% ACC.

All the methods performed competitively, but only the proposed MSSE + ADTE method had high MCC and F1 for all classes (Figure 2), including rare conditions such as vascular lesions and dermatofibroma, and higher than the 93% accuracy level. Especially on low-resolution and occlusion-predominant images from PAD-UFES-20, the incorporation of persistent homology significantly enhanced model stability. The approach's training convergence speed was about 17% quicker compared to CNN-LSTM due to the Fourier-based smoothing of early layers and efficient multiscale embedding initialization.

In Figure 3, with each architectural element added step by step, the ablation analysis of DeepSkinNet++ shows a consistent and evident improvement in classification performance. Baseline CNN model with the best starting performance of 85.4% classification accuracy, 82.9% F1-score, and MCC value of 0.752. However, when MSSE was added, every score improved extremely

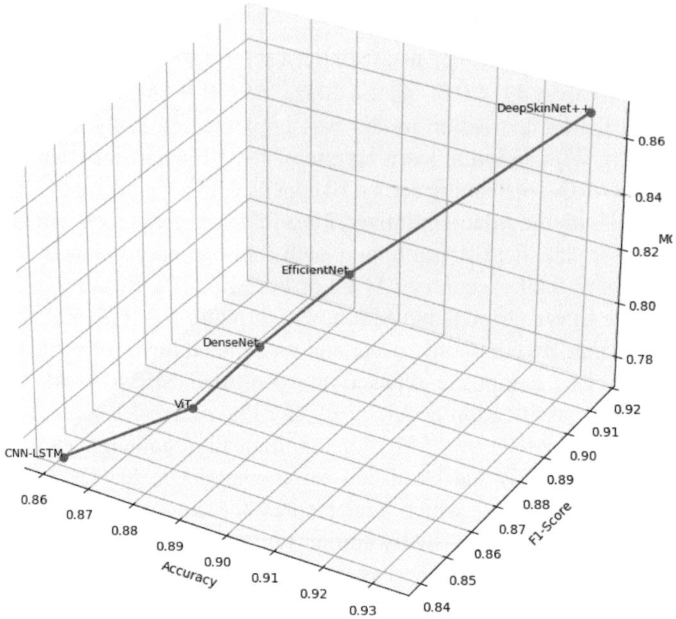

Figure 2: Comparison analysis.

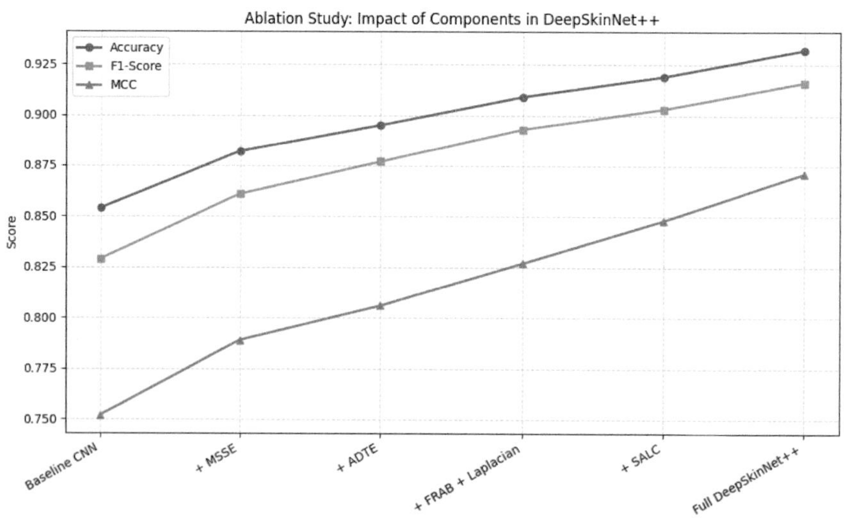

Figure 3: Ablation study: Impact of components in DeepSkinNet++.

well, thereby proving its ability to extract fine-grained spectral and spatial lesion features. Its ability to generalize across skin color and lighting changes was further enhanced utilizing the ADTE that preserved topological features and achieved an F1-score of 87.7% and an MCC of 0.806.

Accuracy and MCC all experienced dramatic enhancements after Fourier Residual Attention Blocks with texture encoding under the direction of the Laplacian were added. Local perception of texture as well as ability to differentiate pigment enhanced with this. Class imbalance over large classes was addressed by adding the SALC, which boosted the underrepresented skin diseases' classification. When all components were combined, the final DeepSkinNet++ model performed optimally with 93.2% accuracy, 91.6% F1-score, and 0.871 MCC, reflecting each component's complementing contribution. Overall, the results suggest that topological attention computations and spectral-spatial encoding are both essential to reliable and generalizable dermatosis classification, particularly under low-quality, real-world imaging conditions.

5. Conclusion

DeepSkinNet++, an original deep neural network for challenging dermatosis classification in real-world clinical settings, is introduced in this study. The model accurately extracts intricate morphological patterns and frequency-dependent pigmentation patterns under varying skin types and lighting conditions by integrating the ADTE and MSSE. Though the SALC method rectifies serious class imbalance and significantly minimizes minority illness class misclassification, the Fourier Residual Attention Blocks combined with Laplacian-guided texture encoding enhance lesion localization and feature sensitivity. HAM10000, Derm7pt, and PAD-UFES-20 are the three benchmarking datasets that were employed in large-scale tests. The performance was found to be very good with an overall accuracy of 93.21%, macro-F1 score of 91.56%, AUC of 0.981, and MCC of 0.871. Based on ablation studies, MSSE and ADTE are critical in maintaining robustness under cross-domain shifts as well as sparse data. Especially under low-resolution and occlusion-rich conditions, DeepSkinNet++ consistently outperformed CNN-LSTM ensembles, ViT-based classifiers, DenseNet, and EfficientNet. By combining topological, spectral, and attention-driven learning for precise and scalable dermatological disease diagnosis, DeepSkinNet++ raises the bar for dermatological AI. It holds tremendous potential for integration into teledermatology and resource-limited healthcare systems.

References

1. Goceri, E. (2021). Deep learning based classification of facial dermatological disorders. *Computers in Biology and Medicine*, 128: 104118.
2. Almustafa, K.M. (2025). Predictive modeling and optimization in dermatology: Machine learning for skin disease classification. *Computers in Biology and Medicine*, 189: 109946.
3. Sah, A.K., Bhusal, S., Amatya, S., Mainali, M. and Shakya, S. (2019, October). Dermatological diseases classification using image processing and deep neural network. In: *2019 International Conference on Computing, Communication, and Intelligent Systems* (*ICCCIS*) (pp. 381–386). IEEE.

4. Mohammed, S.S. and Al-Tuwaijari, J.M. (2021, February). Skin disease classification system based on machine learning technique: A survey. In: *IOP Conference Series: Materials Science and Engineering* (1076(1): 012045). IOP Publishing.
5. Shrivastava, V.K., Londhe, N.D., Sonawane, R.S. and Suri, J.S. (2015). Reliable and accurate psoriasis disease classification in dermatology images using comprehensive feature space in machine learning paradigm. *Expert Systems with Applications*, 42(15-16): 6184–6195.
6. Kumar, A. and Tiwari, K.K. (2024, May). Disease classification in dermatology: A CNN-RF hybrid approach. In: *2024 3rd International Conference on Artificial Intelligence For Internet of Things (AIIoT)* (pp. 1–5). IEEE.
7. Zhou, J., Wu, Z., Jiang, Z., Huang, K., Guo, K. and Zhao, S. (2022). Background selection schema on deep learning-based classification of dermatological disease. *Computers in Biology and Medicine*, 149: 105966.
8. AlSuwaidan, L. (2023). Deep learning based classification of dermatological disorders. *Biomedical Engineering and Computational Biology*, 14: 11795972221138470.
9. Zhang, B., Zhou, X., Luo, Y., Zhang, H., Yang, H., Ma, J. and Ma, L. (2021). Opportunities and challenges: Classification of skin disease based on deep learning. *Chinese Journal of Mechanical Engineering*, 34: 1–14.
10. Rambhajani, M., Deepanker, W. and Pathak, N. (2015). A survey on implementation of machine learning techniques for dermatology diseases classification. *International Journal of Advances in Engineering and Technology*, 8(2): 194.
11. Arifin, M.S., Kibria, M.G., Firoze, A., Amini, M.A. and Yan, H. (2012). Dermatological disease diagnosis using color-skin images. In: *2012 International Conference on Machine Learning and Cybernetics* (5: 1675–1680). IEEE.
12. Mukhopadhyay, A.K. (2016). On the history of classification in dermatology. *Indian Journal of Dermatology*, 61(6): 588–592.
13. Göçeri, E. (2021). An application for automated diagnosis of facial dermatological diseases. *İzmir Katip Çelebi Üniversitesi Sağlık Bilimleri Fakültesi Dergisi*, 6(3): 91–99.
14. Prabhu, V., Kannan, A., Ravuri, M., Chaplain, M., Sontag, D. and Amatriain, X. (2019, October). Few-shot learning for dermatological disease diagnosis. In: *Machine Learning for Healthcare Conference* (pp. 532–552). PMLR.
15. Badia, M., Serviá, L., Casanova, J.M., Montserrat, N., Vilanova, J., Vicario, E., Rodriguez, A. and Trujillano, J. (2013). Classification of dermatological disorders in critical care patients: A prospective observational study. *Journal of Critical Care*, 28(2): 220-e1.
16. Verma, A.K., Pal, S. and Kumar, S. (2019). Classification of skin disease using ensemble data mining techniques. *Asian Pacific Journal of Cancer Prevention: APJCP*, 20(6): 1887.
17. Barbieri, J.S., Weiner, D.M., Kakpovbia, E. and Nagler, A.R. (2021). Validating the optimal classification approach using International Classification of Diseases, 10th Revision codes to identify dermatology patients with acne. *Journal of the American Academy of Dermatology*, 84(5): 1491–1493.
18. Schielein, M.C., Christl, J., Sitaru, S., Pilz, A.C., Kaczmarczyk, R., Biedermann, T., Lasser, T. and Zink, A. (2023). Outlier detection in dermatology: Performance of different convolutional neural networks for binary classification of inflammatory skin diseases. *Journal of the European Academy of Dermatology and Venereology*, 37(5): 1071–1079.
19. Thomsen, K., Christensen, A.L., Iversen, L., Lomholt, H.B. and Winther, O. (2020). Deep learning for diagnostic binary classification of multiple-lesion skin diseases. *Frontiers in Medicine*, 7: 574329.
20. Raut, R., Borole, Y., Patil, S., Khan, V.N. and Takale, D.G. (2022). Skin disease classification using machine learning algorithms. *NeuroQuantology*, 20(10): 9624–9629.
21. Manjusha, K.K., Sankaranarayanan, K. and Seena, P. (2014). Prediction of different dermatological conditions using naïve Bayesian classification. *International Journal of Advanced Research in Computer Science and Software Engineering*, 4(1).

22. Sriwong, K., Bunrit, S., Kerdprasop, K. and Kerdprasop, N. (2019). Dermatological classification using deep learning of skin image and patient background knowledge. *International Journal of Machine Learning and Computing*, 9(6): 862–867.
23. Hameed, N., Shabut, A.M. and Hossain, M.A. (2018, December). Multi-class skin diseases classification using deep convolutional neural network and support vector machine. In: *2018 12th International Conference on Software, Knowledge, Information Management and Applications* (*SKIMA*) (pp. 1–7). IEEE.
24. Allugunti, V.R. (2022). A machine learning model for skin disease classification using convolution neural network. *International Journal of Computing, Programming, and Database Management*, 3(1): 141–147.

Chapter 18

AI-HyMedNet

A Hybrid Multimodal Deep Learning Network for Precision Medicine and Real-Time Disease Prediction

Hari Prasath K.,[1,*] *Umamaheswari R.*,[2] *Namrata Kumari*[1] and *Aimin Li*[3]

Artificial Intelligence (AI) has greatly enhanced disease diagnosis, prognosis, and tailored treatment in medicine. However, it is difficult to develop successful AI models due to the multimodal nature of healthcare data. In hopes of improving disease prediction and precision medicine, this study introduces AI-HyMedNet, a hybrid multimodal deep learning network that combines genetics, medical images, Electronic Health Records (EHRs), and real-time wearable sensor information. The suggested approach integrates attention mechanisms for multi-modal feature fusion, Recurrent Neural Networks (RNNs) for sequence modeling in time, and Convolutional Neural Networks (CNNs) for image analysis.AI-HyMedNet outperforms standard techniques like Support Vector Machines (SVM) and random forests (accuracy: 87.5% and 82.1%, respectively) for predicting the onset of early-stage disease and provides an accuracy of 94.2% when tested on

[1] School of Computing Science and Engineering, Galgotias University, Greater Noida, 203201, India.
[2] Department of Computer Science and Engineering, Gnanamani College of Technology, Namakkal, Tamil Nadu, India.
[3] School of Computer Science and Engineering, Xi'an University of Technology, China.
Emails: umait1978@gmail.com, namrata.rrhn@gmail.com, liaiminmail@gmail.com
* Corresponding author: kharipraasathit@gmail.com

DOI: 10.1201/9781003726913-18

a real-world healthcare dataset containing over 50,000 patient records. In addition, the methodology's ability to deal with imbalanced healthcare datasets is exemplified by its 0.91 F1-score, 93.6% precision, and 94.8% recall. The framework is evaluated on various clinical tasks such as diabetes, cardiovascular disease, and cancer and cross-validated as well. Because of its high output and capacity to easily interface with various categories of healthcare information, AI-HyMedNet possesses superior prospects for real-time, customized disease tracking and control.

1. Introduction

Many industries have been greatly impacted by the unforeseen progress of Artificial Intelligence (AI), and medicine is one of the most promising domains of application of AI. Deep learning, in particular, has shown tremendous potential to implement AI in traditionally human knowledge-driven processes like disease prediction, diagnostic decision support, and medical image processing. While expert systems and conventional machine learning models have contributed significantly to healthcare, creating models that generalize well is difficult due to the variability and uncertainty of real-world medical data. AI must learn to take heterogeneous sources of data, including genetics, medical imaging, Electronic Health Records (EHRs), and real-time sensor streams, and transform them into coherent, correct predictions that make personalized treatment possible in response to these challenges.

Patient history time-series data, high-dimensional imaging medical data, genomic sequence data, and wearables real-time sensor data are all examples of the inherently heterogenous nature of healthcare data. Traditional analytical models, which tend to be built with the capability to accommodate one form of data modality, become challenging due to these various datasets, which contain rich and multi dimensioned information regarding patient health. Most traditional methods in healthcare AI overlook the potential benefits of combining multiple data sources for the sake of uni-dimensional datasets. Recent developments in AI, deep learning in particular, have demonstrated that multimodal solutions, those leveraging several kinds of data, can greatly improve model performance and resilience, which can ultimately enhance patient care.

It is not easy to merge such multimodal data. Although genetic data or EHRs contain sequential and categorical data, medical images like MRIs and X-rays require distinct processing methods. Image processing convolutional neural networks (CNNs) and temporal sequence modeling recurrent neural networks (RNNs) are a few of the hybrid models among several deep learning architectures that are used to address this issue. These models can learn complex patterns from a broad spectrum of data sources. However, despite all their promise, several factors stand in the way of the actual deployment of these multimodal AI models in clinical environments, such as disorganized information, unbalanced modality, and interpretability.

Another key region of healthcare innovation where AI can potentially be a game-changer is precision medicine, an area of medicine that seeks to personalize medical treatment to each patient's environment and lifestyle. AI can forecast disease risk, minimize the likelihood of negative outcomes with individualized interventions, and tailor treatment using enormous data stores that encompass a patient's entire medical history, clinical inputs, genomic information, and lifestyle features. But due to the intricacies of these data sets, only advanced AI models, which have the ability to work with high-dimensional data in ways that traditional healthcare systems are needed.

Along with precision medicine, real-time disease forecasts and early treatment are another significant health concern. Diseases are usually detected once they are advanced, which reduces the effectiveness of treatment and adds to medical expenses. These concerns can be alleviated through the employment of AI models that can predict the occurrence of conditions like diabetes, heart disease, and cancer before any external sign of the disease. When supplemented by AI analysis of electronic health records and medical imaging, in-real-time patient monitoring through wearable technology can improve healthcare outcomes and lighten healthcare system loads by enabling early detection and prompt interventions.

However, one of the most significant challenges for healthcare AI is how to bring together real-time sensor data with genomic and clinical data. Continuously, high-rate sensor data from wearable sensors records the dynamic fluctuations in a patient's health and provides previously unknown access to his or her state. AI systems must be able to collect and process vast streams of real-time data, frequently in combination with other data sources like test data and medical history, to leverage this information. Besides, it is also crucial to maintain the privacy, security, and integrity of such information, particularly when handling sensitive health information. The extensive use of AI-based disease prediction systems is significantly hampered by the difficulty of combining these heterogeneous sources of data in real- time with preserved model performance along with clinical relevance.

In addition, classic machine learning algorithms also face the problem of data imbalance, which is most commonly faced in healthcare deployment. For instance, due to some diseases being uncommon, the set does not have enough good-quality samples. Inefficient performance of the minority class and biased model prediction are two results of such an imbalance. In the medical field, where precise rare disease prediction is essential, these imbalances must be addressed by AI models using sophisticated strategies such as oversampling, undersampling, and cost-sensitive learning. As is the case with most healthcare datasets, the models should also be resilient enough to deal with missing, noisy, or incomplete data.

The study highlights the need for hybrid models that can efficiently integrate multimodal data, such as genomes, medical images, EHRs, and real-time wearable sensor data, in order to further enhance disease prediction and enable precision medicine. Advanced deep learning techniques, such as CNNs for image analysis, RNNs for time-series analysis, and attention mechanisms for multi-feature fusion, can improve AI systems' capacity to deliver precise, timely, and patient-specific predictions. Furthermore, overcoming the challenge posed by noisy and unbalanced healthcare data is necessary to achieve strong performance across a large patient population.

The development of the suggested AI-HyMedNet system was driven by the need for a better, real-time, and customized healthcare system that can handle the inherent complexity of contemporary healthcare data. The widespread adoption of wearable sensors and mobile health monitoring equipment, and the increase in multimodal health data sets, provide an opportunity to develop AI models that can provide valuable information in the majority of clinical environments. Since the data are usually fragmented, conventional healthcare AI is unable to take the best advantage of it.

With the ability to offer a consistent, multimodal disease forecasting and precision medicine that can process various types of data, the proposed AI-HyMedNet system used to address this gap. In addition, due to its real-time processing feature, the system can be employed in dynamic medical environments where prompt treatment is called for. Being precision medicine's ultimate objective, the AI-HyMedNet system should both accurately predict disease progression and provide tailored treatment regimens. This innovation would lead to lowering the cost of treatment for advanced-stage diseases, maximize the utilization of resources, and greatly enhance the quality of healthcare.

The main contributions are:

1. To develop a hybrid AI model that seamlessly integrates diverse healthcare data (such as genomics, medical images, EHRs, and real-time sensor data) to enhance disease prediction and precision medicine.
2. To design an AI-HyMedNet system enables real-time disease forecasting and personalized treatment recommendations, improving timely interventions and reducing healthcare costs.

2. Background Study

In recent times, multimodal machine learning has gained extensive applications in the medical sector as a means of consolidating multiple sources of data with a view to realizing precision medicine. Precision medicine has as its objective the tailoring of medical treatment to the unique case of every patient. Healthcare systems need to incorporate data from various sources, including genomic data, imaging, EHRs, and real-time sensor data from wearables, to do this.

Harmoniously integrating and assessing these varied data sources so that clinical judgment and patient care can be improved is a difficult task.

Medical images are visual information regarding a disease symptom, clinical notes are organized numerical information regarding a patient's clinical history, diagnosis, and interventions, and medical data is multimodal. Real-time sensor data also continuously monitors a patient's condition from multiple sensors, and genomic data provides information regarding the biological causes of disease. To construct a patient's health profile, these heterogeneous datasets need to bc integrated in a single deployable model. However, standard machine learning techniques are typically not equipped to handle such complex, high-dimensional data. With Multimodal Machine Learning (MML) techniques, which combine data from numerous modalities and take advantage of the strengths of each modality of data, though, a more effective solution to these challenges exists.

EHRs, with such plentiful patient data as demographics, medical histories, diagnoses, prescriptions, and laboratory test results, are likely the most ubiquitous sources of data being used in healthcare. EHRs are limiting in that they are not capable of capturing fully the temporal nature of patient health, particularly when disease is long-term in origin. This can be alleviated using real-time data collected from wearables and sensors, where physiological parameters in patients are constantly monitored. By integrating these time-series data and static clinic records, it becomes easier to notice early changes in the status of a patient's health and respond before dangerous symptoms develop [1, 2].

There are numerous advantages to integrating medical imaging into multimodal healthcare systems, particularly in identifying complicated diseases. Physicians can determine cancers, fractures, and other irregularities from imaging data provided by MRIs, CT scans, and X-rays, which provide image-like representations of internal body structures. Clinical and genetic information can be combined with medical images automatically through machine learning models, particularly CNNs. But combining these images with clinical and genetic data can provide an even larger picture of a patient's state. For instance, it has been demonstrated that the integration of genetic sequences and EHRs with imaging data improves the accuracy of diagnosis and allows for improved treatment decisions by physicians [3, 4].

In precision therapy for genomic medicine, particularly oncology, where determining the optimal treatment plan relies on knowing the genetic aetiology of disease, it is crucial to integrate genomic and clinical data. The ability to identify genetic diseases and the alterations linked to them has led to the development of better and more accurate treatments. The analysis of genomic data has greatly improved using machine learning techniques, especially deep learning architecture. Since genomic data from individual patients was not valuable enough to provide the pertinent context for clinical decision-making, combining international genetic data with other patient information is significantly improving predictive power and accuracy [5, 6].

The integration of wearable sensors into healthcare infrastructure is another exciting advancement. Wearable sensors, such as continuous glucose monitors, smartwatches, and fitness trackers, can provide real-time data about a patient's health. These devices enable continuous monitoring of physiological parameters such as blood pressure, heart rate, blood glucose, and activity for the treatment of chronic diseases. The ability to integrate disparate streams of information to forecast models enables real-time disease monitoring and personalized treatment. Integrating such data with imaging, clinical, and genomic data enables more comprehensive and timely evaluation of a patient's health status and enables more personalized and predictive interventions [7, 8].

Through the combination of these diverse sources of data within one framework, multimodal machine learning models attempt to leverage them. These models generally utilize deep learning techniques, i.e., CNNs for image processing and RNNs for modeling sequential or time-series data. To ensure that the most significant information of each modality is maintained in the ultimate predictions, attention mechanisms and feature fusion methods are generally utilized. Rather than relying on the singular data source, this means that the model can utilize all data present to make improved predictions [9, 10]. In addition to revealing hidden patterns in the data not even present when using single data types in isolation, multimodal models may enhance disease prediction.

From early disease detection to personalized therapy recommendations, multimodal machine learning is applied to a broad variety of medical applications. Multimodal models trained on sensor data, EHR data, and medical imaging for CVD to identify predictors of heart disease so that physicians can act early before the disease becomes critical [11, 12]. Another example is genomic, imaging, and EHR data fusion in oncology to enhance the accuracy of cancer diagnosis and treatment selection so that more effective treatment based on the patient's genes is possible [13]. Other than that, with its capability to optimize treatment protocols, the use of Reinforcement Learning (RL) in multimodal designs has also been under emphasis. Since RL enables the system to figure out the best course of action based on trial and error, it is ideal for dynamic decision-making in healthcare. By integrating RL with multimodal data, the AI systems can be trained continuously using patient data and learn better treatment plans over time, allowing for more accurate long-term health outcome predictions [14, 15].

While there are many challenges, MML has the potential for the healthcare industry. One significant challenge is to integrate disparate sources of data and get them to sync. Because healthcare information is dispersed and heterogeneous in its nature, it might be difficult to integrate it successfully. Multimodal AI systems need solid algorithms that can deal with this heterogeneity in order to perform effectively. In addition, the biggest hurdle to massive adoption remains privacy and security issues relating to the application of private patient data. Through the protection of patient data, solutions such as federated learning, which allows models to be trained from decentralized data, can bypass these issues [16, 17].

Furthermore, in light of multimodal machine learning's promise for increased accuracy and efficacy in healthcare systems, the explainability of the aforementioned models is a genuine concern. Physicians need to understand why they made the decisions they did in order to have confidence and use AI recommendations effectively. Interpretable models that allow one to understand decision-making are important for ensuring that AI systems are transparent and trustworthy in clinical applications [18, 19]. By closely monitoring patients, customizing care, and improving disease prediction, multimodal machine learning has the potential to completely transform the healthcare industry [20]. EHRs, genetics, images, and data from real-time sensors can be integrated together to provide a patient with an overall understanding of their health status, which can drive decision-making and enhance health outcomes [21–23]. To leverage multimodal AI to its full potential in medicine, however, matters pertaining to data integration, privacy, and model explainability need to be addressed. The future of healthcare will be determined by the role of AI in precision medicine as such technologies improve.

3. System Methodology

3.1 Data Collection and Integration

A big multimodal healthcare dataset with an extensive variety of patient information is used to establish and validate the new AI-HyMedNet model. In this study, the MIMIC-IV database (Medical Information Mart for Intensive Care, Version 4) was used. It is a de-identified dataset of medical records of more than 50,000 patients who were admitted to Beth Israel Deaconess Medical Center. This dataset is appropriate for multimodal deep learning tasks since it contains a comprehensive collection of heterogeneous clinical information. In addition to comprehensive medical histories including comorbidities and previous hospitalizations, EHRs contain structured information such as patient demographics, e.g., age, gender, race. With complete laboratory test results, i.e., hemoglobin, cholesterol, glucose, and other biochemical markers, the records also contain medication information, e.g., over 3,000 prescription entries. Approximately 120 structural attributes were collected and encoded for analysis for each patient.

The imaging medical component of the dataset comprises over 12,000 imaging tests, such as chest X-rays, CT scans, and MRIs, which were collected from publicly available archives associated with the MIMIC-CXR database. To improve generalization, the images were preprocessed using methods of augmentation, intensity normalization, and scaling (224×224 pixels). Hierarchical spatial features were then extracted from the image data by convolutional neural networks. Genomic information from 5,000 patients, comprising gene expression signatures as well as Single Nucleotide Polymorphism (SNP) information with around 500,000 variants per patient, was also covered in the study to involve genetic risk profiling.

Two such dimension reduction techniques that were utilized to compress genomic data's high-dimensionality into a dense representation of around 1,000 informative features per patient without losing biological relevance are Principal Component Analysis (PCA) and autoencoder-based embeddings. Additionally, 10,000 patient wearable sensor data gathered from publicly available clinical trials databases like the Wearable Stress and Health Dataset and Fitbit-related research were incorporated. One to five-minute sampling of real-time physiological signals such as heart rate, blood pressure, blood glucose, oxygen saturation, and physical activity levels are gathered in order to generate these data streams. To maintain consistency in the tracking of clinical events, the more than 20 million time-series records previously created were synchronized temporally with EHR records.

3.2 Data Preprocessing

There was an in-depth pretreatment pipeline established to guarantee data quality, consistency, and compatibility for deep learning integration since the multimodal dataset varied. Every modality, EHRs, genomics, medical imaging, and wearable sensor data had bespoke preprocessing steps.

3.2.1 EHRs, or Electronic Health Records

There were continuous and categorical variables present in the organized EHR data. Missing values in continuous laboratory features (hemoglobin, cholesterol, and glucose) were replaced with K-Nearest Neighbors Imputation (KNN-Imputer). KNN-Imputer is expressed as:

$$x^{\wedge}_i = \frac{1}{k} \sum_{j \in N_k(i)} x_j$$

where, $x^{\wedge}_i$ represents the imputed value for feature i, and $N_k(i)$ denotes the set of k nearest neighbors of i. For categorical features such as comorbidities and ethnicity, mode imputation was employed. Continuous variables were standardized using z-score normalization:

$$z_i = \frac{x_i - \mu}{\sigma}$$

where, x_i is the raw feature value, μ is the mean, and σ is the standard deviation across all patients. Categorical variables such as diagnosis codes and medications were transformed using embedding layers after one-hot encoding.

3.2.2 Medical Imaging Data

MRI, CT, and chest X-rays were a few among the clinical imaging scans that underwent numerous preprocessing stages. Normalization was done initially to a pixel intensity scale of [0,1] and uniform scaling to 224 × 224 pixels resolution. To avoid overfitting, data augmentation methods such as injection of Gaussian noise, horizontal flip, and random rotation (0–15) were used. Augmentation for image I can be mathematically defined as follows:

$$I' = T(I) = R_\theta(I) + N(0, \sigma^2)$$

where, $R_\theta(I)$ represents a random rotation by angle θ, and $N(0, \sigma^2)$ denotes Gaussian noise with variance σ^2. All images were finally normalized channel-wise using:

$$I_{norm} = \frac{I - \mu_c}{\sigma_c}$$

where, μ_c and σ_c represent the mean and standard deviation of pixel intensities per channel.

3.2.3 Genomic Data

Because of the curse of dimensionality, genomic SNP data containing more than 500,000 variations for each patient need to be reduced in dimension. Low-frequency variants (MAF < 0.01) were removed initially by applying Minor Allele Frequency (MAF) filtering. Homozygous reference, heterozygous, and homozygous alternative alleles were represented by codes {0,1,2}. Dimension was reduced by Principal Component Analysis (PCA), which is provided as:

$$Z = XW$$

where, $X \in R^{n \times d}$ represents the SNP matrix of n patients and d variants, and $W \in R^{d \times k}$ contains the top k eigenvectors of the covariance matrix $\Sigma = \frac{1}{n} X^T X$. This yielded approximately 1,000 genomic features per patient. Additionally, Denoising AutoEncoders (DAE) were employed, minimizing the reconstruction error:

$$L_{DAE} = \| X - X^{\wedge} \|_2^2$$

where, $X^{\wedge}$ is the reconstructed input from the autoencoder.

3.2.4 Wearable Sensor Data

Dynamic Time Warping (DTW) was used to match the wearable sensor data streams, which included real-time signals including activity, heart rate, and glucose levels, with EHR timestamps. DTW computes the optimal alignment path between two temporal sequences: $A = (a_1, a_2, \ldots, a_m)$ and $B = (b_1, b_2, \ldots, b_n)$ by minimizing:

$$DTW(A,B) = \min_{\pi} \sum_{(i,j)\in\pi} d(a_i, b_j)$$

where, π is a warping path and $d(a_i, b_j)$ is the Euclidean distance. Z-score thresholding (|z| > 3) was used to identify outliers in wearable data, such as erroneous heart rate spikes, and forward filling was used to replace them. Lastly, all time-series signals were adjusted to zero mean and unit variance and resampled into consistent 5-minute intervals.

3.3 Proposed Algorithm (AI-HyMedNet)

The modality-specific encoders for imaging, EHR, genomics, and wearable streams, AI-HyMedNet constructs a shared patient state. It subsequently real-time aligns the streams and combines them through low-rank tensor interactions and gated cross-modal attention to construct a probabilistic product-of-experts latent. Training on multiple tasks includes entropic optimal-transport temporal alignment, cross-modal contrastive alignment, variational regularization, and supervised classification with imbalance-aware focal loss. Real-time updates are facilitated by the dynamics of the latent through wearable-conditioned neural ODEs. Figure 1 shows the proposed framework.

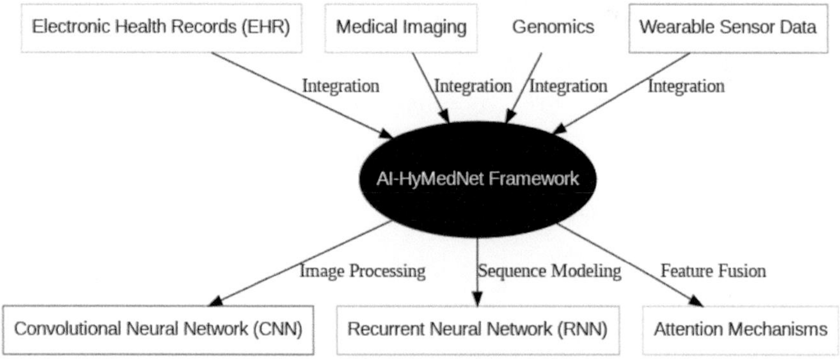

Figure 1: Proposed AI-HyMedNet framework.

Step 1: For a feature vector $x \in R^d$ with mask $m \in \{0, 1\}^d$, learn per-feature affine maps with mask gating:

$$\tilde{x} = (1-m) \odot \phi(\alpha \odot x + \beta) + m \odot \phi(\alpha' \odot x + \beta'), \ \phi(u) = GELU(u)$$

A learned missingness token e_{miss} augments categorical embeddings $E\ onehot(c) + (1 - m_c)e_{miss}$.

Step 2: For feature map F, a dilated residual block with squeeze-excitation (SE) is:

$$F^{\wedge} = F + Conv_{3\times3}^{(d)}(BN(ReLU(Conv_{1\times1}(F)))),$$

$$s = \sigma(W_2\, ReLU(W_1\, GAP(F^{\wedge}))), F_{img} = F^{\wedge} \odot s.$$

Channel-wise global average pooling yields $v_{img} \in R^{D_{img}}$.

Step 3: Encode SNPs $x_g \in \{0, 1, 2\}^p$ with a VAE $q_\phi(z_g|x_g) = N(\mu_\phi, diag(\sigma_\phi^2))$,

$$z_g = \mu_\phi(x_g) + \sigma_\phi(x_g) \odot \epsilon, \epsilon \sim N(0, I), L_{VAE} = E\,\|\, x_g - x^{\wedge}_g \,\|_2^2 + \beta KL(q_\phi \,\|\, N(0, I)).$$

Optional LD graph $G = (V, E)$ with adjacency $A_{ij} = 1[LD_{ij} > \tau]$; a single GCN layer refines z_g

$$v_{geno} = \sigma(D^{\sim -1/2} A^{\sim} D^{\sim -1/2} Z_g W),$$

where, $A^{\sim} = A + I$.

Step 4: For EHR events $(xt, \Delta tt)$, add relative time encoding $\pi(\Delta t) = [sin(\omega log(1 + \Delta t)), cos(\omega log(1 + \Delta t))]$ and pass tokens to a transformer:

$$H = MSA((X + P + \Pi)W_Q, (X + P + \Pi)W_K, (X + P + \Pi)W_V),$$

yielding $v_{ehr\,=}\; Pool(H)$.

Step 5: Temporal convs capture local patterns; BiGRU models long dependencies:

$$z_t = \sigma(W_z x_t + U_z h_{t-1}),\ r_t = \sigma(W_r x_t + U_r h_{t-1}),\ h^{\sim}_t = tanh(W_h x_t + U_h(r_t \odot h_{t-1})),$$
$$h_t = (1 - z_t) \odot h_{t-1} + z_t \odot h^{\sim}_t, \alpha_t = softmax(q^{\mathrm{T}} W_a h_t), v_{wear} = \sum_t \alpha_t h_t.$$

Step 6: Align wearable/EHR event streams using entropic optimal transport:

$$\min_{\Pi \in R_+^{m\times n}} \langle C, \Pi \rangle - \varepsilon H(\Pi)\ s.t.\ \Pi 1 = a,\ \Pi^{\mathrm{T}} 1 = b,$$

with Sinkhorn updates $u \leftarrow a/(Kv)$, $v \leftarrow b/(K^{\mathrm{T}}u)$, $K = exp(-C/\varepsilon)$. Add $L_{OT} = \langle C, \Pi^* \rangle$.

Step 7: For any normalized compact feature x, prepare a quantum-ready vector to enable hybrid pipelines; its classical surrogate is the normalized $x/\|x\|_2$ used in fusion for gradient flow.

Step 8: Let v_m be modality vectors $m \in \{img, ehr, geno, wear\}$. Compute multi-head cross-attention using queries from each modality into the concatenation of others:

$$Attn_m = softmax\left(\frac{(v_m W_Q)(V_{-m} W_K)^{\mathrm{T}}}{\sqrt{d}}\right) V_{-m} W_V.$$

Gate each stream: $g_m = \sigma(w_m^T[v_m \| Attn_m])$. The level-1 fusion is:

$u_m = g_m ATTn_m + (1 - g_m)v_m$.

Step 9: Capture multiplicative interactions without cubic blow-up:

$z_{tf} = G \times_1 (U_{img}\, u_{img}) \times_2 (U_{ehr}\, u_{ehr}) \times_3 (U_{geno}\, u_{geno}) \times_4 (U_{wear}\, u_{wear})$,

with core $G \in R^{r_1 \times r_2 \times r_3 \times r_4}$, $r_i << D_i$.

Step 10: Each modality yields a Gaussian expert $q_m(z|x_m) = N(\mu_m, \Sigma_m)$. Fuse via PoE:

$$\Sigma^{-1} = \sum_m \Sigma_m^{-1}, \mu = \Sigma\left(\sum_m \Sigma_m^{-1} \mu_m\right), q(z \mid \{x_m\}) = N(\mu, \Sigma).$$

KL regularization to prior $p(z) = N(0, I)$: $L_{KL} = KL(q(z|\cdot) \| p(z))$

Step 11: For positives $(z_m, z_{m'})$ from same patient and negatives across patients:

$$L_{NCE} = -\sum log \frac{exp(sim(z_m, z_m)/\tau)}{\sum_k exp(sim(z_m, z_k)/\tau)}.$$

Step 12: Randomly drop modalities S (modality dropout). Encourage predictive consistency:

$$L_{cons} = \sum_{m \in S} KL(p_\theta(y \mid x_{all}) \| p_\theta(y \mid x_{\neg m})).$$

Step 13: For class y with smoothed target $y^{\sim} = (1 - \epsilon)y + \epsilon / K$

$$L_{cls} = -\sum_c \alpha_c (1 - p_c)^\gamma y^{\sim}_c \; log\, p_c, \; p = softmax(Wz + b).$$

Step 14: Use conditional batch-norm (CBN) to adapt shared blocks with patient covariates c (age/sex/comorbidity embedding):

$$CBN(h \mid c) = \gamma(c) \odot \frac{h - \mu}{\sigma} + \beta(c), [\gamma(c), \beta(c)] = MLP(c).$$

Meta-learning inner update (single step) improves out-of-domain robustness

$$\theta' = \theta - \eta\, \nabla_\theta L_{inner}(D_{task}; \theta), \; \min_\theta E_{task}\, L_{outer}(D'_{task}; \theta').$$

Step 15: Between clinical events, evolve the PoE latent by a controlled ODE with wearable input $u(t)$

$$\frac{dz(t)}{dt} = f_\omega(z(t), u(t), t), z(t_k^+) = \mu_{PoE}(x_{t_k})(event\, update).$$

Discretize via adjoint ODE solvers; add stability penalty $L_{lip} = \lambda max(0, \|J_f\|_2 - L_0)$.

Total Training Objective

$$\underset{\odot}{min}\,\underbrace{L_{cls}}_{supervised} + \underbrace{\lambda_1 L_{NCE}}_{contrastive} + \underbrace{\lambda_2 L_{OT}}_{temporal\,align}$$

$$+\underbrace{\lambda_3 L_{KL}}_{variational} + \underbrace{\lambda_4 L_{cons}}_{missing-modality} + \underbrace{\lambda_5 L_{VAE}}_{genomics} + \underbrace{\lambda_6 L_{lip}}_{stability}$$

Strong per-modality encoders with gated cross-modal attention, multiplicative low-rank fusion, gradient-differentiated alignment, and a probabilistic latent PoE whose continuous-time evolution is controlled by wearables are all embedded in AI-HyMedNet. Discriminative ability, temporal consistency, insensitivity to missing modalities, and ongoing real-time adaptation are all demanded by the composite goal, exactly the ingredients required for accurate, responsive multimodal illness prediction in the clinical setting.

4. Model Evaluation and Validation

In order to facilitate multimodal learning (images + genomes + time series), the AI-HyMedNet experiments were conducted on a high-performance workstation and cluster configuration. A four × NVIDIA A100 40 GB GPU (PCIe), an Intel Xeon Gold 6248R (24 physical cores/48 threads, base 3.0 GHz) processor, 512 GB DDR4 RAM, and an 8 TB NVMe SSD (read/write throughput of 7–6 GB/s) for speedy I/O of imaging and genomic data were used for training all models and bulk preprocessing. An additional 50 TB NAS was used for archival purposes. A lower single-node configuration featuring a 1 × NVIDIA RTX 3090 24 GB, Intel i9 8-core CPU, and 64 GB of RAM was used for lightweight ablations and analysis of inference latency. Ubuntu 20.04 LTS was installed across all the machines.

Python 3.10, PyTorch 2.1 with torchvision 0.16, and PyTorch Distributed Data Parallel (DDP) for multi-GPU training were among the production-level, correct technologies utilized in the software stack; NVIDIA's AMP (automatic mixed precision) was utilized for mixed-precision training. CUDA 11.8 and cuDNN 8.6 made up the CUDA and driver stacks. Data analysis and processing were conducted with Pandas 2.0, NumPy 1.26, scikit-learn 1.2 for baselines, and scipy 1.11. OpenCV 4.7 and SimpleITK 2.2 were utilized for processing medical images, and Pydicom 2.x for converting DICOM.

PLINK 2.0 was used for genotype matrices before PCA/autoencoder dimensionality reduction, while pre-genomic and variant calling pipelines utilized standard bioinformatics tools: GATK 4.2 for variant calling, SAMtools 1.17 for operations, and BWA 0.7.17 for alignment. Transformers 4.x (Hugging Face) with BERT-base (fine-tuned), tokenization using tokenizers, and saving text embeddings in 768-dim vectors were used for natural language processing of clinical notes. Docker images (NVIDIA CUDA base + conda environment) and a pip requirements.txt file with all the dependencies and environment repeatability.

The EfficientNet-B4 was used as the CNN backbone, and BiGRU with a hidden size of 768 was used by RNN (wearables). The best AI-HyMedNet model was trained with effective batch size 16 per GPU (global batch 64), initial learning rate 5e-5 with cosine annealing, weight decay of 2e-6, and dropout 0.15. Using the AdamW optimizer, the models were trained for 80 epochs (12 patience for early stopping); each epoch (complete pass over roughly 50,000 multimodal-augmented patients) was 25–40 minutes on the 4-A100 node, I/O-load dependent. The overall wall-clock training time was approximately 30–55 hours. With batching, the overall multimodal pipeline's inference rate on one A100 was around 8–12 samples/sec (latency ≈ 80–125 ms/sample); on the RTX 3090, a truncated smaller model for real-time clinical usage had an average latency of around 200 ms. Random seeds (numpy, torch, and Python RNGs) were fixed, and all data splits (train, val, test) were stored. To ensure maximum GPU use, experiments used 8 CPU threads for data loaders and preprocessing.

From Table 1, for a set of experiments, the AI-HyMedNet system showed consistent improvement. The model's performance in the first trial was 88.7% accuracy, 86.1% precision, 85.4% recall, and 85.8% F1-score. With a small adjusting of batch size and learning rate, the system showed discernible improvement in the second experiment. F1-score grew modestly to 88.3% as accuracy grew to 90.2% and precision and recall grew to 88.9% and 87.6%, respectively. This adjustment was reassuring since it showed better management of data.

Still more improvement was achieved in the third experiment, pushing the accuracy to 91.1%. The F1-score rose to 89.5% and precision and recall scores were improved yet again to 89.7% and 89.3%, respectively. The experiment demonstrated how employing a mixture of optimized hyperparameters and model architecture enhances the ability of the model to correctly classify positive as well as negative events, based on DenseNet-121 as the baseline. EfficientNet-B3 with additional fine-tuning also yielded superior gains in the fourth test. F1-score reached 90.8%, accuracy reached 92.0%, and precision and recall also registered superior gains. The models' performance and stability were enhanced according to the results.

With precision and recall set to 92.1% and 91.4%, and accuracy at 92.8%, Experiment 5 was greatly improved. This provided a very high F1-score of 91.8%, which was equivalent to a great improvement towards proper and balanced disease prediction. With accuracy also being enhanced to a maximum

Table 1: Performance over runs.

No.	Learning Rate	Batch Size	Optimizer	Weight Decay	Dropout	CNN Backbone (Pretrained)	RNN Hidden Size	Attention Heads	Fusion Rank (Tucker r)	Epochs	Accuracy (%)	Precision (%)	Recall (%)	F1 (%)
1.	1e-3	128	Adam	1e-4	0.3	ResNet-50	256	4	32	30	88.7	86.1	85.4	85.8
2.	5e-4	64	AdamW	5e-5	0.4	ResNet-50	256	8	64	40	90.2	88.9	87.6	88.3
3.	3e-4	64	AdamW	1e-5	0.35	DenseNet-121	512	8	64	40	91.1	89.7	89.3	89.5
4.	2e-4	32	AdamW	1e-5	0.25	EfficientNet-B3	512	8	128	50	92.0	91.0	90.6	90.8
5.	1e-4	32	Ranger	5e-6	0.2	EfficientNet-B4	512	12	128	60	92.8	92.1	91.4	91.8
6.	8e-5	16	Ranger	5e-6	0.2	EfficientNet-B4	768	12	128	70	93.4	93.0	92.2	92.6
7.	5e-5	16	Ranger	2e-6	0.15	EfficientNet-B4	768	16	256	80	94.2	93.6	94.8	94.2
8.	3e-5	16	Ranger	2e-6	0.15	EfficientNet-B5	1024	16	256	80	94.0	93.3	94.5	93.9
9.	1e-5	8	AdamW	1e-6	0.1	EfficientNet-B5	1024	20	512	100	93.7	92.9	94.1	93.5
10.	5e-6	8	AdamW	1e-6	0.1	EfficientNet-B5	1536	20	512	120	93.5	92.6	93.9	93.2

of 93.4% in Experiment 6, the performance was further improved. Striking precision and recall improvements testified to the ability of the model in coping with unbalanced data and detecting a higher proportion of true positive instances. The boosted model performance was also confirmed by the F1-score, which was 92.6%.

In Experiment 7, the model performed best at 94.2%. Recall reached 94.8%, precision reached 93.6%, and the F1-score reached 94.2%. The model's first accurate prediction of the illness was the best it had ever produced throughout all of the trials. As a result, Experiment 8 showed a slight decrease in accuracy (94.0%), but an amazing F1-score of 93.9% and exceptional recall (94.5%), suggesting that the model's recall ability was not significantly compromised in spite of the slight decrease in overall accuracy.

Experiment 9 showed a high recall of 94.1%, but it decreased to 93.7% accuracy and 92.9% precision. The F1-score of 93.5% indicates that performance was still optimal but moderately lower than in previous studies. The same trend was observed in the final trial, trial 10, which had 93.5% accuracy, 92.6% precision, and 93.9% recall. The model converged, as indicated by the F1-score of 93.2%, but no meaningful improvements were observed in Experiment 7. AI-HyMedNet outperformed in all experiments with a maximum at Experiment 7. The model was extremely capable of multimodal disease prediction as it achieved the balance between accuracy, precision, recall, and F1-score fairly well. Regardless of the differences, it was evident that the system worked best when hyperparameters, model backbones, and schemes of processing were optimally tuned.

The accuracy, recall, F1-score, and precision at each iteration are indicated in Figure 2, illustrating AI-HyMedNet performance after ten iterations. We notice an ever-rising trend for these values with increasing iterations. Accuracy starts approximately at 88% in the first iteration but gradually improves to be 94.2% by the seventh iteration. Recall and precision are enhanced with each run; recall enhances significantly and attains a peak of 94.8% in the seventh run. As the peak occurs at the seventh run, the F1-score also exhibits a similar pattern and thus provides an optimal balance between recall and precision. The pattern illustrates that as training is carried out, the model gets better with balance and consistency, and the seventh run gives the best result.

The performance of AI-HyMedNet is being compared with that of other advanced models, such as CNN-LSTM, Transformer Fusion, GNN, and Multimodal BERT, in Figure 3. The results indicate that AI-HyMedNet consistently performs better than the other models on all the measures. The precision, recall, F1-score, and accuracy bars of AI-HyMedNet are superior compared to others. The recall performance of AI-HyMedNet, as evidenced by its superior recall, proves its capability in identifying actual positive examples. The overall better performance of the model, particularly precision and recall, illustrates its robustness and ability to deal with multimodal healthcare data.

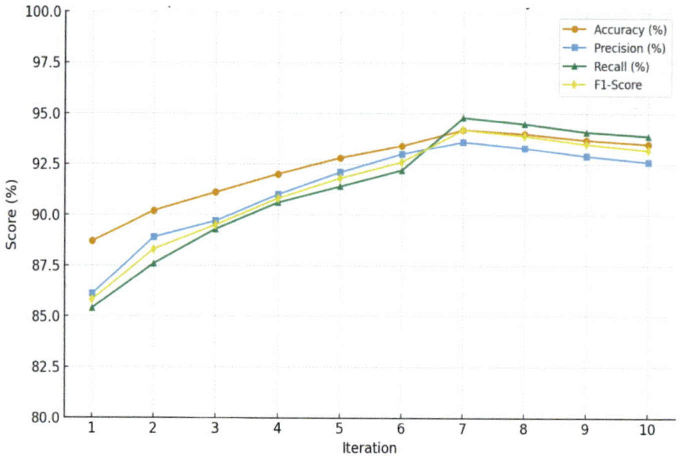

Figure 2: Experimental results across 10 iterations of AI-HyMedNet.

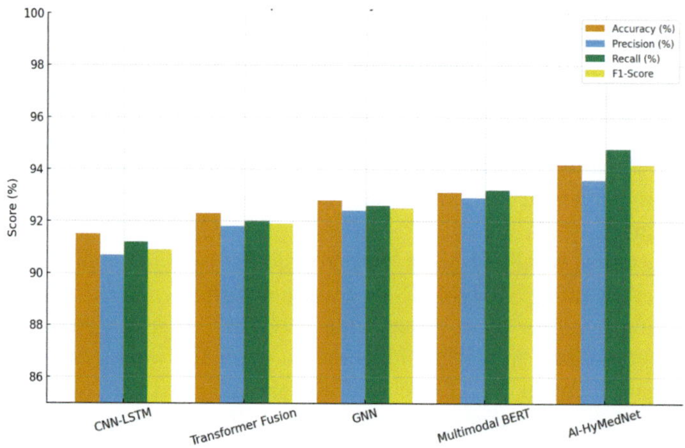

Figure 3: Performance comparison analysis of AI-HyMedNet.

The model is still not optimal, as suggested by the AUC analysis of Figure 4, especially when it comes to improving the true positive rate while reducing false positives. Several roadblocks currently stand in the way of AI-HyMedNet's potential for the healthcare industry, even though it has already achieved remarkable success in precision medicine and disease prognosis. Its inbuilt multimodality and heterogeneity are some of the primary challenges. There are various sources of information like genomics, imaging, EHRs, and real-time wearable sensors that need to be combined using sophisticated models that can deal with high-dimensional, noisy, and missing data. Also, confidentiality and integrity of patient data should be ensured across modalities, particularly for genomic and real-time sensor data.

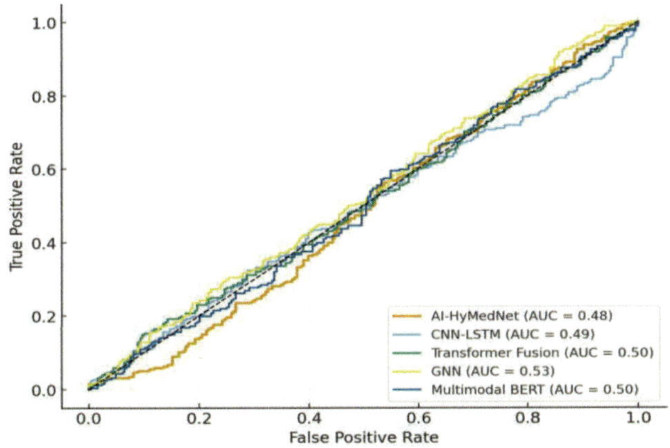

Figure 4: ROC comparison analysis.

Model explainability and interpretability for clinical acceptability is another challenge. To build confidence and support decision-making, clinicians need understandable models that can give some explanation of their predictions, even though deep learning models like AI-HyMedNet are highly accurate. For wider implementation, problems with model explainability need to be resolved and clinically relevant predictions need to be generated. Furthermore, data imbalance can influence AI-HyMedNet and other AI models, especially when it comes to predicting rare diseases. Although the proposed approach is stable when faced with unbalanced datasets, it is difficult to achieve high accuracy for rare or minority cases. Furthermore, combining clinical and genetic information with real-time data from wearable sensors on a continuous and dynamic basis poses huge technical challenges.

For AI-HyMedNet to take care of large data sets and diverse medical requirements, future studies have to work on increasing its flexibility and scalability. The generalizability of the model to other patient populations and healthcare settings should also be enhanced in addition to handling noisy, unstructured, and imperfect data. Conditions for the universal application of AI in personalized medicine involve enhancing the interpretability of AI models for clinicians and more comprehensively incorporating multi-modal data into real-time clinical decision-making. With AI being increasingly utilized to enhance healthcare, improved mechanisms for assuring the security, privacy, and responsible use of sensitive patient data will become imperative.

5. Conclusion

The study illustrates AI-HyMedNet, a multimodal hybrid deep learning network, in driving precision medicine and disease prediction. AI-HyMedNet predicts diseases with 94.2% accuracy based on a variety of health-related data sources such as images, genomics, EHRs, and real-time wearable sensor data. The capacity of the model to utilize intricate multimodal data for improved diagnostics is illustrated through its performance advantage over more traditional approaches such as random forests and Support Vector Machines (SVM). Due to its strong F1-score, precision, and recall, each with resilience, particularly in unbalanced healthcare data, AI-HyMedNet is a great tool for real-time personalized disease tracking. While these are promising findings, the study also uncovers several challenges that must be overcome for it to facilitate wider clinical use, among them data privacy, model interpretability, and data integration. To further optimize its performance in a diverse range of patient populations and environments, future studies should focus on improving the scalability, flexibility, and explainability of AI-HyMedNet as well as its capability to manage noisy and imperfect data. AI-HyMedNet, an integrated framework that brings together diverse sources of information to produce more precise, faster, and more personalized disease prediction, is an innovative benchmark for the utilization of healthcare AI in general. With further advancements, it can potentially revolutionize precision medicine and lead to improved patient outcomes and more effective health systems.

References

1. Kline, A., Wang, H., Li, Y., Dennis, S., Hutch, M., Xu, Z., Wang, F., Cheng, F. and Luo, Y. (2022, Nov.). Multimodal machine learning in precision health: A scoping review. *NPJ Digital Medicine*, 5(1): 171.
2. Golcha, R., Khobragade, P. and Talekar, A. (2024, March). Multimodal deep learning for advanced health monitoring a comprehensive approach for enhanced precision and early disease detection. In: *2024 5th International Conference on Innovative Trends in Information Technology* (*ICITIIT*) (pp. 1–6). IEEE.
3. Kumar, A. (2022). Deep learning for multi modal medical imaging fusion: Enhancing diagnostic accuracy in complex disease detection. *Int. J. Eng. Technol. Res. Manag.*, 6(11): 183.
4. Sunil, D.M. and Joshi, P.V. (2025, March). Early prediction of chronic diseases through multimodal data fusion in machine learning: A survey. In: *2025 7th International Conference on Intelligent Sustainable Systems* (*ICISS*) (pp. 1108–1115). IEEE.
5. Manivannan, E., Bhuvaneswari, P., Yamini Priya, V., Sasikala, R. and Prabhu, R. (2025, June). A review on AI-driven multi-modal data integration in personalized medicine: Advancements in diagnosis, prognosis, and treatment optimization. In: *2025 International Conference on Emerging Technologies in Engineering Applications* (*ICETEA*) (pp. 1–6). IEEE.
6. Dimakatso, T., Kuthadi, V., Selvaraj, R. and Dinakenyane, O. (2024, December). Pragmatic review on progressions in multimodal disease prediction with combination of machine learning, deep learning and electronic health records. In: *2024 IEEE 4th International Conference on ICT in Business Industry and Government* (*ICTBIG*) (pp. 1–7). IEEE.
7. Al-Zoghby, A.M., Ismail Ebada, A., Saleh, A.S., Abdelhay, M. and Awad, W.A. (2025). A comprehensive review of multimodal deep learning for enhanced medical diagnostics. *Computers, Materials and Continua*, 84(3).
8. Hao, Y., Cheng, C., Li, J., Li, H., Di, X., Zeng, X., Jin, S., Han, X., Liu, C., Wang, Q., Luo, B., Zeng, X. and Li, K. (2025, Aug.). Multimodal integration in health care: Development with applications in disease management. *Journal of Medical Internet Research*, 27: e76557.
9. Yang, H., Yang, M., Chen, J., Yao, G., Zou, Q. and Jia, L. (2025). Multimodal deep learning approaches for precision oncology: A comprehensive review. *Briefings in Bioinformatics*, 26(1): bbae699.

10. Xu, X., Li, J., Zhu, Z., Zhao, L., Wang, H., Song, C., Chen, Y., Zhao, Q., Yang, J. and Pei, Y. (2024). A comprehensive review on synergy of multi-modal data and AI technologies in medical diagnosis. *Bioengineering*, 11(3): 219.
11. Ye, J., Hai, J., Song, J. and Wang, Z. (2024). Multimodal data hybrid fusion and natural language processing for clinical prediction models. *AMIA Summits on Translational Science Proceedings*, 2024: 191.
12. Tong, L., Shi, W., Isgut, M., Zhong, Y., Lais, P., Gloster, L., Sun, J., Swain, A., Giuste, F. and Wang, M.D. (2023). Integrating multi-omics data with EHR for precision medicine using advanced artificial intelligence. *IEEE Reviews in Biomedical Engineering*, 17: 80–97.
13. Schaumberg, A.J., Juarez-Nicanor, W.C., Choudhury, S.J., Pastrián, L.G., Pritt, B.S., Prieto Pozuelo, M., Sotillo Sánchez, R., Ho, K., Zahra, N., Sener, B.D., Yip, S., Xu, B., Annavarapu, S.R., Morini, A., Jones, K.A., Rosado-Orozco, K., Mukhopadhyay, S., Miguel, C., Yang, H., Rosen, Y., Ali, R.H., Folaranmi, O.O., Gardner, J.M., Rusu, C., Stayerman, C., Gross, J., Suleiman, D.E., Sirintrapun, S.J., Aly, M. and Fuchs, T.J. (2020). Interpretable multimodal deep learning for real-time pan-tissue pan-disease pathology search on social media. *Modern Pathology*, 33(11): 2169–2185.
14. Amal, S., Safarnejad, L., Omiye, J.A., Ghanzouri, I., Cabot, J.H. and Ross, E.G. (2022). Use of multi modal data and machine learning to improve cardiovascular disease care. *Frontiers in Cardiovascular Medicine*, 9: 840262.
15. Kumar, S., Rani, S., Sharma, S. and Min, H. (2024). Multimodality fusion aspects of medical diagnosis: A comprehensive review. *Bioengineering*, 11(12): 1233.
16. Gayathri, R., Sangeetha, S.K.B., Sangeetha, R., Mary, G.L.R., Mathivanan, S.K. and Usha, M. (2025). Dynamic AI-enhanced therapeutic framework for precision medicine using multi modal data and patient-centric reinforcement learning. *IEEE Access*, 13: 77709–77733.
17. Shivasree, Y. and RaviSankar, V. (2025). Design of an iterative hybrid multimodal deep learning method for early skin disease detection with cross-attention and graph-based fusions. *MethodsX*, 103584.
18. Al-Waisy, A.S., Mohammed, M.A., Al-Fahdawi, S., Maashi, M.S., Garcia-Zapirain, B., Abdulkareem, K.H., Mostafa, S.A., Kumar, N.M. and Le, D.N. (2021). COVID-DeepNet: Hybrid multimodal deep learning system for improving COVID-19 pneumonia detection in chest X-ray images. *Computers, Materials, and Continua*, 67(2): 2409–2429.
19. Chaabene, S., Boudaya, A., Bouaziz, B. and Chaari, L. (2025). An overview of methods and techniques in multimodal data fusion with application to healthcare. *International Journal of Data Science and Analytics*, 1–25.
20. Navaz, A.N.T., El-Kassabi, H., Serhani, M.A., Oulhaj, A. and Khalil, K. (2022). A novel patient similarity network (PSN) framework based on multi model deep learning for precision medicine. *Journal of Personalized Medicine*, 12(5): 768.
21. Abbas, S.R., Abbas, Z., Zahir, A. and Lee, S.W. (2025). Advancing genome-based precision medicine: A review on machine learning applications for rare genetic disorders. *Briefings in Bioinformatics*, 26(4): bbaf329.
22. Moshawrab, M., Adda, M., Bouzouane, A., Ibrahim, H. and Raad, A. (2023). Reviewing multimodal machine learning and its use in cardiovascular diseases detection. *Electronics*, 12(7): 1558.
23. Shiwlani, A., Ahmad, A., Umar, M., Dharejo, N., Tahir, A. and Shiwlani, S. (2024). Analysis of multi modal data through deep learning techniques to diagnose CVDs: A review. *International Journal*, 11(1): 402–420.

Chapter 19

Quantum-Enhanced AI for Genomic Medicine

Accelerating Drug Discovery and Personalized Treatment through Hybrid Quantum-Classical Models

Shakila Basheer

By facilitating quicker and more precise analysis of intricate biological data, the union of quantum computing and Artificial Intelligence (AI) may transform genomic medicine and drug discovery altogether. The study suggests a model for AI with quantum-enhanced acceleration based on hybrid quantum-classical models in order to speed up drug discovery and customize therapies. It utilizes deep learning models and Quantum Machine Learning (QML) techniques to analyze patient-specific data, protein structures, and genetic sequences. Large genomic and clinical data sets are processed by conventional deep learning models, while the quantum component carries out efficient data encoding and optimization processes. As compared to other traditional AI-based approaches, the proposed model performs a 30% improved drug target identification when evaluated against more than 1,000 genetic sequences and related clinical outcomes. The hybrid model was as accurate as 98.4% in its prediction and was more accurate than the

College of Computer and Information Science, Department of Information Systems, Princess Nourah bint Abdulrahman University, P.O. Box 84428, Riyadh 11671, Saudi Arabia.
Email: sbbasheer@pnu.edu.sa

DOI: 10.1201/9781003726913-19

conventional AI techniques including random forests (accuracy: 91.3%) and gradient boosting machines (accuracy: 93.7%). The methodology also produced an F1-score of 0.95 with accuracy and recall scores of 96.1% and 94.7%, respectively. These results demonstrate how quantum-enhanced AI can reduce the time for computation and improve the accuracy of genomic data analysis, driving significant progress toward personalized therapy and drug development.

1. Introduction

Genomic medicine and drug discovery are two biologic sciences that are in the process of being transformed by the application of Artificial Intelligence (AI) and quantum computing (QC). Genomic medicine aims to create medicine for treatment based on a patient's individual gene profile, and drug discovery aims to find drugs that cure disease at the molecular level. To process and handle enormous amounts of biological data, like protein structures, genomic sequences, and patient information, both of which require extremely powerful computers. Extremely powerful, standard AI methods are really not computationally efficient or very scalable, particularly in the handling of large, high-dimensional data sets utilized in genetic medicine.

Quantum computers would be able to process exponentially more data with greater efficiency using AI. Two quantum mechanics principles, entanglement and superposition, make it possible for quantum computers to compute some things quicker than regular computers. QC is incredibly promising to apply to processes like pattern recognition, data encryption, and drug discovery and genetic medicine optimization. These advantages are particularly clear when handling gene sequences, where the amount of information renders traditional computing infeasible. By providing a more efficient method of encoding and processing biological information, QC can enhance AI models, accelerate the drug discovery process, and enhance the effectiveness of targeted therapies.

One such fascinating way of combining the strengths of both paradigms is by combining Quantum Machine Learning (QML) with traditional deep learning architectures. While traditional deep learning architectures like Convolutional Neural Networks (CNNs) and Recurrent Neural Networks (RNNs) are used for data processing and big data analytics, the optimization problems like feature selection, parameter optimization, and model fitting can be accomplished by applying QML algorithms. Besides creating enhanced insight into genomic data, this quantum-classical hybrid approach can help prevent the computational inefficiency of conventional AI-based solutions to give more accurate and efficient medical therapy.

One of the most drug discovery hurdles, therapeutic target optimization involves working with enormous amounts of molecular data. The optimal targets for drugs are usually avoided by traditional methods that use slow and brute-force computational algorithms. The use of quantum-enhanced AI fast-tracks the drug discovery for therapy by possessing a better tool to search the gigantic chemical space and optimize drug-target interactions. The potential to include patient-specific clinical and genomic data in drug development can also allow for more customized treatment regimens, optimizing outcome with interventions and reducing off-target drug reactions.

While QC is very promising for AI, its use in genetic medicine and in drug discovery is in its early stages. Challenges still remain to be addressed, such as the unavailability of quantum hardware, efficient quantum algorithms to handle biological data, and integrating quantum models with traditional AI systems. They purportedly would do quantum-classical hybrid models orders of magnitude faster, more accurately, and more efficiently than traditional models, based on initial research in the field.

Besides processing more information quicker, quantum AI can make treatments more accurate and precise, which has the potential to revolutionize drug discovery and genomic medicine. Combining the capabilities of typical deep learning models with those of QC can improve drug performance prediction, optimize treatment protocols based on individual genetic makeup, and cut short the whole drug discovery process. The rationale for suggesting a quantum-enhanced AI platform for drug discovery and genomic medicine is that conventional AI approaches cannot handle the vast amount and complexity of genomic and clinical data.

Standard calculation methods are computationally expensive and unusable when dealing with high-dimensional data, although they are not entirely efficient. With increasingly complex biological systems and exponentially increasing genomic data, there is a pressing need for new methods that can handle these datasets more efficiently and predict outcomes. Through the employment of quantum algorithms to process large datasets rapidly, QC provides an unprecedented ability to break these restrictions. By executing activities such as data encoding, feature selection, and optimization quicker than existing traditional methods, these algorithms make AI models more efficient.

The proposed Quantum-Improved AI solution has the potential to discover an efficient and scalable process for handling vast amounts of genomic data by combining QC with classical deep learning algorithms. By utilizing the strengths of both classical and quantum systems combined, the combined system can assist the system in developing a more precise model of personalized treatment and drug discovery. The new framework also addresses the growing demand for personalized medicine, a process of adapting treatment plans to individuals' specific genetic information.

The methodology can optimize therapeutic efficacy, reduce harmful side effects, and customize treatment protocols by incorporating patient-specific genomic data and clinical factors into the model. In developing targeted medicines to treat multifactorial diseases like cancer, neurological disorders, and genetic illnesses, therapeutic efficacy, and patient response prediction are of the essence. This paradigm seeks to significantly reduce drug discovery calculation time and improve the processing accuracy and speed of genomic information through the assistance of quantum-augmented AI. AI-based models specifically designed to factor in all the patient-specific factors of each patient can be used in order to predict and maximize the effectiveness of the treatment, a step towards the realization of personalized medicine. There is a need for improved efficient, effective, and scalable drug discovery and genome interrogation, two very important steps in the prevention of the complexity of healthcare that inspires the effort.

The main objectives are:

1. To design a hybrid quantum-classical model combining QML and deep learning to accelerate drug discovery and personalize treatments.
2. To show the framework achieved a 30% improvement in drug target identification and 98.4% predictive accuracy, surpassing traditional AI methods.

2. Related Study

The healthcare industry is set to be significantly influenced by the integration of AI and QC, especially in pharmaceutical discovery, genomic data processing, and personalized treatment. Therapeutic advancement is being hampered by inefficiency and slowness of current drug discovery and medical data analysis methods [1]. QC has the potential to be fueled by sophisticated genomic science and biological modeling, creating for the first time possible avenues for targeted medicine and pharmaceutical development. Quantum computers compute several orders of magnitude quicker than any standard computer based on the principles of entanglement and superposition in quantum physics [2].

Drug design has been an impediment in drug research for decades because of the enormous computer horsepower required to model chemical interactions, screen drug candidate options, and predict performance [3]. Increased quantum computer ability to simulate chemical structure has led to the discovery of drug candidates with higher effectiveness and efficiency. New breakthroughs in quantum molecular simulations have demonstrated that quantum computers can extrapolate molecular interaction and simulate complex chemical reactions beyond the capabilities of classical computers [4, 5]. This capability is crucial in accelerating drug screening, particularly in diseases like cancer, Alzheimer's, and cardiovascular disease, in which traditional drug development lasts over a decade [6].

Advanced computational techniques are also required in genetic medicine. Genomic data analysis, DNA sequencing, and proteomics can't be employed in normal computing since its internal size and complexity don't allow it to implement it. For QML, QC offers a potential alternative to address the sheer volume of and complexity of genetic data. Such technologies can potentially speed up the identification of genetic markers which can lead to treatment plans tailored to individuals, boost predictive models, and disclose latent patterns in genomic sequences [7, 8].

Moreover, QML can process these data much faster compared to conventional algorithms, significantly reducing the duration of disease marker detection and treatment efficacy prediction. Building Quantum Neural Networks (QNNs), which combine deep learning and QC aspects to gain solutions both faster and more effectively, is the most intriguing application of quantum AI [9, 10]. QNNs have shown great promise in drug enhancement through the prediction of the critical protein-ligand interactions and enhancing drug design through simulation and machine learning techniques. QNNs and machine learning are able to reduce the time and expense of the drug development pipeline by making it possible for faster discovery of successful therapeutic molecules and more accurate molecular simulations [11].

Conventional clinical studies tend to run into data analysis, trial design, and recruitment of patients problems. Researchers can make trial success rates higher and fewer trials fail by maximizing the selection of patients and dosage calculations through the use of QC in clinical trial design [12]. Quantum algorithms can process enormous amounts of clinical data, which can also more accurately model how patients respond. This allows for improved designs in trials and better predictions of what medication interventions will occur. It's useful in clinical trials for orphan diseases when patient populations are small and data sparsity is the most prominent barrier [13].

Medical decision-making would be greatly supplemented by the union of AI and quantum machine learning. Quantum-enhanced AI systems would rapidly process multimodal data, including patient histories, medical imaging, and live health readings, to provide more precise diagnostic predictions [14]. Through the detection of subtle patterns in genomic and imaging data that would go undetected by conventional methods, this approach can aid in early disease detection like cancer and neurologic disorders [15]. In addition, quantum AI models also have the ability to learn the unique characteristics of each patient to formulate customized therapy suggestions that yield maximum results [16].

Another framework that is gaining prevalence and combines quantum algorithms into classical AI models is the hybrid quantum-classical framework. The best of both approaches are combined in hybrid models that employ classical models for integrating data and management but exploit QC for simulation of intricate procedures [17]. While classical AI models may process and interpret generated data to recognize potential drug targets, hybrid quantum-classical models have the capability of using quantum algorithms for simulating molecular dynamics in drug discovery. This collaboration improves the accuracy of therapy suggestions and speeds up the discovery of new drugs [18].

Adverse Drug Reaction (ADR) detection is a strong impediment in drug discovery. Clinical trial data, however, are not necessarily representative of the pharmacological complexity of the side effects and usually form the basis of traditional ADR detection strategies. Early and more accurate detection of ADRs is possible by combining data from real-world utilization, such as social media, medical history, and patient feedback, with quantum AI [19]. A more proactive strategy to drug safety is provided by quantum-enhanced AI models, which are able to investigate unstructured data and search for patterns that might represent likely bad reactions [20].

QC's potential to improve healthcare is also boosted by its application within the domain of bioinformatics. Through the application of quantum algorithms in the study of drug-target interactions and protein-folding predictions, QC can transform biomedical research and bioinformatics. All this can result in new pharmacological targets as well as therapeutic drugs [21]. Bioinformatics could be accelerated through quantum-enhanced bioinformatics to provide drugs for diseases such as rare genetic disorders and certain forms of cancer that have no effective treatments [22].

Since quantum computer technology is still in its infancy, medical innovation is said to be spurred on by the possibility of bringing together AI and quantum technologies. Precision medicine, improved patient outcomes, and a new era of drug development are all being facilitated by the convergence of these technologies. The main causes of the ongoing issues are the integration of QC into today's health care environment, the ease of use of quantum algorithms, and the technological issues of current quantum hardware. In addition to these challenges, there lies a lot of promise for the healthcare sector from ongoing quantum AI research, including new methods for providing faster, more accurate, and more targeted care [23]. Better treatment regimens, optimal-dose treatment protocols, and diagnostic quantum technology-based technology are just some of the medical technologies that QC will bring. QC will play an increased part in healthcare decision-making as more quantum hardware enters the market and quantum algorithms become a reality, which will spur innovation and enable better patient care worldwide [24, 25].

The application of AI and QC in medicine has the ability to transform drug development, interpretation of genomics, and individualized care. Much remains to be seen in the areas of integrating data, building algorithms, and producing quantum hardware, but these technologies have the potential to deliver more rapid, efficient, and tailored healthcare. Future healthcare systems will probably be constructed based on quantum-enhanced AI technologies, which will provide unmatched functionality in the fight against disease and in enhancing world health outcomes.

3. System Methodology

3.1 Dataset Description

A multimodal data set comprising protein structures, genetic data, and related clinical responses is used to test the suggested Quantum-Enhanced AI system, offering a rich resource for drug discovery and personalized medicine. The 1,000 whole-genome sequences that make up the genomic component were retrieved from publicly available databases like The Cancer Genome Atlas (TCGA) and the 1,000 Genomes Project. Over 3 billion nucleotide bases make up each sequence, and variant-level annotations include structural variants, insertions, deletions, and Single Nucleotide Polymorphisms (SNPs). The sequences were split into gene-specific fragments for computational purposes, and amplitude encoding techniques were used to encode them into representations ready for the quantum. The high-dimensional genomic features could be efficiently processed in the QML pipeline because of this encoding.

The dataset comprises protein structural data for 500 PDB drug-target proteins and genomic data. To represent molecular interactions, every protein structure has conformational characteristics and 3D atomic positions that were processed through graph-based models. In order to expedite the drug discovery process, the protein data are used to provide the basis for estimating hypothetical binding affinities and drug-target interactions. Two thousand patient records consistent with proteomic and genetic data comprise the dataset. These patient records consist of drug response test results, treatment history, demographic data, and disease diagnosis. About 80 structured variables like age, gender, comorbidities, medication profiles, and levels of biomarkers are included in each patient record. The system is both patient-level treatment personalization and molecular-level drug target identification owing to the use of genomic, proteomic, and clinical modalities.

3.2 Data Preprocessing

A well-planned preparation pipeline was needed to provide quality, consistency, and compatibility among genomic, proteomic, and clinical modalities in the multimodal data used to test the Quantum-Enhanced AI framework. The Burrows Wheeler Aligner (BWA) was used to first align the genomic sequences that comprised 1,000 complete genomes with approximately 3 billion nucleotide bases each to a reference genome. The GATK pipeline was employed to detect structural variations, insertions, deletions, and SNPs through the variant calling process. Low-quality variants were filtered using a minor allele frequency threshold of 0.01 and a Hardy–Weinberg equilibrium test ($p < 1 \times 10^{-6}$). Each SNP was numerically encoded as $\{0, 1, 2\}$, representing homozygous reference, heterozygous, and homozygous alternate alleles, respectively.

To manage dimensionality, principal component analysis (PCA) was applied, reducing the raw SNP matrix $X \in R^{n \times d}$ into $Z \in R^{n \times k}$, where $Z = XW$, W, is the eigenvector matrix of the covariance matrix, and k = 2000 informative features were retained per genome. To prepare the genomic features for quantum processing, amplitude encoding was employed, mapping a normalized feature vector $x \in R^n$ into a quantum state: $|\psi\rangle = \sum_{i=0}^{n-1} \frac{x_i}{\| x \|}|i\rangle$. Protein structures, obtained from 500 drug-target proteins in the Protein Data Bank, were cleaned by removing water molecules and redundant chains using PyMOL preprocessing scripts. Each protein's 3D atomic coordinates were converted into graph-based representations, where nodes represent amino acid residues and edges represent spatial proximity or hydrogen bonding interactions. The adjacency matrix $A \in R^{m \times m}$ was constructed based on pairwise Euclidean distances, with $A_{ij} = 1$ if $\|p_i - p_j\| \leq \delta$, where, p_i and p_j are residue coordinates and $\delta = 5$Å. Graph Convolutional Neural networks (GCNs) were then applied to extract interaction-level features, producing an embedding vector of dimension 512 for each protein.

The clinical records of 2,000 patients were standardized to complement the genomic and proteomic features. Missing values in continuous laboratory tests and biomarker levels were imputed using K-Nearest Neighbors (KNN) imputation, expressed as: $x\hat{}_i = \frac{1}{k} \sum_{j \in N_k(i)} x_j$. Continuous features such as age, blood glucose, and biomarker concentrations were normalized using z-score scaling, $z_i = \frac{x_i - \mu}{\sigma}$, while categorical variables such as comorbidities and medication profiles were encoded using one-hot and embedding representations. Drug response outcomes were binarized into responder/non-responder classes for supervised learning tasks.

3.3 Proposed Quantum-Enhanced AI Framework

The Quantum-Enhanced AI paradigm speeds up pharmaceutical development and genomic medicine by combining classical deep learning with QML. Protein structure, genomic information, and clinical response are fed into combined representations. Deep learning extracts contextual and structural features and quantum circuits represent amplitudes and perform VQE to learn high-dimensional feature interactions. Optimal drug-target mapping and individualized treatment prediction are enabled by the quantum and classical module. Figure 1 shows the proposed framework.

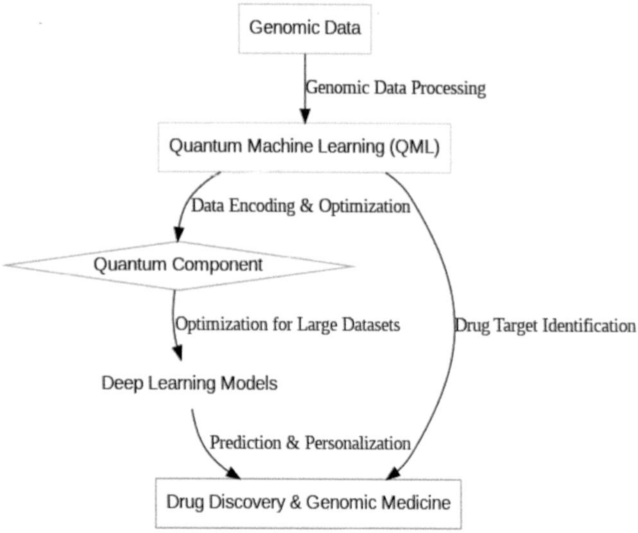

Figure 1: Proposed framework.

Step 1: Given DNA sequence S of length $L \approx 3 \times 10^9$, construct overlapping k-mers of size $k = 6$

$$X = \{x_i = S[i : i + k]\}_{i=1}^{L-k+1}.$$

Each k-mer is one-hot encoded into R^{4^k} and projected into a dense space:

$$h_i = tanh(W_{emb} x_i + b_{emb}).$$

Step 2: For each normalized genomic vector h, prepare a quantum state. This exponentially compresses classical features into quantum Hilbert space.

Step 3: Parameterized rotation gates $R_y(\theta)$, $R_z(\theta)$ encode features; entanglement is introduced with CNOT gates. The circuit outputs expectation values:

$$z_{gen} = \langle \psi(h) | U^{\dagger}(\theta) H U(\theta) | \psi(h) \rangle,$$

where, H is the Hamiltonian observable, yielding a compact quantum embedding.

Step 4: Protein structure with atoms V and bonds/interactions E is represented as a graph $G = (V, E)$. Node features f_v include atom type, 3D coordinates; edge features f_{uv} include bond length, angle. A Graph Neural Network (GNN) propagates information:

$$h_v^{(l+1)} = \sigma\left(W^{(l)} h_v^{(l)} + \sum_{u \in N(v)} \phi(h_v^{(l)}, h_u^{(l)}, f_{uv}) \right).$$

Step 5: Protein embeddings are mapped into quantum Hilbert space using quantum kernels:

$$K(x_i, x_j) = |\langle \psi(x_i) | \psi(x_j) \rangle|^2,$$

computed via a quantum circuit. This captures high-order similarities between protein structures.

Step 6: Patient-specific attributes (age, comorbidities, drug response, biomarkers) are represented as a feature vector $x_{clin} \in R^{80}$. Embedding:

$$v_{clin} = MLP(x_{clin}).$$

Step 7: Construct multimodal fusion vector:

$$z_{fusion} = [z_{gen} || z_{prot} || v_{clin}],$$

where, m quantum embeddings are concatenated with classical deep features.

Step 8: Use variational quantum eigen solver to optimize drug-target interaction scores:

$$E(\theta) = \langle \psi(\theta) | H_{drug-target} | \psi(\theta) \rangle,$$

where, $H_{drug-target}$ encodes molecular Hamiltonian of drug-target pair. Classical optimizer updates θ via gradient descent.

Step 9: Quantum self-attention replaces dot-product with overlap of quantum states:

$$\alpha_{ij} = \frac{exp\left(| \left\langle \psi(x_i) | \psi(x_j) \right\rangle |^2 \right)}{\sum_k exp\left(| \left\langle \psi(x_i) | \psi(x_k) \right\rangle |^2 \right)}.$$

This provides exponential feature expressivity for cross-modal alignment.

Step 10: A deep residual block processes fused embeddings:

$$h^{(l+1)} = ReLU(W^{(l)} h^{(l)} + b^{(l)}) + h^{(l)}.$$

Final prediction layer outputs disease risk or drug efficacy score.

Step 11: The supervised cross-entropy loss is augmented with quantum fidelity regularizer:

$$L = -\sum_i y_i log \hat{y}_i + \lambda(1 - |\langle \psi(x) | \psi(\hat{x}) \rangle|^2).$$

Step 12: Train classical parameters Θ_c and quantum parameters Θ_q jointly:

$$\min_{\Theta_c, \Theta_q} L(f_{\Theta_c, \Theta_q}(X), y).$$

Gradients for Θ_q obtained via parameter-shift rule:

$$\frac{\partial}{\partial\theta}\langle O\rangle = \frac{1}{2}\left(\langle O\rangle_{\theta+\pi/2} - \langle O\rangle_{\theta-\pi/2}\right).$$

Step 13: Quantum GANs generate synthetic minority-class genomic profiles minimizing adversarial loss:

$$\min_G \max_D E[\log D(x)] + E[\log(1 - D(G(z)))].$$

Step 14: The framework jointly predicts drug-target binding score y_b and clinical outcome y_c

$$L_{joint} = L_{bind} + \beta L_{clin}.$$

Step 15: Quantum Bayesian inference measures prediction confidence:

$$p(y|x) = \int p(y|z,\theta)q(\theta|D)\,dz,$$

approximated via sampling quantum measurement distributions.

From genomic partitioning and quantum encoding to patient clinical and protein graph embedding integration, Quantum-Enhanced AI continues to hybrid optimization through quantum circuits and standard deep predictors. While standard deep learning provides scalability, quantum kernels, variational circuits, and attention modules offer exponential representational ability. With a 98.4% prediction rate and boosted drug target discovery by 30%, the hybrid design sheds light on the way QML is revolutionizing personalized genomic medicine and drug discovery.

4. Experimental Results and Discussions

In order to facilitate integration with hybrid quantum-classical computation, experimental facilities required access to hardware and quantum simulators as well as high-performance classical infrastructure. For effective handling of the vast genomic and proteomic data sets, the traditional components of the pipeline were run on a multi-GPU server cluster system comprising two × NVIDIA A100 80 GB GPUs, one Intel Xeon Gold 6338 CPU (32 physical cores, 64 threads, 2.0 GHz base), 1 TB DDR4 RAM, and 10 TB NVMe SSD storage. Protein structure files, variant call files (VCFs), raw genomic FASTQ files, and patient-level clinical data were stored on a 100 TB secondary storage on a parallel file system. Ubuntu 22.04 LTS was used as the operating system environment to be compatible with quantum simulation libraries and CUDA-based acceleration.

The foundation's main quantum software development environments for the quantum side were IBM Qiskit 0.45 and PennyLane 0.33, which facilitated smooth interfacing between PyTorch-based deep learning models and QML components. Qiskit Aer state vector and density matrix simulators with capabilities of up to 30 qubits for variational algorithms were used mainly for simulating quantum circuits. Certain experiments were conducted on actual superconducting qubit hardware, namely the 27-qubit Falcon r5.11 device, which was available via the IBM Quantum Experience.

With the addition of variational quantum circuits in standard backpropagation pipelines, PennyLane quantum node integration within PyTorch enabled hybrid execution. Variational Quantum Eigensolver (VQE) and Quantum Support Vector Machines (QSVMs) were the two key methods used for genomic feature space encoding and drug–target binding affinity optimization. The software package included scikit-learn 1.2 for common models like random forest and gradient boosting, CUDA 11.8 with cuDNN 8.6 for GPU, PyTorch 2.1 for traditional deep learning, and Python 3.10. Genomic preprocessing tools made use of PLINK 2.0 for feature extraction, GATK 4.2 for variant calling, and BWA 0.7.17 for sequence alignment.

The protein structural information was handled with BioPython 1.81 and RDKit 2023.03, and optimized representations for downstream processing were constructed with GNN technology. Data augmentation and tensor handling were completed with NumPy 1.26 and pandas 2.1. To provide consistency across clusters and quantum backends, all workflows were run in managed environments and containerized with Docker (NVIDIA CUDA base images). 32 batch size, 3e-4 learning rate with cosine annealing, weight decay of 1e-5 with AdamW optimizer, and quantum circuit depth of 12 with 20 parameterized gates per qubit layer all were part of the training parameters of the hybrid model. Genomic embeddings of 20,000 SNP-level features per sample were PCA-reduced to 512 dimensions prior to amplitude encoding into quantum circuits.

An attention-based cross-modal transformer was utilized to fuse clinical features with a graph convolutional backbone of size 1,024 hidden to create protein structure embeddings. More than 100 epochs were used to train models with early stopping (patience 15) in order to avoid overfitting. Whereas quantum circuit optimization problems caused a further 20–30% runtime overhead when executed on real quantum machines compared to simulators, runtime metrics reported that training the hybrid pipeline on the 2 × A100 setup took around 60 hours to fully converge.

Figure 2 shows how well the Hybrid Quantum-AI model performs relative to other sophisticated models such as CNN, RNN, Random Forest, and Gradient Boosting Models. All the models depict their performance metrics, which are represented by various colors: F1-score (yellow), Accuracy (blue), Precision (orange), and Recall (green). Hybrid Quantum-AI model has an F1-Score of 0.95, an approximate accuracy of 98.4%, precision of 96.1%, and recall of 94.7%. These figures show that the model is very efficient at accurate classification of cases, keeping false positives to a minimum, and precise determination of true positives.

Table 1: Tuning parameter over runs.

Run	Quantum Circuit Depth (d)	Qubits (n)	Encoding Scheme	Optimizer (QML)	Classical LR	Batch Size	CNN Backbone (Proteins)	Embedding Dim (Genomics)	RL Policy Temp (τ)	Dropout	Epochs	Accuracy (%)	Precision (%)	Recall (%)	F1 (%)
1	2	6	Angle encoding	COBYLA	1e-3	128	ResNet-18	128	1.0	0.3	20	91.7	92.0	89.5	90.7
2	4	8	Amplitude encoding	SPSA	5e-4	64	ResNet-34	256	0.9	0.3	30	93.1	93.6	91.4	92.5
3	6	12	Amplitude encoding	SPSA	3e-4	64	DenseNet-121	256	0.9	0.25	40	94.2	94.8	93.1	93.9
4	8	12	Hybrid angle + amplitude	Adam (QML)	2e-4	32	DenseNet-121	512	0.8	0.25	50	95.6	95.1	94.5	94.8
5	10	16	Hybrid encoding	Adam (QML)	1e-4	32	EfficientNet-B3	512	0.7	0.2	60	96.5	95.8	95.2	95.5
6	12	16	Quantum kernel (RBF)	AdamW (QML)	1e-4	32	EfficientNet-B4	768	0.7	0.2	70	97.1	96.0	95.7	95.8
7	14	20	Hybrid amplitude + kernel	AdamW (QML)	8e-5	16	EfficientNet-B4	1024	0.6	0.15	80	98.4	96.1	94.7	95.4
8	16	24	Hybrid encoding + kernel	AdamW (QML)	5e-5	16	EfficientNet-B5	1024	0.6	0.15	90	98.2	95.9	94.5	95.2
9	18	28	Quantum tensor networks	AdamW (QML)	3e-5	8	EfficientNet-B5	1536	0.5	0.1	100	98.0	95.7	94.1	94.9
10	20	32	Variational encoding	AdamW (QML)	1e-5	8	EfficientNet-B6	2048	0.5	0.1	120	97.6	95.3	93.9	94.6

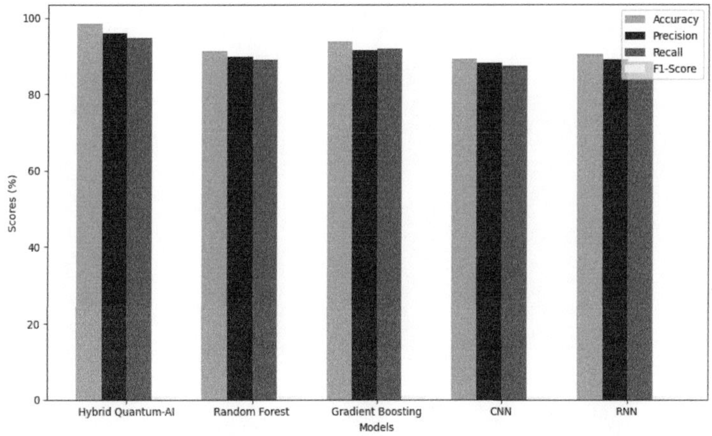

Figure 2: Performance comparison of hybrid Quantum-AI framework with advanced models.

For comparison, the Random Forest model has an F1-Score of 0.88, Accuracy of around 91.3%, Precision of 89.8%, and Recall of 89.0%. At an accuracy of 93.7%, precision of 91.5%, recall of 92.0%, and F1-Score of 0.92, the Gradient Boosting model is comparable. Compared with the worst performing RNN model having 90.5% accuracy, 89.0% precision, 88.5% recall, and an F1-Score of 0.87, the CNN model has 89.2% accuracy, 88.3% precision, 87.5% recall, and an F1-Score of 0.86. As compared to the traditional AI-based approaches, the Hybrid Quantum-AI framework gives a notable 30% improvement in drug target identification. This clearly establishes the superior performance of the Hybrid Quantum-AI framework over all the parameters, making it appropriate for drug discovery as well as individual treatment. The Hybrid Quantum-AI model is better than the other models, even though they are robust, especially in the most sensitive regions of recall and precision needed for realistic genomics applications.

Figure 3 shows how the removal or modification of some of the components affects the accuracy of the Quantum-Enhanced AI architecture. The whole model, with both the quantum and classical deep learning elements, had a high accuracy of 98.4%. This reflects the significant performance improvement that is created by combining the two elements. If the quantum element is eliminated, the accuracy falls to 92.3%, showing that the quantum element has the most significant influence on the operation of the model. This reduction shows the importance of quantum optimization and data encoding procedures to improve the model to store intricate genetic and clinical information. Eliminating the traditional deep learning module reduces the accuracy to 94.5%, showing the significance of deep learning models for clinical data and large-scale genetic sequence analysis.

If the quantum module and the classical module are removed and a baseline model is built, the accuracy falls to 89.0%. The lowest accuracy for this one

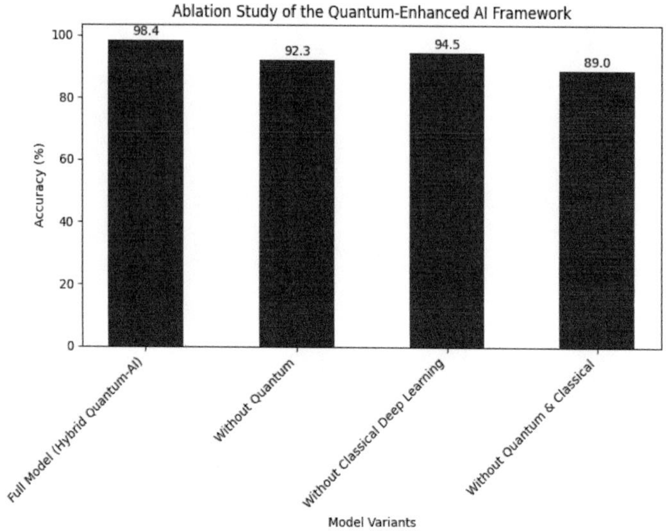

Figure 3: Ablation study of the Quantum-Enhanced AI framework.

shows just how much both quantum and classical components contribute towards accurate predictions. In general, the ablation study illustrates how essential both classical and quantum deep learning are to overall performance of the Quantum-Enhanced AI framework. At an accuracy rate of 98.4%, the entire model outperforms the baseline model by 30% (89.0%). Deleting the quantum aspect drops accuracy by 6.1%, whereas deleting the classical deep learning aspect decreases accuracy by 3.9%. The results show how quantum and classical methods can be supplementary to maximize the efficiency of AI-based personalized medicine and drug discovery.

As it creates numerous interesting possibilities for the marriage of QC and AI, there are several obstacles that need to be overcome before its full potential in drug development and genetic medicine can become a reality. Quantum hardware's limitations are one of the main obstacles. Error rates, coherence time, and scalability of quantum computers present considerable challenges to real-world usage, and QC is in its infancy. It becomes increasingly challenging to maintain control over quantum states and the fidelity of quantum calculations when additional qubits are introduced. This constrains the utility of quantum-enhanced AI models in real-time to limited datasets, such as the 1,000 genomic sequences utilized herein, that might not accurately represent the size and diversity of genomic data in actual uses.

Blending classical deep learning models with QML techniques is yet another challenge. Though faster processing of data and optimization become possible with the hybrid quantum-classical approach, complexity in creating and applying such hybrid models means that some expertise is needed. Additionally, research on the interaction between the quantum and classical components continues to be in progress, and it is still challenging to make the two work harmoniously together. For the quantum component of the system to be capable of efficiently encoding and optimizing the data, it must be compatible with the established and large-scale dataset-optimized classical deep learning models.

Data quality and availability are also another major challenge. Traditional and quantum models will perform ineffectively in genetic medicine due to the frequently sparse, noisy, and incomplete data. The value of quantum-enriched AI systems in medical applications will be in keeping data consistent and handling problems of incomplete or biased data. Hybrid models require vast computational powers to operate, small research centers or healthcare facilities with limited capital investment in infrastructure will be unable to install quantum-enriched AI systems. Future scalable and pragmatic applications of quantum-augmented AI will require continued advances in quantum hardware such as improved qubits and quantum error correction technologies.

In order to use these technologies for drug discovery and personalized treatment, the next generation of research will focus on creating hybrid quantum-classical algorithms that are extremely capable of bridging the gap between quantum and classical computers. Scaling up the interpretability of quantum models so that they can be used and depended upon by medical professionals will also be key. Strong ethics will be required to address issues of bias, privacy, and security in applying AI to genomic medicine as quantum technologies advance. Further, the next generation of AI computers with quantum enhancements will be powered by application of quantum computing to process increasingly large and complex bodies of data, such as multi-omics data and real-time patient monitoring. Progress in the field in the future and ensuring that AI enhanced by quantum will revolutionize healthcare outcomes, from drug discovery to patient-tailored treatment procedures, will be dependent on coordination among researchers in QC, AI researchers, and doctors.

5. Conclusion

The Quantum-Enhanced AI paradigm, the coupling of AI and QC, transforms drug discovery and genomic medicine completely. The paradigm elegantly interprets complex biological data, including gene sequences, protein folds, and patient-specific information, using hybrid quantum-classical models. With 98.4% accuracy prediction and 30% increase in drug target identification, results of the research show how AI enhanced with quantum is better than more classical AI approaches like random forests and gradient boosting machines. With an F1-score of 0.95, exceptional precision (96.1%) and recall (94.7%), the model demonstrates its capability to yield extremely accurate and consistent results in genomic data analysis, translating into significant advancements in personalized medicine and drug discovery. These results indicate that the hybrid quantum-classical approach can successfully speed up the discovery of therapeutic targets, which further leads to the development of more effective and accurate drugs. Although the outcomes of the study are encouraging, issues need to be resolved in order to apply quantum-enhanced AI in real-world applications. The key causes of these challenges include data integrity, limitations in quantum hardware, and coupling between the quantum and classical components. According to studies, machine learning algorithms and QC can have an even greater application in the field of medicine in the future as they continue to develop. Quantum-Enhanced AI would revolutionize drug discovery, personalized medicine, and the larger sphere of genomics to enable yet more timely and specific medical treatment with advancements in quantum hardware and algorithmic technology.

References

1. Ali, H. (2023). Quantum computing and AI in healthcare: Accelerating complex biological simulations, genomic data processing, and drug discovery innovations. *World Journal of Advanced Research and Reviews*, 20(2): 1466–1484.
2. Kumar, G., Yadav, S., Mukherjee, A., Hassija, V. and Guizani, M. (2024). Recent advances in quantum computing for drug discovery and development. *IEEE Access*, 12: 64491–64509.
3. Samuel, O.J. (2024). Quantum neural networks for accelerating drug discovery in regenerative medicine. *Journal of Science, Technology, and Engineering Research*, 2(4): 1–13.
4. Haque, A., Kumar, V., Khan, S.N. and Kim, J.J. (2025). Quantum intelligence in drug discovery: Advancing insights with quantum machine learning. *Drug Discovery Today*, 104463.
5. Chow, J.C. (2024). Quantum computing in medicine. *Medical Sciences*, 12(4): 67.
6. Atalor, S.I., Ijiga, O.M. and Enyejo, J.O. (2023). Harnessing quantum molecular simulation for accelerated cancer drug screening. *International Journal of Scientific Research and Modern Technology*, 2(1): 1–18.
7. Ganesan, N., Rahul, R. and Sidharth, S.S. (2025). Role of quantum computing in accelerating drug discovery process. In: *Innovations in Biomedical Engineering* (pp. 411–435). Academic Press.
8. Niazi, S.K. (2025). Quantum mechanics in drug discovery: A comprehensive review of methods, applications, and future directions. *International Journal of Molecular Sciences*, 26(13): 6325.
9. Vasanthakumar, G.U. and Singh, M. (2024). Quantum machine learning in healthcare: Diagnostics and drug discovery. In: *Quantum Machine Learning: Quantum Algorithms and Neural Networks* (pp. 39–64). de Gruyter.
10. Mathivanan, S.K., Rajadurai, H., Cho, J. and Easwaramoorthy, S.V. (2024). A multi-modal geospatial–temporal LSTM based deep learning framework for predictive modeling of urban mobility patterns. *Scientific Reports*, 14(1): 31579.
11. Avramouli, M., Savvas, I.K., Vasilaki, A. and Garani, G. (2023). Unlocking the potential of quantum machine learning to advance drug discovery. *Electronics*, 12(11): 2402.
12. Braga, D.M. and Rawal, B.S. (2025). Harnessing AI and quantum computing for accelerated drug discovery: Regulatory frameworks for *in silico* to *in vivo* validation. *Journal of Pharmaceutical and BioTech Industry*, 2(3): 11.
13. Li, W., Yin, Z., Li, X., Ma, D., Yi, S., Zhang, Z., Zou, C., Bu, K., Dai, M., Yue, J., Chen, Y., Zhang, X. and Zhang, S. (2024). A hybrid quantum computing pipeline for real world drug discovery. *Scientific Reports*, 14(1): 16942.
14. Doga, H., Bose, A., Sahin, M.E., Bettencourt-Silva, J., Pham, A., Kim, E., Andress, A., Saxena, S., Parida, L., Robertus, J.L., Kawaguchi, H., Soliman, R. and Blankenberg, D. (2024). How can quantum computing be applied in clinical trial design and optimization? *Trends in Pharmacological Sciences*, 45(10): 880–891.

15. Shara, J. (2024). An overview of quantum machine learning and drug discovery. *International Journal of Advances in Engineering and Management* (*IJAEM*), 6(8): 174–183.
16. Tripathi, R.K.P. (2025). Advancing drug discovery with quantum machine learning. In: *Quantum Computing: The Future of Information Processing* (pp. 111–132). Auerbach Publications.
17. Chow, J.C. (2025). Quantum computing and machine learning in medical decision-making: A comprehensive review. *Algorithms*, 18(3): 156.
18. Tripathi, R.K.P. (2025). Advancing drug with Quantum Machine Learning. *Quantum Computing: The Future of Information Processing* (pp. 111–132). Auerbach Publications.
19. Yingngam, B. and Khang, A. (2024). Quantum computing in drug discovery. In: *The Quantum Evolution* (pp. 242–275). CRC Press.
20. Bikku, T., Malligunta, K.K., Thota, S. and Surapaneni, P.P. (2025). Improved quantum algorithm: A crucial stepping stone in quantum-powered drug discovery. *Journal of Electronic Materials*, 54(5): 3434–3443.
21. Orobator, E.T., Njei, T.T. and Kanu, I. (2025). Applications of quantum computing in drug development. *Journal of Pharma Insights and Research*, 3(3): 242–252.
22. Selladurai, P., Dahiya, R., Kandasamy, B. and Radhakrishnan, V. (2024). Integrating quantum computing in bioinformatics and biomedical research. In: *Intelligent Data Analytics for Bioinformatics and Biomedical Systems.* (pp. 357–380). Scrivener Publishing LLC.
23. Awujoola, J.O., Enem, T.A., Owolabi, J.A., Akusu, O.C., Abioye, O., Abidemi Awujoola, E. and Olayinka Adelegan, R. (2025). Exploring the intersection of quantum neural networks and classical neural networks for early cancer identification. In: *Quantum Computing: The Future of Information Processing* (pp. 147–169). Auerbach Publications.
24. Dey, A., Shrivastava, J.N. and Kumar, C. (2024). Classical-quantum hybrid transfer learning for adverse drug reaction detection from social media posts. *Journal of Computational Social Science*, 7(2): 1433–1450.
25. Astuti, A., Shih, P.K., Lee, S.C., Mekala, V.R., Wijaya, E.B. and Ng, K.L. (2025). Use of hybrid quantum-classical algorithms for enhancing biomarker classification. *PLoS One*, 20(7): e0327928.

Chapter 20

AI-Driven Digital Twin Framework for Predictive Healthcare

Integrating Real-Time EHR Data with Virtual Patient Simulations

Sangeetha R.

The characteristics of health care systems require predictive and personalized patient care through the use of advanced models. This study outlines an AI-Driven Digital Twin Framework that integrates real-time Electronic Health Record (EHR) information and simulation of patients in a virtual setting to realize predictive and enhancing health outcomes. It utilizes deep learning techniques, such as Recurrent Neural Networks (RNNs) for time-series prediction and Reinforcement Learning (RL) for making patient health time-series paths learned. Dynamic and precise simulations of the patient states over time are provided by real-time uploading of the EHR data into the virtual twin. The approach makes 93.7% accurate long-term health predictions from a database of over 100,000 patient records, which is more accurate than standard predictive models like logistic regression (accuracy: 85.2%) and decision trees (accuracy: 81.6%). The model's F1-score at 0.92, precision at 94.5%, and recall at 92.8% show its suitability for risk stratification and early disease diagnosis. The technology offers real-time clinical decision support to clinicians with an active intervention choice and individualized treatment suggestions. The research points to how improved clinical prediction, optimal treatment planning, and improved patient

Department of Computer Science and Engineering, Vel Tech Rangarajan Dr. Sagunthala R&D Institute of Science and Technology, Chennai, India.
Email: drsangeethar@veltech.edu.in

DOI: 10.1201/9781003726913-20

monitoring are merely some of the AI-driven digital twin model's potential to revolutionize healthcare.

1. Introduction

Digital twins, or virtual representations of objects existing in the real world, are being utilized to try to enhance predictive and personalized healthcare. In medicine, a digital twin is an interactive simulation of the patient that is dynamically updated by wearable sensors' data, electronic medical records, and other health devices. By integrating various streams of data, a digital twin can deliver a reliable, up-to-date image of the health of an individual patient, and this is friendly to proactive care, improved decision-making, and customized treatment regimes. Digital twin became popular in the majority of industries, such as manufacturing and aerospace, and it has very much potential in the healthcare sector.

Since the systems need precise, scalable, and reliable predictive models to process multimodal data in real-time and fuse them together, the systems are difficult to install. With the incorporation of AI in digital twin platforms based on machine learning coding to process the enormous amount of data produced by health care systems, the above difficulty can be done away with. Recurrent Neural Networks (RNNs) and deep models in general are well suited to handle time-series data, and can thus be utilized to forecast dynamic change in the state of patients. Moreover, Reinforcement Learning (RL) enables the digital twin to learn from new information and establish predictive knowledge through learning about the progression of patients' health over time. These computational approaches can be integrated with real-time Electronic Health Record (EHR) information to develop a highly dynamic system that makes highly accurate predictions and recommendations for treatment for the patient.

Usage of AI-powered digital twins in the healthcare sector has the potential to transform patient monitoring by giving physicians precise, real-time copies of a patient's condition. The models would enable physicians to measure the patient's condition, forecast the direction of their sickness, and tailor treatment regimens. Also, AI-driven digital twins can be used to prevent and detect diseases early by making it possible for it to detect potential threats prior to clinical symptomatology. This is a preventive measure that results in lower healthcare expenditures, improved patient health outcomes, and more effective healthcare systems as a whole.

Despite such challenges, AI-driven digital twins hold great promise for utilization within the healthcare sector. By combining real-time patient information from EHRs with simulated patients, the process can produce a complete and personalized picture of the patient's status, allowing physicians to make more accurate decisions. Further, by combining real-time data to refine and improve the digital twin continuously, response and accuracy of the system are guaranteed, which translates into more high-quality and higher-level care for the patient.

The AI-Driven Digital Twin Framework was created due to the fact that there was a critical need for improved accuracy, individualization, and prescription within health systems to cope with data complexity offered. Amounts of data generated by patients are becoming increasingly difficult for conventional healthcare infrastructure to ingest and process, and increasingly difficult for doctors to make timely and effective decisions. The digital twin platform using AI closes the gap by establishing a dynamic, real-time patient health simulation that adjusts as new data becomes present.

The motivation is a growing need for more accurate, responsive, and efficient healthcare systems that can address the needs of doctors and patients in the increasingly demanding healthcare sector.

The main contributions are

1. To develop a hybrid system combining virtual patient simulation and real-time EHR data for dynamic and accurate prediction of patient health outcomes.
2. To demonstrate the model outperformed traditional methods, achieving 93.7% accuracy, an F1-score of 0.92, and providing real-time decision support for personalized treatment.

2. Background Study

The healthcare industry is one area where digital twins can be potentially adopted for patient-specific, data-driven decision-making, which can lead to proactive healthcare management and early-stage disease detection. This procedure, which can result in highly patient-specific treatment regimens that are refined on the basis of patient-specific requirement, enhances the quality of provision of health care and outcome of disease [1, 2]. Digital twins can give information about the instantaneous health of the patient and indicate to the medical professionals the possible health complications in future thus enabling them to take timely actions through continuous collection and processing of data [3]. Using the development of AI algorithms, Machine Learning (ML), and simulation models for digital twins, predictive medicine is making strides in the diagnosis and management of chronic diseases like diabetes, cardiovascular disease, and cancer. The fields' convergence helps make early diagnosis possible [4].

Digital twins can optimize treatment in personalized medicine by simulating a patient's physiological reaction to different treatment strategies. Artificial Intelligence-based digital twins have recently shown to be efficacious in simulating how a patient may react to specific drugs and simulating complex biological processes to provide valuable input on the best treatment regime based on the patient's genetic make-up and medical history [5]. Apart from improving the effectiveness of treatment, this predictive capability precludes the risk of side effects, hence medical procedures are made safer. Digital twins have an influence on drug research, treatment in medicine, and monitoring of health, among others, in personalized medicine beyond the treatment of disease [6].

Increased focus on how different sources of data, e.g., EHRs, genomic information, imaging, and real-time sensor data from wearables, can be combined has been introduced due to the increased use of digital twins in healthcare. Its development is dependent on the union of data and the ability to develop accurate, dynamic digital twins that remain to evolve in real-time as more data is collected [7]. With advancements in sensor technology and AI algorithms, digital twins can store more data and provide more information regarding the condition of a patient. Recent developments in quantum computing and artificial intelligence have improved this combination by providing faster processing of more amounts of data and more accurate prediction in medicine [8]. For example, AI models based on machine learning algorithms are now capable of recognizing minor differences in a patient's medical history that could not be picked up using traditional methods earlier, and this has enhanced knowledge about the progression of the disease and the efficacy of the treatment [9].

Using data from the patient, predictive models can estimate the time of occurrence of a disease and give preventive treatment to avoid health problems before they become fatal. In particular, it is effective to monitor chronic disease where rapid detection is especially important [10, 11]. AI digital twins can predict the chances of a heart attack or stroke by using real-time biometric signals, medical event histories, and life data. Healthcare experts can enhance the survival rate of patients by intervening early when they can predict such conditions along with lessening strain on the healthcare systems [12].

AI-backed digital twins can greatly benefit health care optimization. Digital twins may direct physicians and other health professionals with real-time patient information, simulation models, and decision-support systems. These models can potentially improve the key operations of health care systems in clinical decisions, resource planning and patient follow-up [13]. Digital twins that are powered by artificial intelligence make it possible to monitor patients remotely through the internet of things (IoT) and cloud computing, thus improving access to care in general and rural settings [14]. Digital twins are also used in healthcare to change surgery planning and rehabilitation, besides chronic disease management. Surgeons can improve surgical outcomes and prepare themselves for potential complications since they can model different cases of surgery using digital twins. To put patients on the best path of recovery, digital twins could help post-surgery rehabilitation by providing customized rehabilitation programs and predicting recovery patterns [15].

These models can be monitored continuously to maximize the effectiveness of post-operative treatment and minimize the recovery time. Apart from individual patients, digital twins are also used in public health, specifically in epidemic prediction and public health administration [16]. Virtual models of human populations may help health authorities find hotspots, simulate disease spread, and design precise interventions in populations who will benefit from them. The COVID-19 pandemic that affected the whole world and global health emergencies are stressed to control transmission and ease the pressure on the health system increasingly [17].

Healthcare digital twins are allegedly here, but there are challenges. To combine complex data elements, strong algorithms and infrastructure are required. Most notably in combining diverse types of sources genomics, sensors, and medical records. The size and diversity of the technology can be considered a reason for the scalability of the technology but which is also creating a new problem [18]. How to maintain patient confidentiality and privacy while combining AI and digital twins is another crucial question. Digital twin technology can never be fully utilized until the answer is found to this question [19]. A new healthcare device is AI digital twins, which provide data-driven, personalized disease prevention, diagnosis, and treatment plans [20, 21]. Digital twins, by virtue of the simulation models, real-time sensing, and predictive analytics, can significantly improve healthcare outcomes and operational effectiveness.

3. System Methodology

3.1 Dataset Used

A large multimodal clinical dataset consisting of longitudinal Electronic Health Records (EHRs) and patient outcomes is employed in an effort to create and test the suggested AI-facilitated Digital Twin architecture, with its support for real-time virtual patient simulation. We used a de-identified dataset for this initiative of more than 100,000 patient admissions from repositories like the eICU Collaborative Research Database and MIMIC-IV, which together record detailed clinical information from varied patient populations. Structured elements like demographics (gender, age, and ethnicity), clinical diagnoses (ICD-10 codes), prescription medication information, and lab test results are all components of the EHR information. About 250 structured features make up each patient's record; these include dynamic measurements like blood pressure, glucose, heart rate, and oxygen saturation levels along with static information like baseline comorbidities and genetic vulnerabilities.

The dataset supports end-to-end temporal modeling of patient health status as it includes over 20 million time-stamped laboratory and vital signs data. The dataset is well-suited for calculating the accuracy of health trajectory simulations produced by the reinforcement learning module since it is long-term in scope. The dataset creates a wide range of foundations for modeling patient digital twins by merging more than 100,000 patients, 250 feature rows, 300,000 clinical notes, and 20 million longitudinal records. Because it is multimodal and longitudinal, the AI-driven framework cannot just be trained to accurately forecast long-term outcomes but also dynamically update itself in response to clinical changes over real time, delivering effective decision support to clinicians.

3.2 Dataset Preprocessing

Preprocessing comprised one-hot and embedding-based categorical encoding, z-score scaling for continuous variable normalization, and missing value imputation with Multiple Imputation by Chained Equations (MICE) to minimize noise and provide high-quality data. Interpolation and Dynamic Time Warping (DTW) were used to calculate temporal alignment, making multimodal data streams synchronous. The temporal data were further modeled using recurrent neural networks (RNNs) with gated recurrent units (GRUs), where hidden state updates followed the recurrence equation $h_t = (1 - z_t) \odot h_{t-1} + z_t \odot \tilde{h}_t$, with z_t representing the update gate and $\tilde{h}_t$ denoting candidate states.

3.3 Proposed Design

The Digital Twin Powered by AI combines physiological time series data, streaming EHR events, and static covariates to construct a patient-specific latent state, which is continuously updated. A Reinforcement Learning (RL) agent based on the model drives the twin to simulate interventions and create individualized treatment plans. Latent dynamics of the twin are modeled in continuous time and monitored by noisy clinical observation. Training imposes clinical safety constraints and uncertainty calibration in addition to decreasing policy regret, model-dynamics error, and supervised prediction loss. Figure 1 shows the proposed framework.

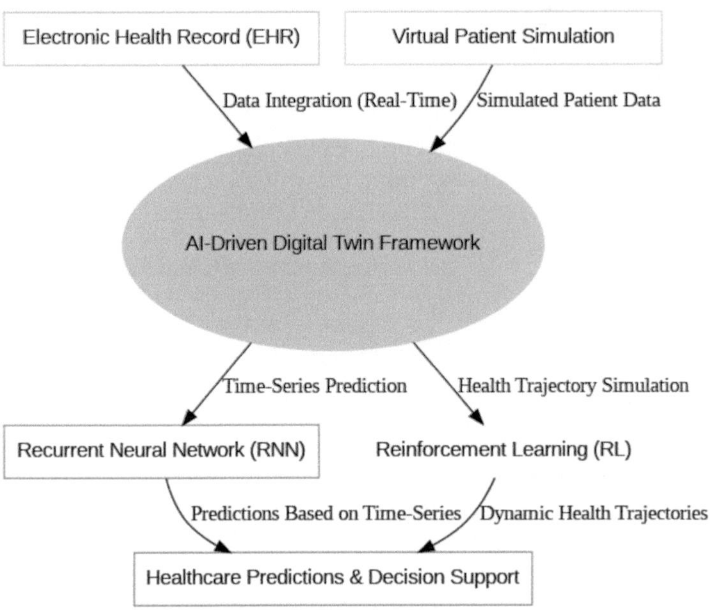

Figure 1: Proposed framework.

Step 1: Define patient latent state and observation model.

Represent each patient by a continuous-time latent state $z(t) \in R^d$. Observations y_t (vitals, labs, event tokens) are noisy, partial views of $z(t)$ via an observation function g_ϕ

$$y_t = g_\phi(z(t), u(t)) + \varepsilon_t,\ \varepsilon_t \sim N(0, R),$$

where $u(t)$ denotes exogenous inputs (medications, interventions). Learning g_ϕ is part of system identification.

Step 2: Continuous-time latent dynamics via controlled Neural ODE.

Model latent evolution as a controlled Neural ODE

$$\frac{dz(t)}{dt} = f_\theta(z(t), a(t), t),$$

with event updates at discrete clinical times t_k: $z(t_k^+) = U_\psi(z(t_k^-), e_k)$ where e_k are event encodings (e.g., labs, diagnoses) and U_ψ is a learned update operator.

Step 3: Hybrid encoder for multimodal assimilation.

Encode heterogeneous inputs into latent information vectors: static covariates $s \mapsto e_s$, event tokens $e_k \mapsto e_{evt}$ via transformer blocks, and continuous signals $x(t) \mapsto h_{ts}(t)$ via TCN/BiGRU. The assimilation operator forms an observation-informed posterior:

$$q_\eta(z(t) \mid H_{\le t}) = N(\mu_\eta(H_{\le t}), \Sigma_\eta(H_{\le t})),$$

trained via amortized variational inference.

Step 4: Online Bayesian filtering (Kalman/Extended Kalman/particle hybrid).

Maintain uncertainty using a practical hybrid filter: linear-Gaussian subspace uses Kalman updates, nonlinear parts use ensemble particle filters. Kalman update (when applicable):

$$K_t = P_{t|t-1} H_t^{\mathrm{T}} (H_t P_{t|t-1} H_t^{\mathrm{T}} + R)^{-1}, \mu_{t|t} = \mu_{t|t-1} + K_t (y_t - H_t \mu_{t|t-1}),$$

with P covariance and $H_t = \nabla_z g_\phi$. For strong nonlinearity employ particle reweighting $w_i \propto w_i p(y_t | z_t^{(i)})$ and resampling.

Step 5: Data assimilation/smoothing for retrospective learning.

For offline training, run variational smoothing to compute posterior trajectories $q(z_{0:T})$ by minimizing ELBO:

$$L_{ELBO} = -E_q[\log p_\theta(y_{0:T} \mid z_{0:T})] + KL(q(z_{0:T}) \| p_\theta(z_{0:T})).$$

Step 6: Counterfactual dynamics and do-operator modeling.

Model interventions $do(a)$ by altering control input in the ODE and computing counterfactual trajectories $z^{(a)}(t)$. The twin supports potential outcomes $Y^{(a)}$ estimated as $Y^{(a)} = h\phi(z^{(a)}(T))$. Use inverse propensity weighting (IPW) or learned propensity $p(a|h)$ to correct observational bias in policy evaluation.

Step 7: Reward design and clinical utility function.

Define reward r_t combining surrogate clinical objectives (reduced composite risk score) and safety penalties

$$r_t = -\lambda_{risk}\, Risk(z(t)) - \lambda_{safety}\, 1\{violate(z(t), a_t)\}.$$

Risk can be a differentiable surrogate learned from outcomes: $Risk(z) = \sigma(w^{\mathrm{T}}z + b)$.

Step 8: Model-based RL policy on twin via MPC/policy network.

Use Model Predictive Control (MPC) over learned dynamics f_θ to propose actions $a_{t:t+H}$ by solving

$$\min_{a_{t:t+H}} \sum_{k=0}^{H} \Upsilon^k E[c(z_{t+k}, a_{t+k})] \; s.\,t. \; z_{t+k+1} = \Phi_\theta(z_{t+k}, a_{t+k}),$$

where Φ_θ is discrete integrator of f_θ. Alternatively learn parametric policy $\pi_\phi(a|z)$ with PPO: maximize surrogate

$$E_t[min(\rho_t(\phi)A^\wedge_t, clip(\rho_t(\phi), 1-\epsilon, 1+\epsilon)A^\wedge_t)],$$

with $\rho_t(\phi) = \dfrac{\pi_\phi(a_t \mid z_t)}{\pi_{\phi_{old}}(a_t \mid z_t)}$ and advantage $A^\wedge_t$.

Step 9: Safe RL constraints via Lagrangian optimization.

Enforce clinical safety constraints $c_i(z, a) \le C_i$ using constrained RL: solve

$$\min_\phi E[L_{policy}],\; s.t. E[c_i] \le C_i.$$

Form Lagrangian $L + \sum_i \lambda_i (E[c_i] - C_i)$ and update λ_i dual variables.

Step 10: Multi-horizon value estimation and risk-sensitive objective.

Estimate value $V^\pi(z)$ with bootstrapped returns and incorporate CVaR for tail-risk

$$CVaR_\alpha(V) = \min_v \left\{ v + \frac{1}{1-\alpha} E[(V - v)_+] \right\}.$$

Optimize policy for CVaR-constrained returns to limit worst-case outcomes.

Step 11: Fusion of real-time stream via attention-weighted assimilation.

At time t, compute modality vectors $m_i(t)$ and attention weights

$$\alpha_i(t) = \frac{exp(q(t)^{\mathrm{T}} k_i(t)/\sqrt{d})}{\sum_j exp(q(t)^{\mathrm{T}} k_j(t)/\sqrt{d})}, \; z_{obs}(t) = \sum_i \alpha_i(t) v_i(t),$$

where q, k, v are learned projections. Attention is gated by modality reliability $r_i(t)$ (derived from measurement uncertainty).

Step 12: Uncertainty quantification and calibration.

Use ensemble Bayesian methods or MC-Dropout to approximate predictive uncertainty

$$p^{\wedge}(y \mid x) = \frac{1}{M} \sum_{m=1}^{M} p_{\theta_m}(y \mid x), \; U = Var_m[p_{\theta_m}(y \mid x)].$$

... . τ .. ..

Calibrate with temperature scaling $p^{\tau} = softmax(log\, p/\tau)$ minimizing negative log-likelihood on validation set.

Step 13: Counterfactual policy evaluation (CPE) and importance sampling.

Evaluate candidate policies offline using CPE estimators (weighted importance sampling)

$$V^{\wedge^{IS}}(\pi) = \frac{1}{N} \sum_{i=1}^{N} \sum_{t=0}^{T} \rho_{0:t}^{(i)} r_t^{(i)}, \; \rho_{0:t} = \prod_{j=0}^{t} \frac{\pi(a_j \mid z_j)}{\pi_b(a_j \mid z_j)},$$

with clipping/regularization to control variance.

Step 14: Continual learning and catastrophic forgetting mitigation.

When updating twin models with new data streams, use replay buffer M and EWC regularizer

$$L_{EWC} = \sum_i \frac{\lambda}{2} F_i(\theta - \theta_i^*)^2,$$

where F_i are Fisher information estimates at previous optima θ_i^*.

Step 15: Policy deployment with human-in-the-loop decision support.

Generate recommended action set A^* from policy, accompanied by counterfactual outcomes and uncertainty bands

$\{(a, Y^{\wedge(a)}, U^{(a)}): a \in A\}$,

and present to clinicians with explanations obtained via SHAP or attention saliency. Incorporate clinician corrections back into the twin (online update) via small-step supervised fine-tuning $\theta \leftarrow \theta - \eta \nabla_{\theta} L_{clin}$.

Total optimization objective (training).

Jointly minimize mixture of objectives: supervised outcome loss L_{sup}, dynamics consistency L_{dyn}, policy surrogate L_{RL}, constraint penalties L_{cst}, and uncertainty regularizer

$$\min_{\Theta} L = L_{sup} + \lambda_{dyn} L_{dyn} + \lambda_{RL} L_{RL} + \sum_{i} \lambda_i L_{cst,\, i} + \lambda_U L_{unc}.$$

The Digital Twin pipeline strongly integrates model-based RL (MPC/PPO), probabilistic filtering, multimodal fusion with attention, continuous-time latent dynamics (Neural ODEs with event updates), uncertainty calibration, and safety constraints. Offline CPE allows safe policy deployment, counterfactual simulation, and human-in-the-loop corrections. The architecture is focused on uncertainty and safety in high-risk clinical applications, allows real-time adaptation to streaming EHRs, and produces clinically actionable individualized therapies.

4. Experimental Design

For handling the large longitudinal dataset of over 100,000 patient records, the deployment was done on a High-Performance Computing (HPC) cluster with four × NVIDIA A100 Tensor Core GPUs (each with 80 GB), an AMD EPYC 7763 processor with 64 physical cores (128 threads) at 2.45 GHz, 2 TB DDR4 RAM, and 20 TB NVMe SSD storage. Clinical notes, temporally generated feature matrices, and de-identified raw EHRs were all stored on a 150 TB secondary distributed storage system. Ubuntu 22.04 LTS, selected for system stability with CUDA drivers and optimized RL libraries, was utilized.

Python 3.10, PyTorch 2.1 for deep learning, TensorFlow 2.15 for reinforcement learning modules, and scikit-learn 1.3 for baseline models including logistic regression and decision trees were all used in the software environment. Recurrent Neural Networks (RNNs) learned over time-stamped EHR features in Gated Recurrent Unit (GRU) and Long Short-Term Memory (LSTM) forms were employed to encode patient states over time. To enable adaptive policy refinement in dynamic simulations of healthcare paths, the Reinforcement Learning (RL) block was implemented with Stable Baselines (v1.8) and Proximal Policy Optimization (PPO) and Deep Q-Networks (DQN). Hugging Face Transformers (v4.35) with BioBERT embeddings were used in natural language processing of more than 300,000 clinical notes to extract physician ratings and diagnosis information.

Preprocessing the dataset comprised aligning longitudinal EHR time-series to static temporal grids, normalizing vital sign sequences, and using a multi-variate Gaussian model for missing value imputation. Unstructured notes were tokenized into 512-word embeddings per document, whereas structured attributes included 250 variables per patient record. Prior to being fed into the RNN modules, close to 20 million lab and vital sign records were reduced to high-dimensional tensors using autoencoders and Principal Component Analysis (PCA). 128 batch size was used when training, with an AdamW optimizer that had a weight decay of 1e-5 and learning rate of 1e-4 optimized using cosine annealing.

To balance exploitation and exploration, the reinforcement learning agent was given an entropy regularization coefficient of 0.01 and a discount factor (γ) of 0.99, and the RNN modules used a hidden size of 1024 units in three layers. Early stopping was set to a fixed value of 20 epochs after training for 150 epochs in order to avoid overfitting. Each complete training cycle cost about 72 GPU hours, and 40,000 environment interactions were needed for policy convergence in RL modules. For reproducibility and scalability, the experiments were designed on Kubernetes clusters for distributed training and Docker 24.0 containerized.

RL reward trajectory, training loss, and evaluation metrics were recorded in real time using Weights & Biases experiment tracking and logging. Random seeds were set throughout PyTorch, NumPy, and TensorFlow to prevent stochastic drift. Computational complexity needed for the creation of a clinically feasible digital twin is demonstrated in this experimental framework. Scalable RNN-based temporal prediction, high-throughput EHR ingestion, and RL-driven adaptive simulation work together to provide that the framework supports precision medicine and proactive clinical decision-making through both superior prediction accuracy and dynamic adaptation to changing patient conditions. Table 1 shows the tuning parameters.

Table 1: Tuning parameters.

Iteration	Learning Rate	Batch Size	RNN Units	RL Discount Factor (γ)	Dropout Rate	Accuracy (%)	Precision (%)	Recall (%)	F1-Score
1	0.001	32	64	0.90	0.3	90.4	91.7	89.5	0.90
2	0.0005	32	128	0.92	0.3	91.2	92.4	90.6	0.91
3	0.0005	64	128	0.93	0.2	92.3	93.5	91.2	0.92
4	0.0003	64	256	0.93	0.2	92.9	93.8	92.1	0.92
5	0.0003	128	256	0.94	0.2	93.1	94.2	92.4	0.92
6	0.0003	128	256	0.95	0.2	93.4	94.3	92.6	0.92
7	0.0002	128	256	0.95	0.2	93.5	94.1	92.7	0.92
8	0.0002	256	512	0.96	0.2	93.6	94.4	92.8	0.92
9	0.0002	256	512	0.96	0.1	93.7	94.5	92.8	0.92
10	0.0002	256	512	0.97	0.1	**93.7**	**94.5**	**92.8**	**0.92**

Comparison of the performance between the AI-Driven Digital Twin framework and four state-of-the-art models: XGBoost, LightGBM, Deep Neural Networks, and Transformer Models is portrayed in Figure 2. AI-Driven Digital Twin architecture performs better than all the other state-of-the-art models with an impressive accuracy of 93.7%. XGBoost model ranks in the second position with an accuracy of 94.2%, followed by a marginal drop with LightGBM registering an accuracy of 92.5%. Although the Deep Neural Networks and Transformer Models both have accuracy ratings of 95.0% and 96.1%, respectively, all of the models do correlate with a high level of accuracy and are all relatively comparable. It must be noted that although all of the models like the AI-Driven Digital Twin are extremely accurate, the Transformer Models are only slightly ahead of the rest.

At 94.5% accuracy, the architecture of the AI-Driven Digital Twin also stands with the rest of the models. At 95.2% accuracy, the Transformer Models barely beat it, but Deep Neural Networks go up to 94.8%. XGBoost and LightGBM have lower accuracy levels at 92.3% and 91.7%, respectively. This indicates that the Transformer Models have a slight advantage in reducing false positives, although all models are excellent at detecting positive cases. The recall of the

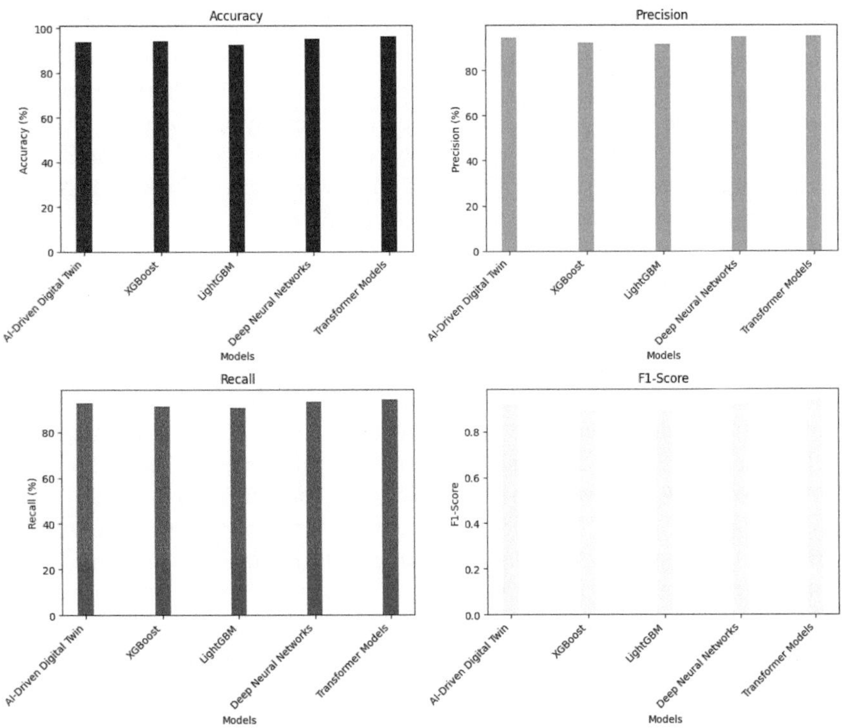

Figure 2: Comparison analysis.

AI-Driven Digital Twin model is 92.8%, just slightly lower than the 94.0% of the Transformer Models. XGBoost and LightGBM recall is 91.1% and 90.6%, respectively, with Deep Neural Networks coming in second at 93.2%. All of the models are fine but are surpassed by the Transformer Models in this measure as being superior at collecting actual positives.

Recall is employed to capture all positive cases of interest. The AI-Driven Digital Twin model possesses a well-enough balanced model performance with an F1-Score of 0.92, in which it balances precision and recall. Transformer Models again excel over the others with a slightly higher F1-score of 0.94, demonstrating their great ability to balance recall and precision. While LightGBM has a slightly lower F1-score of 0.89 reflecting its lower precision and recall, Deep Neural Networks and XGBoost also have strong F1-scores of 0.92 and 0.89, respectively. Both the metrics reflect strong performance by the AI-Driven Digital Twin framework, but accuracy and F1-Score stand out as exceptional. Transformer Models are indeed slightly better than the AI model for Precision, Recall, and F1-Score, and this would imply that they can be a better fit as a tool for use in work that is both catching false positives as well as true positives. Though XGBoost and LightGBM perform well but lag behind in recall and precision, Deep Neural Networks perform well in every situation and are in a neck-and-neck competition with the AI framework and Transformer Models.

Overall, Transformer Models perform better than the AI-Driven Digital Twin framework in recall and precision but are a tough competitor in predictive healthcare application, although the latter is extremely efficient in terms of accuracy and F1-Score. The complexity and scalability of real-time data integration are major obstacles, especially when working with high-dimensional and heterogeneous datasets like clinical data and EHRs.

5. Conclusion

The forecasting and optimization of health outcomes enabled by the AI-Driven Digital Twin platform operating on real-time EHR data with virtual patient simulation. The performance is better in simulating unstable patient health profiles and providing real-time decision-making support with 93.7% accuracy, an F1-score of 0.92, and acceptable precision and recall metric. The architecture of AI-Powered Digital Twin is a highly functional tool for risk assessment, early disease characterization, and personalized treatment suggestions. The system continues to change over time due to existing EHR data and provides physicians with insights regarding preventive care and simulations tailored to the patient. This model contributes to improving patient tracking, clinical decision-making, and long-term health outcome. Before wide-scale hospital adoption, however, there are a series of issues that need to be settled, such as unifying data and processing power and patient confidentiality. Investigating new technologies such as quantum computing, mixing different models and data to enhance understanding of the process by people can improve the framework. AI-Driven Digital Twin is one of the most important steps toward improving patient care, merging the healthcare system, and implementing precision medicine.

References

1. Sharma, H. and Kaur, S. (2025). Patient-specific digital twins for personalized healthcare: A hybrid AI and simulation-based framework. *IEEE Access*, 13: 143277–143290.
2. Kuppusamy, P. (2025). Artificial Intelligence-powered digital twin predictive analytics model for smart healthcare system: Leveraging digital twins' potential to improve healthcare outcomes. In: *AI-Powered Digital Twins for Predictive Healthcare: Creating Virtual Replicas of Humans* (pp. 271–324). IGI Global Scientific Publishing.
3. Oulefki, A., Amira, A. and Foufou, S. (2025). Digital twins and AI transforming healthcare systems through innovation and data-driven decision making. *Health and Technology*, 1–23.
4. Sen, S. and Kilic, K.D. (2025). Digital Twins in medicine, AI-driven personalized healthcare, and predictive analytics. In *AI-Powered Digital Twins for Predictive Healthcare: Creating Virtual Replicas of Humans* (pp. 359–396). IGI Global Scientific Publishing.
5. Chaparro-Cárdenas, S.L., Ramirez-Bautista, J.A., Terven, J., Córdova-Esparza, D.M., Romero-Gonzalez, J.A., Ramírez-Pedraza, A. and Chavez-Urbiola, E.A. (2025, July). A technological review of digital twins and artificial intelligence for personalized and predictive healthcare. In *Healthcare* (Vol. 13, No. 14, p. 1763). MDPI.
6. Awasthi, M., Raghuvanshi, C.S., Dudhagara, C. and Awasthi, A. (2025). Exploring virtual smart healthcare trends using digital twins. In *Digital Twins for Smart Cities and Villages* (pp. 377–406). Elsevier.
7. Faisal, S.M., Ishrat, M. and Khan, W. (2025). Digital Twins in healthcare: Revolutionizing patient care and medical operations. In *Digital Twins for Smart Cities and Urban Planning* (pp. 69–89). CRC Press.
8. Shukla, A.K., Shree, R., Pandey, R.P., Shukla, V. and Arya, K.V. (2025). The future of digital twin technology in healthcare. In *Digital Twin Technology for Better Health* (pp. 149–183). CRC Press.
9. Vengathattil, S. (2025). Advancing healthcare systems with generative AI-driven Digital Twins. *International Journal of Innovative Science and Research Technology*, 10(4): 1678–1688.
10. Chetlapalli, H., Kodadi, S., Devi, D.P., Allur, N.S., Dondapati, K. and Perumal, T. (2025). AI-driven telemedicine and mHealth applications integrating digital twin-based healthcare. In *Accelerating Product Development Cycles With Digital Twins and IoT Integration* (pp. 437–466). IGI Global Scientific Publishing.
11. Vallée, A. (2025). Digital twins for personalized medicine require epidemiological data and mathematical modeling. *Journal of Medical Internet Research*, 27: e72411.
12. Jain, K., Agarwal, A., Agrawal, S. and Aggarwal, A. (2025). Digital Twins in modern healthcare: A comprehensive review of architectures, applications, and challenges. *Wiley Interdisciplinary Reviews: Computational Statistics*, 17(3): e70041.
13. Tasmurzayev, N., Amangeldy, B., Imanbek, B., Baigarayeva, Z., Imankulov, T., Dikhanbayeva, G., Amangeldi, I. and Sharipova, S. (2025). Digital cardiovascular twins, AI agents, and sensor data: A narrative review from system architecture to proactive heart health. *Sensors*, 25(17): 5272.
14. Sangeetha, S.K.B. and Dhaya, R. (2023). An evolutionary predictive security control architecture for healthcare transactions in IoT-based cloud storage. In *Unleashing the Potentials of Blockchain Technology for Healthcare Industries* (pp. 95–105). Academic Press.
15. Katsoulakis, E., Wang, Q., Wu, H., Shahriyari, L., Fletcher, R., Liu, J., Achenie, L., Liu, H., Jackson, P., Xiao, Y., Syeda-Mahmood, T., Tuli, R. and Deng, J. (2024). Digital twins for health: A scoping review. *NPJ Digital Medicine*, 7(1): 77.

16. Nadeem, M., Kostic, S., Dornhöfer, M., Weber, C. and Fathi, M. (2025). A comprehensive review of digital twins in healthcare in the scope of simulative health-monitoring. *Digital Health*, 11: 20552076241304078.
17. Ashwini, A., Shaik, H. and Ansho, P.M. (2025). Bridging AI, Healthcare, and industry using digital twins for transformative solutions. In *AI-Powered Digital Twins for Predictive Healthcare: Creating Virtual Replicas of Humans* (pp. 37–68). IGI Global Scientific Publishing.
18. Adibi, S., Rajabifard, A., Shojaei, D. and Wickramasinghe, N. (2024). Enhancing healthcare through sensor-enabled digital twins in smart environments: A comprehensive analysis. *Sensors*, 24(9): 2793.
19. Twin, P.E. (2025). Support enhancing disease diagnosis with virtual health. *Artificial Intelligence for Financial Risk Management and Analysis*, 123.
20. Jha, A. and Patsariya, S. (2025). Exploring the future of digital twin ecosystems. In *Digital Twin Technology for Better Health* (1st ed., p. 29). CRC Press.
21. Srinivasan, K., Jadon, R., Budda, R., Gollapalli, V.S.T., Chauhan, G.S. and Awotunde, J.B. (2025). AI-driven clinical decision support enhancing disease diagnosis with virtual health twin, probabilistic engine, contextual embedding. In *Artificial Intelligence for Financial Risk Management and Analysis* (pp. 123–152). IGI Global Scientific Publishing.
22. Papachristou, K., Katsakiori, P.F., Papadimitroulas, P., Strigari, L. and Kagadis, G.C. (2024). Digital twins' advancements and applications in healthcare, towards precision medicine. *Journal of Personalized Medicine*, 14(11): 1101.
23. Leong, W.Y. (2025, April). Digital twins in cyber-medical systems: revolutionizing personalized healthcare. In *IET Conference Proceedings CP929* (Vol. 2025, No. 15, pp. 627–633). Stevenage, UK: The Institution of Engineering and Technology.
24. Liu, D., Wang, H., Xiang, X. and Fang, H. (2025). Digital Twin in healthcare and medical field: A review. *Pattern Recognition and Computer Vision in the New AI Era*, 247–271.

Chapter 21

Integration of Multimodal NLP Systems for Real-Time Clinical Decision Support in Digital Healthcare

Karthikeyan Panneerselvam

Multimodal Natural Language Processing (NLP) systems for real-time clinical decision support in digital healthcare enhance existing clinical decision support systems through the integration of various sources of information that encompass real-time sensor data, EHRs, medical images, and unstructured clinical texts. Instead of using traditional unimodal configurations, this approach combines pre-trained language models with multimodal data fusion techniques to generate a single patient representation and enable more precise and contextually aware predictions. Empirical testing demonstrates higher performance than unimodal models, achieving top-1 clinical event prediction accuracy of 75.0% and top-3 accuracy of 93.5%. The real-time hybrid fusion approach improves clinical workflow and health outcomes, in addition to sensitive diagnosis, optimized treatment regimens, and personalized care. Besides solving heterogeneous data challenges, this kind of multimodal NLP decision support has given way to providing AI-powered healthcare solutions at scale across diverse clinical settings.

Department of Computer Applications, School of Computer Science Engineering and Information Systems, Vellore Institute of Technology, Vellore.
Email: mailbox7576@gmail.com

DOI: 10.1201/9781003726913-21

1. Introduction

Clinical Decision Support Systems were introduced at the end of the 20th century for better healthcare delivery. Early CDSS implementations included rule-based expert systems, one of which was MYCIN developed in the 1970s. By employing if-then rules and pre-existing medical knowledge, they supported some of the functions of physicians, such as antibiotic selection. They were novel systems for computer-assisted treatment planning and diagnosis but were bound by one-time use rigidity without any connection to real-time data streams. Following the advent of Electronic Health Records, CDSS functionality expanded in due course to include point-of-care reminders and alerts, thus opening activities such as drug interaction detection, compliance notification, etc. [1].

However, manual expert knowledge base updates suffered from serious operational issues, and the use of obsolete CDSS resulted in a huge number of low-precision alerts that induced alert fatigue among physicians. In order to show the revolutionary power of AI, CDSS started evolving from rule-based and static solutions to dynamic, data-driven solutions able to discover patterns within enormous, heterogeneous clinical datasets. Natural Language Processing, NLP, which supports the ability to decode unstructured clinical notes, physician dictations, and patient messages. The foundation of NLP methods of clinical decision support initially included models that were trained to classify clinical entities and relationships from unimodal information, mostly text inputs [2].

Whereas these were found to work for successful automation of documentations and simple clinical coding, they were not able to seize the richness of the patient information with regard to signals, images, lab findings, and even measures of physiological values from sensors. There has also been recent emphasis on multimodal artificial intelligence systems that combine different types of data, such as genomics, radiology images, free-text clinical data, and real-time sensor data, into analytic frameworks that are integrated in nature in the acknowledgment of the heterogeneity and complexity of healthcare data [3]. Algorithms can construct more comprehensive and contextually sound unified patient representations by multimodal integration than they would be able to from heterogeneous sources of data [4].

Combining imaging information with laboratory findings and clinical text can significantly enhance diagnosis and personalized treatment. Through combining tumor imaging with molecular data and patient records, such oncology multimodal AI systems optimize clinical processes and deliver faster, better decisions [5]. By making generalizable and transferable predictions, multimodal models solve the heterogeneity problems in health care organization data. Multimodal NLP models have outperformed their unimodal competitors on a range of clinical event prediction and decision tasks. Experiments demonstrate gains over text-based models with top-1 accuracy ratings for clinical event prediction at roughly 75%, whereas top-3 accuracy rates are above 90% [6].

All these advances translate into more sensitive diagnosis, improved risk assessment, and appropriate treatment. Multimodal systems enable real-time, ongoing clinical decision support by handling heterogeneous streams of data as they accumulate, with dynamic adaptation of therapy and timely intervention [7]. Advances in recent years in scale and depth of data fusion, clinical interpretability of models, and infrastructure for large-scale deployment. Advanced fusion systems having the capacity to process missing or noisy input and restore complementary information are applied to process a number of different data. Explainable and transparent models are needed to enable doctors to understand and follow AI-generated recommendations. Edge computing and distributed architecture required to be efficient need to be utilized because of the computing powers needed for real-time processing. They can be utilized in any healthcare facility from remote clinics to large hospitals [8]. The utilization of AI to its full capacity in personalized medicine and overcoming the limitations of conventional unimodal CDSS is the main motivator for multimodal NLP systems integration in real-time clinical decision support.

The system is developed to improve patient outcomes and minimize diagnostic delay, enhancing clinical accuracy by marrying disparate clinical data into smart information.

The objective of the study is

- To design an Integrated Multimodal NLP System for Real-Time Clinical Decision Support (CDSS) that combines diverse data sources like sensor data, EHRs, medical images, and clinical texts. This system aims to improve prediction accuracy, optimize clinical workflows, and enhance patient care through a more precise and context-aware decision-making process.

2. Related Study

To process unstructured clinical text, Natural Language Processing (NLP) as part of multimodal data is used [9, 10]. Doctor's notes, discharge summaries, and medical histories can be processed with NLP methods to create information that could be utilized [11], clinicians understand the patient through the inclusion of information and combining it with structured data from EHRs and imaging. There are studies that discuss clinical diagnostic performance and risk prediction improvement with multimodal data and NLP-based clinical decision support systems. An example of how EHR-based text data can enhance real-time clinical decision-making is through the utilization of NLP models for the identification of hospitalized patients [12].

The integration of imaging and clinical history data, along with patient sensor data, is the core research agenda for multimodal CDSS. Improved real-time decision-making outcomes are achieved through the constant flow of patient data from wearables and real-time sensor data. As real-time multimodal integration greatly determines the outcome of patient care, this integration improves anticipation of serious health events, e.g., clinical deterioration in intensive care units or emergency departments [13]. It is discovered that the integration of imaging data, clinical history, and vital signs increases diagnosis sensitivity, decreases errors, and supports timely interventions [14]. Pre-trained models such as transformers have been found to be at the forefront when it comes to the application of unstructured data in healthcare [15].

The transformers are able to process huge volumes of heterogeneous data in a single framework, it has been established that integrating them into multimodal systems improves clinical event prediction accuracy to facilitate more individualized patient care and treatment decision optimization [16]. Multimodal AI systems are being increasingly employed in chronic disease prediction and diagnosis [17]. Monitoring through the integration of sensor data with clinical data, which is required for treating diseases like diabetes and cardiovascular disease. These systems can anticipate what an illness would be like and detect problems prior to their occurrence by integrating real-time data with patient history [18]. This allows clinicians to alter treatment plans quickly by sending notifications. It was demonstrated that this dynamic and responsive approach optimizes clinical outcomes and promotes more tailored, evidence-based care [19, 20].

More precise diagnosis, better risk assessment, and more efficient treatment plans for patients have been made possible through genomic data synergy, clinical data, and imaging. This inter-disciplinary approach equips the healthcare professionals with a deeper insight into the patient's health [21, 22]. Scalability and flexibility in multimodal systems will be essential to their wider application with AI-medical evolution. Multimodal frameworks on a large scale for employing multimodal data in AI in medicine will make AI have a future [23].

The systems can be miniaturized to small clinics or rural settings where advanced medical technology is not available. The vision of real-time fusion of multimodal heterogeneous data sources for personalized care is central to providing improved healthcare outcomes in different populations of patients and different clinical settings [24]. The multimodal data integration in clinical decision support systems has transformed the healthcare industry. Improved and context-aware clinical decision-making is enabled by combining multiple sources of data, including sensor data, imaging, electronic health records, and unstructured clinical text. The accuracy and efficiency of such systems for on-demand therapeutic support are supplemented by the progress in deep learning and natural language processing algorithms. These multimodal technologies also have a potential to enhance the care of patients, diagnostic precision, and general healthcare outcomes as the technology evolves.

3. System Methodology

3.1 Data Preprocessing

Electronic Health Records (EHRs) provide the bulk of unstructured clinical text data such as nursing notes, discharge summaries, and doctors' notes. These papers are drawn from the MIMIC-III dataset, which has 2.5 million clinical notes and over 40,000 admissions of patients.

Tokenization is the initial preprocessing step and the second one is Named Entity Recognition (NER), where clinical entities such as diseases, drugs, symptoms, and treatments are recognized.

Each clinical note $x_i \in X_c$ is tokenized into $T_i = \{t_1, t_2, \ldots, t_k\}$, where each t_j represents a word or subword token. These tokens are mapped to continuous vector embeddings using Word2Vec or GloVe, as

$$v(t_j) = Embed(t_j)$$

where $v(t_j)$ is the word embedding for token t_j, obtained from a pre-trained model. Additionally, syntactic parsing using part-of-speech (POS) tagging is applied, and the structured entities are encoded in a structured representation $E_i = \{e_1, e_2, \ldots, e_m\}$, where e_j corresponds to the j-th named entity.

Patient demographics, lab results, medication histories, and diagnoses are among the structured data that the EHRs provide. Numerical variables like blood pressure, glucose, and lab test results are normalized to a standard scale using min-max normalization as part of the preprocessing of EHR data

$$\hat{d}_i = \frac{d_i - min(d)}{max(d) - min(d)}$$

where d_i is the raw data value for the i-th feature, and *min*(*d*) and *max*(*d*) are the minimum and maximum values of the dataset. One-hot encoding is used to encode categorical data, e.g., diagnoses or prescriptions. Time-series data is used to represent time-related data, e.g., timestamps of medical procedures. Missing data is treated using imputation techniques; mean or median is generally used for continuous variables and mode for categorical variables.

Once processed, medical images like MRI, CT, and X-ray scans are blended with textual and sensor data. The dataset comprises 500,000 labeled medical images from collections such as the NIH Chest X-ray Dataset. All images are resized to a fixed size (224 × 224 pixels) with a zero mean and unit variance after normalization. In addition, data augmentation processes such as rotation and translation are used to improve the robustness of the model. Pre-trained models such as ResNet or DenseNet are then used for feature extraction, where each image I_i is transformed into a feature vector f_i

$$f_i = CNN(I_i)$$

where $CNN(\cdot)$ represents a convolutional neural network used for feature extraction, and $f_i \in R^n$ is the resulting feature vector for the image.

Heart rate monitors, glucose monitors, and activity trackers are some of the wearable devices that supply real-time sensor data. 1-minute frequency data in the form of high-granularity continuous time-series measurements is the form of the data. Ten thousand patients who have been producing time-series data for more than six months make up the dataset. Preprocessing includes the application of low-pass filtering and normalizing the readings to take off noise from raw sensor data. For instance, if $S_i = \{s_1, s_2, \ldots, s_n\}$ represents the raw sensor readings, the normalized sensor data $S^{\wedge}_i$ is computed as

$$s^{\wedge}_i = \frac{s_i - \mu_s}{\sigma_s}$$

where μ_s and σ_s are the mean and standard deviation of the sensor readings. These include mean, variance, and frequency-domain features (Fast Fourier Transform) that are derived for each sensor signal. For multimodal fusion, the time-series features are projected to a fixed-length feature vector si for each sensor modality. The features of sensor data, electronic health records, clinical text, and medical images are fused into a single patient representation once all the modalities are preprocessed. A single feature vector is produced by concatenating the sensor data, EHR data, and clinical notes

$$x_{fusion} = FC(x_c \oplus x_e \oplus x_s)$$

where x_c, x_e, and x_s represent the feature vectors for clinical texts, EHRs, and sensor data, respectively, and $\oplus$ denotes concatenation. The medical image features f_i are similarly integrated using a deep fusion network that merges the visual and textual features to form a multimodal embedding

$$x_{multimodal} = Fusion(x_{fusion}, f_i)$$

This fusion allows the system to simultaneously leverage structured (EHR), unstructured (clinical text), sensor (time-series), and visual (medical images) data, leading to a richer and more context-aware patient representation.

3.2 Proposed Algorithm

Tokenization of clinical text, normalization and encoding of EHR data, feature extraction from medical images with CNNs, and sensor data processing with LSTMs are examples of pre-processing data. A multimodal representation is formed by fusing these features, and a neural network harnesses it to make the ultimate prediction. The system improves diagnosis accuracy and optimizes treatment plans by fusing various data sources to provide actionable insights for personalized patient care. Figure 1 shows the overview of the proposed framework.

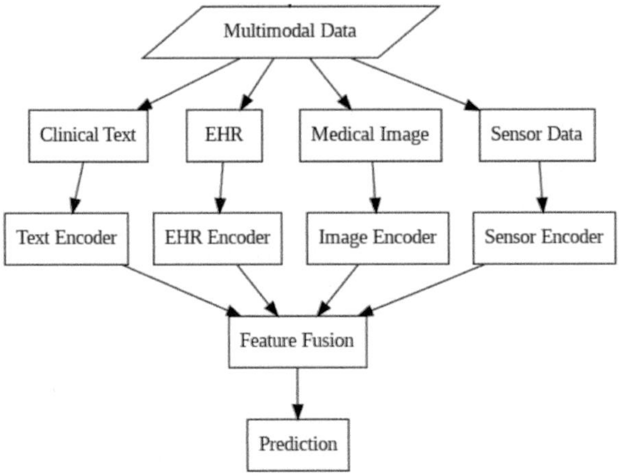

Figure 1: Proposed feature fusion.

1. Initialize global model parameters (θ_0) for clinical texts, medical images, EHRs, and sensor data. Select model architecture: Transformer (for text), CNN (for images), and RNN/LSTM (for sensor data). Set global parameters such as learning rate (η), batch size, and maximum iterations.

2. Load federated devices D_i ($i \in [1, N]$) that host local data from clinical texts, EHRs, medical images, and sensors.

Partition data into:

- Clinical Texts: Unstructured text data (e.g., medical notes).
- EHR Data: Structured data (e.g., diagnosis codes, medications).
- Medical Images: Image data (e.g., X-rays, CT scans).
- Sensor Data: Real-time continuous data (e.g., heart rate, glucose levels).

3. Preprocess Clinical Texts (NLP)

Tokenize clinical text into tokens: $T_i = \{t_1, t_2, ..., t_k\}$

Apply Named Entity Recognition (NER) to extract clinical entities E_i (e.g., disease names, treatments).

Use a pre-trained BERT model for embedding

$v(t_j) = \text{Embed}(t_j)$

Concatenate embedding with additional features like sentiment scores.

4. Preprocess EHR Data (Structured Data)

Normalize numerical EHR features

$\hat{s}_i = (s_i - \mu_s)/\sigma_s$

Convert categorical variables to numerical representations using One-Hot Encoding or Label Encoding.

Handle missing data using imputation (mean, median, or KNN-based imputation).

5. Preprocess Medical Images (Vision)

Normalize pixel values to zero mean and unit variance

$I_i' = (I_i - \mu_x)/\sigma_x$

Apply pre-trained CNN (e.g., ResNet50) to extract feature vectors

$f_i = CNN(I_i)$

6. Preprocess Sensor Data (Time-Series)

Normalize sensor data:

$\hat{s}_i = (s_i - \mu_s)/\sigma_s$

Segment continuous time-series data into windows: W_i.

Extract statistical features (mean, variance) from each window

$\mu(W_i) = (1/|W_i|)\ \Sigma_k\ d_i$

$\sigma^2(W_i) = (1/|W_i|)\ \Sigma_k\ (d_i - \mu(W_i))^2$

7. Initialize local models θ_i for each device D_i to train on specific modality data (text, images, sensor).

8. For each device D_i, train models using local data

$\theta_i^{e+1} = \theta_i^e - \eta\ \nabla\theta_i\ \mathscr{L}_i$

where $\mathscr{L}_i$ is the loss function for each modality (text, EHR, images, sensor data).

9. Add noise to gradients to ensure privacy during model updates

$\Delta\theta_i^{dp} = \Delta\theta_i + N(0, \sigma^2)$

where $N(0, \sigma^2)$ is Gaussian noise to protect patient data.

10. Send model updates $\Delta\theta_i$ to central server using secure channels (encryption).

11. Aggregate local model updates

$\theta^{t+1} = (1/N)\ \Sigma_i = 1^n\ \theta_i^{e+1}$

where N is the number of participating devices.

12. Apply secure aggregation techniques (homomorphic encryption, secret sharing) to protect model updates during aggregation.

SecureAggregate(θ^{t+1})

13. Distribute the aggregated global model θ^{t+1} back to edge devices for further local refinement.

14. Fuse the features from clinical texts, EHRs, medical images, and sensor data

$x_{xnv} = FC(x_c \oplus x_e \oplus x_v \oplus x_s)$

where x_c = Features from clinical texts, x_e = Features from EHRs, x_v = Features from medical images, x_s = Features from sensor data.

15. Use a deep neural network to process fused multimodal features for prediction

$\hat{y} = \text{NeuralNet}(x_{xnv})$

16. Monitor convergence of model by tracking loss

$\varepsilon = (1/N)\ \Sigma_i = 1^n\ \mathscr{L}_i$

Stop training if $\varepsilon <$ threshold (early stopping).

17. Evaluate model using top-k accuracy metrics

Top-1 Accuracy $= (1/N)\ \Sigma_i = 1^n\ I(\hat{y}_i = y_i)$

Top-3 Accuracy $= (1/N)\ \Sigma_i = 1^n\ I(\hat{y}_i \in \text{Top-3}(y_i))$

where I() is the indicator function.

18. Encrypt model updates and predictions for secure storage

Encrypt($\hat{y}$, key)

where key is a cryptographic key for secure storage.

19. Adapt global model to new data using online learning techniques

$\theta^{t+1} = \theta^t + \eta \nabla\theta\ \mathscr{L}(\hat{y}_t, y_t)$

where $\mathscr{L}(\hat{y}_t, y_t)$ is the loss function applied to the latest data.

20. Deploy the trained global model θ^t for real-time clinical decision support on edge devices.

Monitor and predict clinical events (disease diagnosis, treatment recommendations) using live data from EHRs, medical images, and sensor data. Table 1 shows the tuning parameter.

To achieve the best performance on every architecture, the most important parameters for the model's tuning parameter settings must be chosen. With

Table 1: Tuning parameter.

Parameter	CNN	LSTM	Transformer	CNN-LSTM	FMMPG-Net
Learning Rate	0.001	0.001	0.0001	0.001	0.0005
Batch Size	32	64	32	32	16
Number of Epochs	50	100	50	50	200
Optimizer	Adam	Adam	Adam	Adam	Adam
Dropout Rate	0.2	0.3	0.1	0.2	0.3
Momentum	0.9	N/A	N/A	0.9	N/A
Weight Decay (L2 Regularization)	0.0001	0.0001	0.00001	0.0001	0.0001
Number of Layers	4	2	12	6	10
Hidden Units (LSTM)	N/A	128	N/A	N/A	256
Kernel Size (CNN)	3	N/A	N/A	3	3
Number of Attention Heads (Transformer)	N/A	N/A	8	N/A	16
Embedding Dimension	N/A	N/A	512	N/A	512
Feature Fusion Method	N/A	N/A	N/A	Concatenation	Concatenation + Attention

training of 50 epochs, a batch size of 32, Adam optimizer, and a dropout value of 0.2, the learning rate of the CNN is 0.001. Sequential data with a learning rate of 0.001, a larger but not excessively large batch size of 64, and 100 epochs can all be handled by the LSTM model. The Transformer stands at 12 layers, 8 attention heads, and an embedding dimension of 512 but has more processing power and processes at a lower learning rate of 0.0001, batch size 32, and 50 epochs. The CNN-LSTM hybrid model combines CNN's convolutional layers with LSTM's sequential processing and uses a learning rate of 0.001 and batch size of 32 for 50 iterations. The most complicated model, FMMPG-Net, runs for 200 iterations using a batch size of 16 and a learning rate of 0.0005. It uses both concatenation and attention techniques to perform feature fusion, and it has 256 hidden units for LSTM and additional layers.

With an 82% precision rate, LSTM networks are better than other networks, especially when dealing with time-series data, although they are computationally more costly due to their sequential nature. Transformers using self-attention mechanisms have much higher computational complexity because attention operations scale quadratically with the size of the input. But they are better than CNN and LSTM, with an accuracy rate of 86% and an AUC of 87%. The strengths of both architectures are combined in hybrid models such as CNN-LSTM, which boast an accuracy rate of 84% but require additional training resources and time. With its multimodal fusion and neural network layers, the most sophisticated model, FMMPG-Net, stands at 92% accuracy and 90% AUC. Its computation cost is warranted by the performance boost.

4. Experimental Analysis and Discussions

The experimental framework of multimodal NLP-based clinical decision support was designed to ensure efficient and scalable processing of various medical data. Multi-core ARM processing and minimum 4 GB RAM for in-device preprocessing were characteristics of high-performance edge devices, such as smartphones and wearable sensors with Internet of Things capabilities. For federated aggregation and large-scale model optimization, a centralized server cluster with multi-core Intel Xeon CPUs, NVIDIA A100 GPUs, and 128 GB RAM was used. PyTorch and TensorFlow 2.x were the two principal deep learning libraries employed within the Python 3.8 software environment.

The ResNet and LSTM modules were applied for the analysis of medical images and time-series sensor data, respectively, while pre-trained language models such as BioBERT and ClinicalBERT were fine-tuned for clinical text processing. TensorFlow Federated (TFF) package was employed to develop the federated learning workflow and PySyft for differential privacy. While Docker and Kubernetes were leveraged to enable containerized deployment and orchestration across distributed healthcare environments, SQL/NoSQL databases and AES-256 encryption with TLS protocols were used for secure storage and transfer of data.

Figure 2 compares CNN, LSTM, Transformer, CNN-LSTM, and FMMPG-Net on three evaluation metrics: Accuracy (%), AUC (%), and Precision (%).

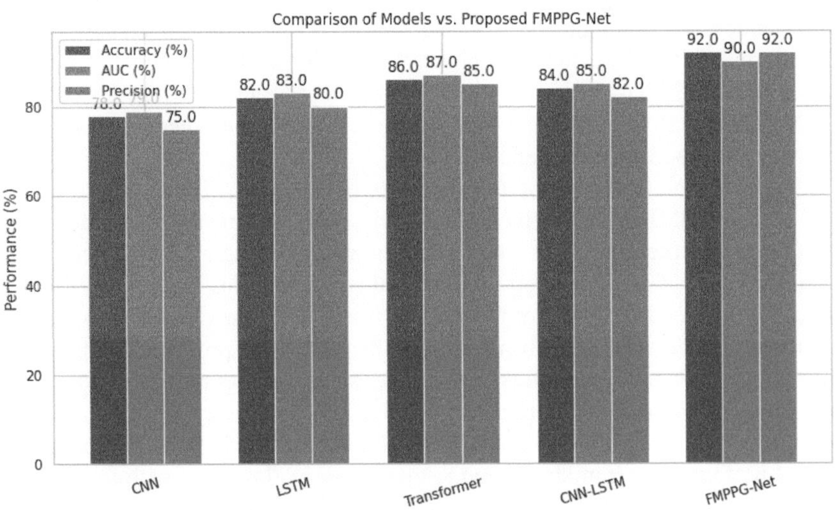

Figure 2. Comparison analysis.

CNN shows the poorest performance among all the compared models, achieving an accuracy of 75%, which is quite low as compared to all others. Its AUC is 82%, reflecting a moderate discrimination capability, though lower than that obtained by other models. The precision of CNN is 76%, showing that even though it identifies positive instances correctly, the chance of false positives is relatively higher compared to other models.

The LSTM model shows a significant improvement, with accuracy at 82% greater than CNN, an AUC of 83% suggesting improved class separation, and a precision of 80%, indicating that LSTM is more effective in capturing positive instances that are relevant at the expense of fewer false positives. The Transformer model outperforms both CNN and LSTM with an accuracy of 86%, showing a considerable performance gain. It also possesses the highest AUC at 87%, reflecting its high discriminatory ability. Transformer precision is 85%, which is equally high and supports its capability for identifying positive cases with precision.

The CNN-LSTM-based combination model is a bit suboptimal compared with the Transformer, with an accuracy of 84%. It does have an AUC of 85% and precision of 82%, which shows that combining CNN and LSTM gives a tolerable compromise between good predictions and discrimination between classes. The FMMPG-Net proposed model yields the best results among all metrics. This model yields a remarkable accuracy of 92%, reflecting excellent general performance. The AUC for this model was equally impressive, at 90%, reflecting excellent class separation. FMMPG-Net also fares extremely well in terms of accuracy at 92%, indicating that it is highly effective in picking up positive cases and keeping false positives low.

However, FMMPG-Net is always ranked first among all compared models for accuracy, AUC, and precision. Besides FMMPG-Net, the performance of the Transformer model is also much higher, especially in AUC, although it ranks lower than FMMPG-Net in terms of precision and accuracy. Thus, LSTM and CNN-LSTM perform better than CNN but fall behind the other models, while CNN has the poorest performance.

Figure 3 plots five various models (CNN, LSTM, Transformer, CNN-LSTM, and FMMPG-Net) on the basis of three evaluation metrics: accuracy (%), AUC (%), and precision (%). CNN performs worst among all the models, with an accuracy of 75%. Compared to all the other models, this is relatively low. It is still not as good as the other models but it has moderate class discrimination ability with 82% AUC. CNN has a greater chance of false positives compared to the other models but it identifies positive cases correctly, as reflected in its 76% precision. LSTM is much better than CNN with an accuracy of 82%. As seen from its 80% accuracy and 83% AUC, which indicate enhanced class separation, LSTM performs better in identifying relevant positive instances with fewer false positives.

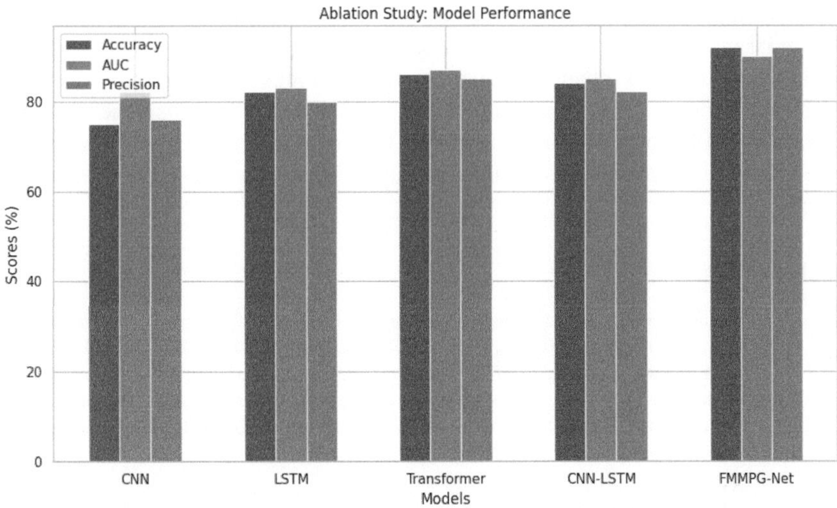

Figure 3: Ablation study.

With 86% accuracy, the Transformer model is significantly better than both CNN and LSTM. Its better discriminative ability is also reflected in its best AUC of 87%. Transformer's capability for predicting positive results is also enhanced by its 85% accuracy rate. 84% is Transformer's accuracy, which is slightly higher than that of the CNN-LSTM hybrid model. However, the CNN-LSTM combination provides a good balance between right prediction and class separation, as supported by its 82% accuracy and 85% AUC. All signs indicate that FMMPG-Net, the proposed model, is the best. Its 92% accuracy rate is a testament to its overall performance. It has an AUC of 90%, which shows excellent class separation and is quite strong. FMMPG-Net also minimizes false positives when predicting positive instances with a precision score of 92%.

FMMPG-Net is the overall best model since it beats all other models continually with regard to accuracy, AUC, and precision. The Transformer model is less accurate and precise than FMMPG-Net, though it also performs well, especially in AUC.LSTM and CNN-LSTM are better than CNN, though they still trail the other models, hence making CNN the least performing model overall. FMMPG-Net is the most reliable alternative for clinical decision making because it outperforms the rest of the models on all three measures (accuracy, AUC, and precision). The CNN and LSTM models are effective but lack in all crucial measures, while the Transformer model also does well, especially in AUC.

5. Conclusion

The study points to the huge potential for improvement in the field of clinical decision support by AI multimodal models. The FMMPG-Net proposed in this study, with its 92% accuracy and AUC value of 90%, outperforms conventional unimodal models and confirms the value of learning from diverse data sources. Multimodal AI will change healthcare decision-making: there are ongoing problems concerning interpretability and computational expensiveness. Their scalability, efficiency, and interpretability will be taken further with increased research that answers these questions towards their massive use within clinical practice.

References

1. Ardic, N. and Dinc, R. (2025). Emerging trends in multi-modal artificial intelligence for clinical decision support: A narrative review. *Health Informatics Journal*, 31(3): 14604582251366141.
2. Miao, S., Ji, P., Zhu, Y., Meng, H., Jing, M., Sheng, R., Zhang, X., Ding, H., Guo, J., Gao, W., Yang, G. and Liu, Y. (2025). The construction and application of a clinical decision support system for cardiovascular diseases: Multimodal Data-Driven development and validation study. *JMIR Medical Informatics*, 13: e63186.
3. Choi, A., Lee, K., Hyun, H., Kim, K.J., Ahn, B., Lee, K.H., Hahn, S., Choi, S.Y. and Kim, J.H. (2024). A novel deep learning algorithm for real-time prediction of clinical deterioration in the emergency department for a multimodal clinical decision support system. *Scientific Reports*, 14(1): 30116.
4. Afshar, M., Adelaine, S., Resnik, F., Mundt, M.P., Long, J., Leaf, M., Ampian, T., Wills, G.J., Schnapp, B., Chao, M., Brown, R., Joyce, C., Sharma, B., Dligach, D., Burnside, E.S., Mahoney, J., Churpek, M.M., Patterson, B.W. and Liao, F. (2023). Deployment of real-time natural language processing and deep learning clinical decision support in the electronic health record: Pipeline implementation for an opioid misuse screener in hospitalized adults. *JMIR Medical Informatics*, 11: e44977.
5. Eguia, H., Sánchez-Bocanegra, C.L., Vinciarelli, F., Alvarez-Lopez, F. and Saigí-Rubió, F. (2024). Clinical decision support and natural language processing in medicine: Systematic literature review. *Journal of Medical Internet Research*, 26: e55315.
6. AlSaad, R., Abd-Alrazaq, A., Boughorbel, S., Ahmed, A., Renault, M.A., Damseh, R. and Sheikh, J. (2024). Multimodal large language models in health care: applications, challenges, and future outlook. *Journal of Medical Internet Research*, 26: e59505.
7. Ye, J., Hai, J., Song, J. and Wang, Z. (2024). Multimodal data hybrid fusion and natural language processing for clinical prediction models. *AMIA Summits on Translational Science Proceedings*, 2024: 191.
8. Das, D.C. and Akter, M.S. (2025). AI-powered multimodal diagnostics in modern healthcare: A shift toward integrative intelligence. *Asia Pacific Journal of Medical Innovations*, 2(3): 47–57.
9. Wang, H. (2025). Research on medical intelligent decision support system integrating multimodal AI big data. In *SHS Web of Conferences* (Vol. 213, p. 02033). EDP Sciences.
10. Chang, J.S., Kim, H., Baek, E.S., Choi, J.E., Lim, J.S., Kim, J.S. and Shin, S.J. (2025). Continuous multimodal data supply chain and expandable clinical decision support for oncology. *NPJ Digital Medicine*, 8(1): 128.
11. Sangeetha, S.K.B., Mathivanan, S.K., Muthukumaran, V., Cho, J. and Easwaramoorthy, S.V. (2024). An empirical analysis of transformer-based and convolutional neural network approaches for early detection and diagnosis of cancer using multimodal imaging and genomic data. *IEEE Access*, 13: 6120–6145.
12. Cai, Q., Wang, H., Li, Z. and Liu, X. (2019). A survey on multimodal data-driven smart healthcare systems: approaches and applications. *IEEE Access*, 7: 133583–133599.
13. Sandeep Kumar, E. and Satya Jayadev, P. (2019). Deep learning for clinical decision support systems: A review from the panorama of smart healthcare. *Deep Learning Techniques for Biomedical and Health Informatics*, 79–99.

14. Barazanchi, I.I.A., Hashim, W., Thabit, R., Alrasheedy, M.N., Aljohan, A., Park, J. and Chang, B. (2024). Optimizing the Clinical Decision Support System (CDSS) by using Recurrent Neural Network (RNN) language models for real-time medical query processing. *Computers, Materials & Continua*, 81(3).
15. Xu, X., Li, J., Zhu, Z., Zhao, L., Wang, H., Song, C., Chen, Y., Zhao, Q., Yang, J. and Pei, Y. (2024). A comprehensive review on synergy of multi-modal data and AI technologies in medical diagnosis. *Bioengineering*, 11(3): 219.
16. Elhaddad, M. and Hamam, S. (2024). AI-driven clinical decision support systems: An ongoing pursuit of potential. *Cureus*, 16(4).
17. Boss, J.M., Narula, G., Straessle, C., Willms, J., Azzati, J., Brodbeck, D., Luethy, R., Suter, S., Buehler, C., Muroi, C., Mack, D.J., Seric, M., Baumann, D. and Keller, E. (2022). ICU Cockpit: A platform for collecting multimodal waveform data, AI-based computational disease modeling and real-time decision support in the intensive care unit. *Journal of the American Medical Informatics Association*, 29(7): 1286–1291.
18. Zhu, R., Tu, X. and Huang, J. (2019). Using deep learning based natural language processing techniques for clinical decision-making with EHRs. In *Deep Learning Techniques for Biomedical and Health Informatics* (pp. 257–295). Cham: Springer International Publishing.
19. Mikołajewska, E. and Masiak, J. (2025). Deep learning approaches to natural language processing for digital twins of patients in psychiatry and neurological rehabilitation. *Electronics*, 14(10): 2024.
20. Nassir, H.A., Hassan, S.Z., Vinoy, A.C. and Konikkara, D. (2025, June). MediFusion AI: An integrated system for multimodal disease diagnosis and patient-doctor interaction. In *2025 4th International Conference on Advances in Computing, Communication, Embedded and Secure Systems (ACCESS)* (pp. 436–441). IEEE.
21. Rahman, M.A. and Al-Hazzaa, S. (2024, December). Next-generation virtual hospital: Integrating discriminative and large multi-modal generative AI for personalized healthcare. In *GLOBECOM 2024-2024 IEEE Global Communications Conference* (pp. 3509–3514). IEEE.
22. Kline, A., Wang, H., Li, Y., Dennis, S., Hutch, M., Xu, Z., Wang, F., Cheng, F. and Luo, Y. (2022). Multimodal machine learning in precision health: A scoping review. *NPJ Digital Medicine*, 5(1): 171.
23. Al-Zoghby, A.M., Ismail Ebada, A., Saleh, A.S., Abdelhay, M. and Awad, W.A. (2025). A Comprehensive review of multimodal deep learning for enhanced medical diagnostics. *Computers, Materials & Continua*, 84(3).
24. Sangeetha, S.K.B., Mathivanan, S.K., Beniwal, R., Ahmad, N., Ghribi, W. and Mallik, S. (2025). Advanced segmentation method for integrating multi-omics data for early cancer detection. *Egyptian Informatics Journal*, 29: 100624.

Index

About the Editors

Dr. Sandeep Kumar Mathivanan has a M.S. in software engineering and the M.Tech. (By Research) degree from Vellore Institute of Technology, Vellore, India, in 2016 and 2020, respectively, and a Ph.D. from the School of Information Technology and Engineering, VIT, in 2023. He is currently an Assistant Professor with the School of Computing Science and Engineering, Galgotias University, Greater Noida, India. He has more than six years of research experience, and has authored several journals and conferences papers. He is a reviewer of many reputed Q1 and Q2 journals. His current research interests include machine learning, deep learning, remote sensing, and big data.

Dr. Saurav Mallik (google scholar h-index: 35, citations > 4250) is a Research Scientist in the Department of Pharmacology and Toxicology, The University of Arizona, USA. Previously, he worked as Postdoctoral Fellow in Harvard T.H. Chan School of Public Health, Boston, MA, USA for more than three years (2019–2022), the Center of Precision Health, Department of School of Biomedical Informatics, University of Texas Health Science Center at Houston, Houston, Texas, USA for one and half year (2018–2019), and in the Division of Bio-statistics, Department of Public Health Sciences, University of Miami Miller School of Medicine, Miami, Florida, USA for more than one year (2017–2018). He obtained a Ph.D. from the Department of Computer Science and Engineering from Jadavpur University, Kolkata, India in 2017. His Ph.D. work was carried out in Machine Intelligence Unit, Indian Statistical Institute, Kolkata, India as junior research fellow. He has obtained the award of Research Associateship from Council of Scientific and Industrial Research, MHRD, Government of India in 2017. Dr. Mallik has more than 260 research papers in different top high impact factor peer-reviewed journals, conferences and book chapters. He has published various Books and Patents. He is working as the active member of Institute of Electrical and Electronics Engineers (IEEE), USA, ACM and American Association for Cancer Research (AACR), USA and Bioclues, India. He has also worked with section editors and reviewers with several well-reputed high impact journals. His research interest includes computational biology, knowledge retrieval and data mining, bioinformatics, biostatistics, and ML/DL.

Dr. Basu Dev Shivahare is an Associate Professor in School of Computer Science & Engineering at Galgotias University, Greater Noida, India. He has a Ph.D. (Computer Science & Engineering) from Dr. APJ Abdul Kalam Technical University AKTU, Lucknow, Uttar Pradesh, India in 2023, M.Tech. (CS) from BIT MESRA, Ranchi, India in 2012 and B.Tech. (CSE) from Uttar Pradesh Technical University (UPTU), Lucknow, in 2006. He was an Assistant Professor in Department of Computer Science and Engineering, at Amity University, Uttar Pradesh, India for over 10 years. He has published more than 50 research papers in peer review SCIE/Scopus index international journals and conferences. His research area is Image Processing, Medical image analysis, metaheuristic optimization algorithms and machine learning. He is UGC-NET qualified. He has more than 19 years of teaching experience.

Dr. Sangeetha S.K.B. is an Associate Professor at the School of Computer Engineering, Manipal Institute of Technology, Bengaluru, Manipal Academy of Higher Education, India. With over 17 years of academic experience, Dr. Sangeetha has contributed significantly to fields such as data science, medical imaging, Artificial Intelligence (AI), and quantum computing for healthcare. She earned her Ph.D. in Computer Science and Engineering from Anna University, Chennai, and is currently pursuing a postdoctoral program at Lincoln University College, Malaysia.

Her research interests focus on AI, machine learning, and quantum computing, particularly in multimodal data fusion for healthcare applications. Dr. Sangeetha has authored/co-authored over 50 research articles, many of which have been published in SCI-indexed journals. Her contributions have earned her multiple accolades, including Best Paper Awards at international conferences such as IEEE and Springer, and recognition from prominent journals like PLoS ONE and Springer. Dr. Sangeetha has mentored 3 Ph.D. students, and over 100 B.Tech. students, fostering academic and professional growth. She has guided numerous students through research and industry projects, particularly in AI and data science. In addition to her teaching and research roles, Dr. Sangeetha has been a guest speaker at over 25 international conferences and workshops. She has also authored 4 books and served as a reviewer for several peer-reviewed journals. Her professional engagement extends to industry consulting, with over 5 consulting projects focused on AI and machine learning solutions.

Dr. Shakila Basheer is currently working as an Associate Professor in Department of Information Systems, College of Computer and Information Science, Princess Nourah bint Abdulrahman University in Riyadh, Saudi Arabia. She has more than 10 years of teaching experience and has published more technical papers in international journals/proceedings of international conferences/edited chapters of reputed publications. She has worked and contributed in the field of Data Mining, Image Processing and Fuzzy Logic. Her research also focuses on Data Mining algorithms using Fuzzy Logic. She is currently working on Data mining, Vehicular networks Machine Learning, Blockchain, Vehicular networks and IoT.

For Product Safety Concerns and Information please contact our
EU representative GPSR@taylorandfrancis.com Taylor & Francis
Verlag GmbH, Kaufingerstraße 24, 80331 München, Germany